CAMBRIDGE BIOLOGICAL SERIES

PALÆONTOLOGY

INVERTEBRATE

PALÆONTOLOGY

INVERTEBRATE

BY

HENRY WOODS, M.A., F.R.S.

EIGHTH EDITION

CBS

CBS Publishers & Distributors Pvt. Ltd.

New Delhi • Bengaluru • Chennai • Kochi • Kolkata • Mumbai
Hyderabad • Nagpur • Patna • Pune • Vijayawada

PUBLISHED BY
THE SYNDICS OF THE CAMBRIDGE UNIVERSITY PRESS
Bentley House, 200 Euston Road, London, N.W.1
American Branch: 32 East 57th Street, New York 22, N.Y.

ISBN: 81-239-1080-0

First Indian Reprint: 1982
Reprint: 1985, 2001, 2004

This edition has been published in India by arrangement with Cambridge University Press, UK

Published by:
Satish Kumar Jain for CBS Publishers & Distributors Pvt. Ltd.,
4819/XI Prahlad Street, 24 Ansari Road, Daryaganj, New Delhi - 110002
delhi@cbspd.com, cbspubs@airtelmail.in • www.cbspd.com
Ph.: 23289259, 23266861, 23266867 • Fax: 011-23243014

Corporate Office: 204 FIE, Industrial Area, Patparganj, Delhi - 110 092
Ph: 49344934 • Fax: 011-49344935
E-mail: publishing@cbspd.com • publicity@cbspd.com

Branches:
- ***Bengaluru:*** 2975, 17th Cross, K.R. Road, Bansankari 2nd Stage, Bengaluru - 70 • Ph: +91-80-26771678/79 • Fax: +91-80-26771680 E-mail: cbsbng@gmail.com, bangalore@cbspd.com
- ***Chennai:*** No. 7, Subbaraya Street, Shenoy Nagar, Chennai - 600030 Ph: +91-44-26681266, 26680620 • Fax: +91-44-42032115 E-mail: chennai@cbspd.com
- ***Kochi:*** Ashana House, 39/1904, A.M. Thomas Road, Valanjambalam, Ernakulum, Kochi • Ph: +91-484-4059061-65 Fax: +91-484-4059065 • E-mail: cochin@cbspd.com
- ***Kolkata:*** 6-B, Ground Floor, Rameshwar Shaw Road, Kolkata - 700014 Ph: +91-33-22891126/7/8 • E-mail: kolkata@cbspd.com
- ***Mumbai:*** 83-C, Dr. E. Moses Road, Worli, Mumbai - 400018 Ph: +91-9833017933, 022-24902340/41 • E-mail: mumbai@cbspd.com

Representatives:

- Hyderabad: 0-9885175004
- Nagpur: 0-9021734563
- Patna: 0-9334159340
- Pune: 0-9623451994
- Vijayawada: 0-9000660880

Printed at: Neekunj Print Process, Delhi (India)

PREFACE TO EIGHTH EDITION

THE general plan of this work is to give, in each group of the Invertebrata, first, a short account of its general zoological features with a more detailed description of the hard parts of the animals; secondly, its classification and the characters of the important genera, with remarks on the affinities of some forms; and thirdly, a description of the present distribution, and the geological range. The account of each genus is followed by the enumeration of one or more typical species, so as to guide the student in making use of a large collection.

The illustrations are employed mainly for the purpose of explaining structure and terminology, and will not enable the student to dispense with the use of specimens. The list of palæontological works is intended to indicate where further information may be obtained in any branch of the subject; it includes works of general interest in each group, and others dealing especially with British fossils.

Minor alterations and corrections have been made in this edition and the list of palæontological works has been brought up-to-date. In connection with the systematic position of the Graptolites attention must be called to the work of Kozlowski on the Dendroids from the Upper Tremadoc. He maintains that they are not Hydrozoa, but are allied to the Pterobranchia among the Hemichordata. This is made probable by the presence among the Graptolites of a Pterobranch related to the living *Cephalodiscus*.

H. WOODS

March 1946

CONTENTS

INTRODUCTION

From the earliest times it has been known that bodies resembling marine animals occur embedded in the rocks. For several centuries two distinct views were held respecting their nature. By some persons they were thought to have once formed parts of living animals, and consequently to indicate that the spot where they are now found was in past ages covered by the sea. Others, feeling it difficult to account for so much geographical change as would be necessitated by this view, considered that they were not of organic origin at all, but had been formed by some 'plastic force' within the earth—that they were in fact 'Sports of Nature'. Since, however, these bodies resemble in every essential respect the hard parts of animals now existing, we may at once reject this hypothesis.

The remains of animals and plants of past ages preserved in the rocks are known as fossils, the study of which forms the subject of Palæontology.

In order that an animal or plant may become a fossil two conditions are generally necessary: First, it must possess a skeleton of some kind or other, since the soft parts are rapidly decomposed; consequently such animals as jelly-fishes leave no trace of their existence, unless it be a mere imprint. Secondly, the organism must be covered up by some deposit, otherwise it will soon crumble to pieces. Now, since there are comparatively few places on land where material is being deposited to any great extent, it follows that terrestrial animals will stand but little chance of being preserved; the greater number after death will remain on the surface and will in a short time be entirely decomposed. A few may become entombed in peat-bogs, in the dust and ashes thrown out by volcanoes, in the sand of sand-dunes, or by a landslip; some may be sealed up in deposits of carbonate of lime, such as the travertine thrown down by calcareous springs, or the stalagmite formed on the floor of caves; and lastly, others may be transported by running water and ultimately buried in the bed of a river, of a lake, or of the sea. Such instances, however, are of comparatively rare occurrence. In the case of aquatic animals the conditions for fossilisation are much more favourable, since deposition is more universal in water than on land. Of such aqueous deposits, those formed in the sea will enclose by far the larger number of animals on account of the greater area which these deposits cover.

The structure and composition of the hard parts vary considerably in different groups of animals and plants; some are therefore much more readily preserved as fossils than others. Thus in *Argonauta* the skeleton consists of a thin shell which is easily broken up; then again in some sponges it is formed of needles of silica, which are held together by the soft parts only and consequently easily become scattered after the death of the animal. But in other cases, as in most of the molluscs and corals, the skeleton is very strong and not easily destroyed, hence these occur abundantly in the fossil form. Perhaps even more important than the structure, is the composition of the hard parts, which, in the case of insects and some hydroids, consist of a horny substance known as chitin: in diatoms, in most radiolarians, and in many sponges, of silica; in the bones of vertebrates, chiefly of carbonate and phosphate of lime; in corals, echinoderms, molluscs and many other animals and some plants, of carbonate of lime;

in most plants, of woody or corky tissue: a larger or smaller amount of organic matter is always combined with the mineral. Of these substances, chitin is with difficulty dissolved. Silica in its ordinary crystalline condition is one of the most stable of minerals, but when secreted by an animal or plant it is glassy and isotropic (*i.e.* singly refracting and without effect on polarised light), and is dissolved with comparative ease, so that such skeletons may be entirely removed by the action of percolating water. In organisms with calcareous skeletons the carbonate of lime is readily dissolved by water containing carbonic acid, but the degree of solubility varies according to the condition in which the carbonate of lime is present. In some animals it occurs as aragonite, in others as calcite. Of these two minerals, aragonite is the harder and heavier, its specific gravity being 2·93, whilst that of calcite is only 2·72; aragonite crystallises in the rhombic system, calcite in the hexagonal. Fossil calcite shells (*e.g. Pecten opercularis*) are translucent, their surface is compact, but their interior porous; on the other hand the aragonite shells (*e.g. Glycimeris glycimeris*) are opaque, and have a chalky appearance but a compact structure throughout. If a shell of each kind be suspended in water containing carbonic acid, it will be found that the one composed of aragenite will lose, in the same time, a much greater proportion of its weight than the other. Further, the calcite shell remains firm longer than the aragonite, the latter being soon reduced to the consistency of kaolin or china-clay. This difference, however, does not appear to be due directly to mineral composition, for Cornish and Kendall found that when crystals of calcite and aragonite were powdered and placed in carbonic acid solutions of the same strength, the aragonite was *not* acted on more rapidly than the calcite, and the same result was obtained with powdered fossil shells. From all these considerations, it is not surprising to find that in some strata the aragonite skeletons have entirely disappeared, whereas those formed of calcite remain. This will obviously be most likely to occur in pervious beds through which water containing carbon dioxide percolates. A striking instance of the difference in the solubility of calcite and aragonite was furnished by some specimens of the common edible mussel, *Mytilus edulis*, in which the inner layer of the shell is formed of aragonite and the outer of calcite; Sorby found specimens in the raised beach at Hope's Nose, Torquay, which had lost the inner layer but not the outer. Similarly, in specimens of *Spondylus* from the Chalk, the inner layer of the shell has been completely removed, but the outer is left. In some cases aragonite is replaced by calcite, but then the organic structure is entirely destroyed, and we get merely a mass of calcite crystals. Calcite is never replaced by aragonite.

The mineral character of the skeleton of the chief calcareous organisms is as follows:

Foraminifera.—The vitreous forms consist of calcite, the porcellanous probably of aragonite.

Porifera.—Calcareous sponges of calcite.

Anthozoa.—The Alcyonaria are of calcite, except *Heliopora*, which is of aragonite; the Madreporaria are of aragonite.

Echinoderma.—All of calcite.

Polyzoa.—Chiefly of calcite.

Brachiopoda.—All of calcite.

Lamellibranchia.—Many consist entirely of aragonite, but *Anomia*, *Ostrea*, and *Pecten* of calcite. In *Pinna*, *Mytilus*, *Spondylus*, *Unio*, and *Trigonia*, the inner layer is of aragonite, the outer of calcite.

Gasteropoda.—The majority are formed of aragonite, but *Scalaria* and some species of *Fusus* are of calcite. In some (*e.g. Patella, Littorina*) the outer layer is calcite.

Cephalopoda.—*Nautilus*, *Spirula*, and *Sepia* are mainly aragonite, as also were probably the Ammonites. *Argonauta* and the guard of *Belemnites* are calcite.

Crustacea.—The shell consists of chitinous material usually containing calcite, and often some phosphate of lime.

The condition in which fossils occur depends, as we have seen, on their original composition and on the material in which they are embedded. The chief types are the following:

1. *The entire organism preserved.* Occasionally the soft parts of the organism are preserved as well as the skeleton, the whole having suffered very little change. Instances of this are the woolly rhinoceros and mammoth found frozen in the mud and ice in Northern Siberia. Insects encased in fossil resin, known as amber, are found in the Oligocene beds on the Baltic shores of Prussia and in the Tertiary beds near Cromer, but it is only rarely that any of the soft parts are preserved.

2. *The skeleton preserved almost unchanged.* Sometimes when the skeleton alone is preserved, it remains almost in its original condition, except that it has lost its organic matter. Thus the shells in the Pliocene beds of England differ from living ones only in being lighter, more porous and generally colourless. In some instances a certain amount of mineral matter, such as carbonate of lime, has been added to the skeleton, making it heavier and more compact.

3. *Carbonisation.* In some plants, and in animals with chitinous skeletons, such as graptolites, the original material usually becomes carbonised. The organism undergoes decomposition and loses oxygen and nitrogen, the relative percentage of carbon therefore increasing. The changes are

similar to those which occurred during the conversion of vegetable matter into coal.

4. *A mould of the skeleton.* Sometimes the skeleton disappears entirely, a mould only remaining: this is especially the case when it consists of aragonite and is embedded in a porous stratum. After the shell of a mollusc has become covered up with sediment, and the soft parts have been decomposed, the interior becomes filled with the same material. Water containing carbonic acid subsequently percolates through the rock and carries away the shell as bicarbonate of lime, so that there is left only a mould of the interior and of the exterior, the space between the two being that which was originally occupied by the shell and, if filled with wax, will give an exact model of it. Excellent examples of this mode of fossilisation are seen in some molluscs from the Portland Oolite, *e.g.* *Aptyxiella* and *Trigonia*. Sometimes after the shell has been removed the space left becomes filled up with mineral matter carried in by percolating water; this has the form of the original skeleton but obviously not its internal structure.

The interior of the shells of Foraminifera may, soon after the death of the animal, become filled with glauconite (silicate of iron and alumina); subsequently the shell itself often disappears, leaving only the internal cast. Glauconite occurs in this way in the various greensand strata, and also in some of the deep-sea deposits at the present day. Somewhat similarly the shells of sea-urchins occurring in the Chalk are sometimes filled with flint; in such cases the shell when buried did not become filled with Chalk, but remained empty until flint was deposited in it from percolating water containing silica in solution.

5. *Petrifaction.* In some deposits the fossils show the minute structure as well as the form of the organism, but

the original material of the skeleton has been replaced by another mineral. Thus we find fossil wood which shows the cells and vessels just as in existing trees, but in which the walls are formed of silica instead of cellulose. The change has gone on in such a manner that as each particle disappeared its place was taken by a particle of silica. The chief minerals which replace the original substance of organisms in this manner are:

(i) Carbonate of lime; calcite sometimes replaces the silica of sponges.

(ii) Silica, as in the fossils from the Blackdown Greensand, and the Thanet Sands near Faversham; also in the wood of the Purbeck dirt-bed in the Isle of Portland.

(iii) Iron pyrites; *e.g.* Ammonites from the Oxford Clay, Lias, etc., and some graptolites.

(iv) Oxide of iron, in the form of limonite in some fossils from the Dogger (Inferior Oolite) of Yorkshire and the Lower Greensand of Potton, etc., and as hæmatite in fossils from the Carboniferous Limestone of Cumberland.

(v) In rare cases there are other replacing minerals, such as sulphate of lime, barytes, blende, galena, malachite, vivianite, and spathic iron.

6. *Imprints.* The footprints of animals and the impressions of jelly-fishes are sometimes found in the rocks, and these, although forming no part of the animal itself, are nevertheless regarded as fossils.

In endeavouring to discover the changes which have taken place on the earth in past geological times, the evidence furnished by fossils is of primary importance. Each great group of the stratified rocks, known as a system, is characterised by a particular assemblage of genera and species, some of which are confined to it and enable us to identify the system. In a similar manner, the smaller divisions—the series and stages—are each characterised by the presence of certain fossils, which do not occur above or below. Further, it is found that the fauna of the smallest division (stage or group of beds) is not of uniform character throughout; although there may be no change in the nature of the rock, some of the species and varieties which are abundant at one level will become rare or will disappear entirely in passing to higher or lower horizons. Consequently, a set of beds may be divided into belts or zones, the general aspect of the fauna of each zone being somewhat different from that of the others, but between these divisions there will be no break either physical or palæontological. If then we have determined the order of succession of the formations in any one area by means of their relative positions, the newer resting on the older, it is fairly easy in any other district, merely by examining the fossils, to refer any set of beds to its proper position in the geological record. But although this law of the identification of strata by the fossils which they contain is of great value, it must not be applied without some caution, for even if two formations were deposited at exactly the same time, it does not necessarily follow that all the genera and species found in the two will be identical. Thus for instance in the seas at the present day the same forms of life do not occur in all parts; animals which live in water of moderate depth are distributed in provinces which depend largely on climatic conditions, each province possessing some forms peculiar to itself. The organisms now being entombed in deposits formed, say, off the British coasts, will as a whole be different from those off the Canary Islands; but still, some of the species and many of the genera will be common to both areas, and would enable us to identify the two deposits as having been formed within the same general period, though perhaps not to prove them absolutely synchronous. Then again there is a distribution of organisms according to the depth of the sea, and the nature of the sea-bottom; so that the fauna of a deep-water formation will necessarily be different from that of a shallow-water one, and that of a sandy deposit different from that of a mud. But in addition to the animals living on the sea-bottom there are others which live near the surface of the ocean, far from land; such *pelagic* forms have a wider geographical range than those which live on the sea-floor in shallow water, and are consequently of great value in determining, as of the same age, deposits found in widely-separated localities.

In addition to their chronological value, fossils are also important in indicating the conditions under which the formations were deposited. In the case of the later beds, where most of the fossils belong to genera which are still existing, it is easy to distinguish a marine deposit from one formed in freshwater or on land. Even in the rocks of earlier periods, in which most of the genera are extinct, we may recognise a marine deposit by the presence of such animals as radiolarians, corals, echinoderms, brachiopods, pteropods, cephalopods, or cirripeds, which at the present day are found only in the sea.

The depth of the sea in which a formation was deposited can be estimated when the fossils belong to living species; when the species are extinct some idea may be formed if the genera to which they belong are found chiefly at some particular depth at the present day. In attempting such determinations it must be remembered that the sea-bottom down to a depth of nearly 50 fathoms may be disturbed by the action of waves and currents in the sea; consequently the animals living on the bottom in shallow water are liable to be carried from their original home to higher or lower levels. One of the surest indications that a formation was laid down in shallow water and not far from land is furnished by the association of the fossil remains of land animals and plants with marine species; another, by the presence of molluscs such as *Pholas*, *Saxicava* and *Lithophaga*, which bore into rocks, and at the present day are found only in shallow water. The proximity of a shore-line is also indicated when the assemblage of fossil forms resembles in general character the faunas which live in littoral regions at the present day. When evidence of the existence of a shore-line is found it is obviously possible to gain some idea of the distribution of land and sea in past times.

The nature of the climates of past ages may be judged to some extent by the character of the fossils; the evidence furnished by land-plants is particularly valuable, since their distribution is determined largely by temperature and is better marked than in the case of marine animals. As far as the latter are concerned it is only when we are dealing with modern species that we can, as a rule, speak with any degree of certainty on this subject; this is owing to the fact that at the present day the individual species of the same genus have often a very different distribution, some being found in warm, others in cold, regions. Even when all the fossils in a formation belong to extinct species, the assemblage of genera is sometimes such as marks some region at the present day; thus, for example, in the London Clay we find that many of the genera of molluscs are now characteristic of tropical or sub-tropical seas.

The study of fossil animals and plants is of the highest importance to the biologist, not only because they include the ancestors of modern species, but because among fossil forms we find many groups (*e.g.* Graptolites, Cystids, Blastoids, Trilobites, Eurypterids), which are altogether extinct, and which often throw light on the relationship of existing animals and plants. Others (*e.g.* Crinoids, Brachiopods, Nautiloids) are represented at the present day by few forms only, but were, in past ages, very abundant; consequently no adequate knowledge of such groups of animals can be obtained from the study of living examples only. In some cases the ancient forms serve to connect groups which, at the present day, appear to be quite distinct; thus, for example, the earliest known bird (*Archæopteryx*, from the Solenhofen Limestone, Upper Jurassic) shows, in several important characters, affinities to the Reptiles.

From the point of view of the biologist, the greatest interest in Palæontology is found in the bearing it has on the subject of evolution: it is only by a study of the stratigraphical succession of fossil forms that the race-history or phylogeny of animals and plants can be traced with certainty; but in attempting such investigations a great difficulty is presented by the imperfection of the record of the life of past ages, since only a very small proportion of the animals and plants has been preserved, and often in an imperfect manner. We have already seen several reasons why this record must be imperfect; some animals are without hard parts, while others, particularly land animals, frequently do not become covered up with sediment. Further, the remains of animals which were originally present in the rocks have been, in some cases, dissolved by percolating water, or to a great extent obliterated by the metamorphism which the rock has undergone. Then again the record of life is incomplete because of the breaks in the succession of the stratified rocks; these breaks have been caused sometimes by denudation having removed a great thickness of rocks, in other cases by a temporary absence of deposition. Even when there is no break in the succession due to these causes a further difficulty in tracing out phylogeny may be introduced by changes taking place in the physical conditions during the deposition of a series of beds; thus there may have been alterations in the depth of the sea, in the nature of the sediment on the floor, or in the temperature of the water; in each case the physical change would react on the fauna tending to cause the animals living on the sea-floor to migrate to other regions where conditions favourable to their mode of life could be found. When such migrations occurred the descendants of the animals which lived when one stratum was deposited would not be found fossil in the overlying beds of the same area.

Notwithstanding this imperfection of the record and the effects of changing physical conditions, many groups of animals are found to undergo gradual modification when traced through series of strata or formations. For example, in the Pliocene deposits of Slavonia there are numerous shells of pond-snails (*Viviparus* or *Paludina*); and specimens found at the top and bottom of the formation, and also at certain intervening levels, differ so much from one another that they appear to belong to distinct species. When, however, examples are collected from all the beds of the formation, the apparently distinct species are seen to be connected by intermediate forms, and a series, showing a gradual passage from the species found in the lowest bed to that in the highest, can be obtained. Similarly in the English Chalk, during the deposition of which the physical conditions continued more nearly uniform than in most other formations, it is found that the sea-urchins, starfishes, etc. undergo slow and gradual changes in various characters when traced from lower to higher horizons.

In several groups of Tertiary Mammalia there is also evidence of gradual modification in structure; thus the earliest known forerunner of the horse, found in the Eocene beds, possessed five toes, and was succeeded in later times by forms with successively fewer toes, until in the Pliocene, the existing type of horse, with only one toe and splint-bones, appeared; other gradual changes also occurred in the character of the teeth, etc.

In attempting to work out phylogeny, in addition to the stratigraphical method just described, the method of comparative anatomy and often the method of ontogeny (or development of the individual) can be used in the case of fossils. In the course of the development and growth of an animal, various stages, which often present resemblances to the adults of other animals, are passed through. The 'recapitulation theory' supposes that the changes seen during the development of the individual (ontogeny) are, in a general way, a rapid but often incomplete repetition of those which occurred in its race-history (phylogeny). Palæontology has, in some cases, given support to this view, by showing that successive stages, similar to those passed through in the development of an animal, also occurred in the history of its race, as seen in the geological record.

On the whole the evidence of Palæontology favours the view that evolution proceeded by slow and gradual modifications; but there were also times, especially in the early history of various groups, when evolutionary changes went on more rapidly. There is also evidence indicating that evolution was *orthogenetic*—that the evolutionary changes in any one group of animals proceeded in definite directions for considerable periods of time; and further, that allied groups descended from the same ancestral stock, have passed through similar or parallel stages in their evolution quite independently of one another and of external conditions, suggesting that the lines of evolution in the various groups were determined by something inherited from the common ancestor. Examples of this are seen in the Chalk starfishes, in the mode of branching of graptolites (p. 69), in the development of horns in different evolutionary series of Titanotheres, and in the evolutionary history of various other groups of Tertiary mammals.

In a natural classification of animals an attempt is made to place together in the same group those forms which are connected by descent; such a classification, if perfect, would be of the nature of a genealogical tree. Each main division is termed a Phylum and includes animals built on the same fundamental plan and believed to have descended from one ancestral stock. Each Phylum is divided and subdivided into smaller and smaller groups, known as Classes, Orders, Families, Genera, and Species. A species includes a group of individuals very closely related to one another, which have descended from the same ancestors and can give rise to offspring which are fertile among themselves; such individuals usually differ from one another to only about the same degree that offspring of the same parents may differ. One species is generally distinguished from another by such characters as ornamentation, shape, relative proportion of parts, and size. In some species one or more groups termed *varieties* may be recognised, and are distinguished from the other forms included in the species by some slight, but fairly well-marked and constant modification. Varieties are frequently connected with the special physical or biological conditions under which they are living. The varieties in some species pass into one another by intermediate forms; but others appear to be fairly distinct and may be regarded as incipient species.

Sometimes two groups of individuals resemble each other so closely that they might be regarded as belonging to the same genus or even to the same species, but they have descended from different ancestors since they are found to differ in development (ontogeny) or in their palæontological history; this phenomenon, of forms belonging to different stocks approaching one another in character, is known as *convergence* or *heterogenetic homœomorphy*, and may occur either at the same geological period or at widely separated intervals. Thus the form of oyster known as *Gryphæa* has originated independently from oysters of the ordinary type in the Lias, in the Oolites, and again in the Chalk; these forms found at different horizons closely resemble one another and have usually been regarded as belonging to one genus (*Gryphæa*), but they have no direct genetic connection with one another. Similarly in various species of Terebratulids a double fold or biplication has arisen in the front part of the shell, thus giving considerable resemblance to different species which are not closely related to one another. Then again sutures similar to those of *Ceratites* from the Trias are developed in some Chalk Ammonites which have no genetic connection with *Ceratites*.

Also, animals belonging to quite distinct groups may, when living under similar conditions, come to resemble one another owing to the development of adaptive modifications, though they do not really approach one another in essential characters; thus analogous or parallel modifications may occur in independent groups—such are the resemblances between flying reptiles (Ornithosaurs) and birds, and between sharks, ichthyosaurs and dolphins.

PHYLUM PROTOZOA

Classes		*Orders*
1. Gymnomyxa ... (Sarcodina)	...	1. Foraminifera 2. Radiolaria 3. Others not found fossil
2. Flagellata or Mastigophora		
3. Infusoria (not fossil)		
4. Sporozoa (not fossil)		

The Protozoa include the lowest forms of animals, such as *Amœba*, *Vorticella*, and *Globigerina*. The body is usually very small, and consists in many cases of one cell only, in others of more than one, but the cells never form tissues as they do in all other animals. A cell consists of *protoplasm*—a viscid or semi-fluid living substance containing granules; in the centre of the cell is a denser, usually spherical body called the *nucleus*—sometimes more than one is present.

In some Protozoa (the *Gymnomyxa*) the protoplasm is naked, and consists of an inner granular mass and a thin, clear, outer layer; such forms are further characterised by having no definite shape, by being able to take in food at any part of the body, and by possessing the power of throwing out lobes or filaments of protoplasm known as *pseudopodia*. In others (the *Flagellata* and *Infusoria*) the protoplasm is surrounded by a firm membrane or cuticle which gives the animal a definite form; the food is generally taken in at one permanent aperture, and pseudopodia are seldom present, but the surface is provided with *cilia* or *flagella*, which are fine threads of protoplasm having a definite form and a rhythmic movement.

Reproduction in the Protozoa takes place usually by fission (*i.e.* division into two parts) and sometimes by the formation of spores. In some cases conjugation of two or more individuals occurs, representing to some extent sexual reproduction. In some of the Protozoa there is no skeleton, but in others a shell is formed.

The Protozoa can be divided into four main groups: (1) the Gymnomyxa, (2) the Flagellata, (3) the Infusoria, (4) the Sporozoa; no examples of the last two divisions have been definitely recognised in the fossil state.

CLASS I. GYMNOMYXA (SARCODINA)

The members of this group possess no external membrane (cuticle), and are able to throw out pseudopodia, by means of which movement takes place and food is obtained.

The Gymnomyxa or Sarcodina are divided into several orders, of which only two have been found fossil, namely, the Foraminifera and the Radiolaria.

ORDER I. FORAMINIFERA

The Foraminifera are characterised by their thread-like pseudopodia, which frequently branch and anastomose; and by possessing in most cases a shell or test, which may be calcareous, arenaceous, chitinous, siliceous, or gelatinous.

The calcareous forms are by far the commonest, and in these, two kinds of shell may be distinguished, namely, the *vitreous* or *perforate* and the *porcellanous* or *imperforate*. In the vitreous, the shell often has a glassy appearance, and is perforated by innumerable tubes for the passage of the pseudopodia; in some forms (*e.g. Rotalia*) these tubes are $\frac{1}{3000}$ of an inch in diameter, but in others (*e.g. Operculina*) only $\frac{1}{10000}$ of an inch. In the porcellanous forms the shell, when viewed by reflected light, is opaque and white, having the appearance of porcelain; it is not perforated by tubes, but possesses one or two large apertures through which most of the pseudopodia pass out—some, however, are given off from the layer of protoplasm which covers the surface of the shell. In these porcellanous Foraminifera the shell is sometimes pitted, producing at first sight the appearance of perforation.

In the arenaceous forms the shell consists of foreign particles joined together by a cement. The particles are usually grains of sand (commonly quartz), but sometimes sponge-spicules, or the shells of other Foraminifera. The cement may be formed of chitinous, calcareous, or ferruginous material. The shell is often imperforate.

The chitinous forms (*e.g. Gromia*) do not occur as fossils.

The shell of the Foraminifera varies considerably in form and structure; in some genera it consists of a single chamber, when it is said to be *unilocular*, as in *Lagena* (fig. 3 F) which is generally flask-shaped. In other cases it consists of several chambers communicating with one another, either by perforations in the walls (septa) between them, or by larger openings. In these *multilocular* forms the shell grows by the addition of a new chamber at the end of the one last formed; this takes place by the protrusion, through the aperture or mouth of the shell, of a mass of protoplasm, at the surface of which the wall of a new chamber is formed either by the secretion of material or by cementing of foreign particles. The arrangement of the chambers in the multilocular Foraminifera is very varied; they may be placed in a straight line as in *Nodosaria* (fig. 3 H), in a curved line as in *Dentalina*, in a plane spiral as in *Cristellaria* (fig. 3 G), or in a helicoid spiral as in *Rotalia* (fig. 3 L, M). The earlier whorls in some spiral forms are partly or entirely covered by the later ones, so that sometimes the last whorl only is visible on the exterior (*e.g. Cristellaria*); but when the later chambers are merely attached to the extremities of the earlier ones, all the whorls can be seen (*e.g. Operculina*). Some genera, such as *Textularia* (fig. 3 E), have two rows of chambers placed side by side; others (*Tritaxia*) have three. In some cases (*e.g. Orbitolites*) there are numerous chambers arranged in concentric rings instead of in a spiral.

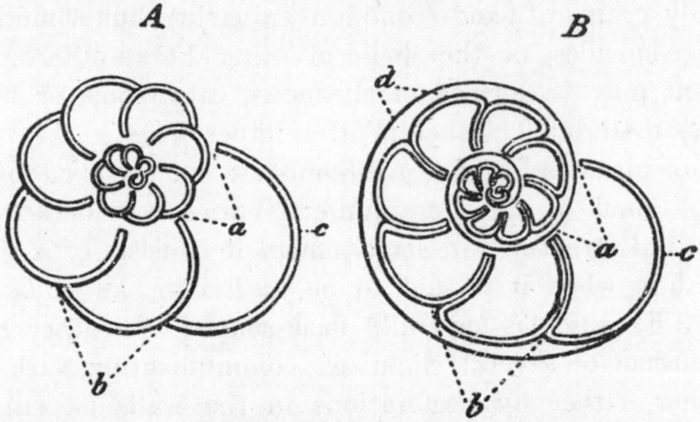

Fig. 1. A, section of a foraminifer in which each septum is formed of a single lamella. B, in which the septum is formed of two lamellæ. *a*, passages between the chambers; *b*, septum; *c*, anterior wall of last chamber; *d*, supplemental skeleton. (After Carpenter.)

In the porcellanous and the simpler vitreous Foraminifera each septum (fig. 1 A, *b*) consists of a single lamella which is really the front wall of the preceding chamber; but in the higher vitreous forms each septum (fig. 1 B, *b*) is formed of two lamellæ, owing to the fact that when a new chamber is added to the shell a new wall is secreted next to the front wall of the last chamber. The shell of the vitreous Foraminifera is at first thin, but may afterwards increase in thickness by the addition of material at the surface; in the higher vitreous forms the outer layers constitute what is known as the *supplemental skeleton* (fig. 1 B, *b*), which is traversed by numerous canals connected with canals in the septa and other parts.

A considerable number of the Foraminifera, especially the higher forms, are *dimorphic*—that is to say, there are two forms of the same species. This fact was first noticed in specimens of *Nummulites* from the Eocene deposits. In one form, the first or initial chamber, which is seen at the

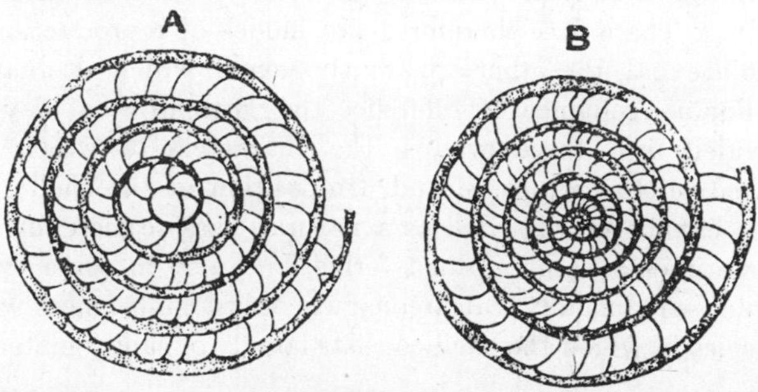

Fig. 2. Dimorphism of *Nummulites lævigatus*, Bracklesham Beds (Eocene), Selsea. A, section of the entire shell of the megalospheric form. × 9. B, section of the *central* part of the microspheric form. × 9.

centre when the shell is split, is large and more or less spherical and is called the *megalosphere* (fig. 2 A); in the other it is much smaller and is known as the *microsphere* (fig. 2 B). These two forms are found associated together and were, at one time, described as different species. In the microspheric type the shell commonly, but not always, grows to a larger size than in the megalospheric type, and individuals of the former are much less numerous than of the latter; in other respects the two are similar. The relationship of the microspheric and megalospheric shells has been elucidated by a study of the life-history of

Polystomella and other living Foraminifera. When reproduction takes place in the microspheric form all the protoplasm passes out of the shell and divides into spherical masses, each of which secretes a shell and develops into a megalospheric individual. In the reproduction of the megalospheric form the protoplasm divides into small rounded portions which pass out of the shell as moving spores—zoospores; it is believed that two zoospores from different individuals conjugate and give rise to a microspheric individual. There are, therefore, two modes of reproduction—one asexual, the other apparently sexual, which alternate.

For convenience of reference the Foraminifera may be divided into three groups, the characters of which are based on the structure and composition of the shell; but this cannot be regarded as a natural classification since it sometimes separates allied forms, and also in some types which are usually calcareous we occasionally meet with species in which the shell consists largely of sandy material.

I. *Porcellanous Forms*

Shell calcareous, porcellanous, not perforated by canals, but provided with one or two large apertures through which the pseudopodia pass out.

Miliola (fig. 3 A—D). Shell multilocular, the early chambers spiral, the later chambers coiled on an elongated axis, each chamber forming half a convolution. In some cases all the chambers are visible externally on both sides of the shell (fig. 3 D); in others, owing to the lateral prolongations of the chambers, only the last one or two are seen (fig. 3 A—C); or it may be that more chambers are shown on one side than on the other. The external features of the shell consequently vary considerably, and on account of this and changes in the plane of coiling, the forms included under the term *Miliola* are now regarded as constituting a number of distinct genera to which

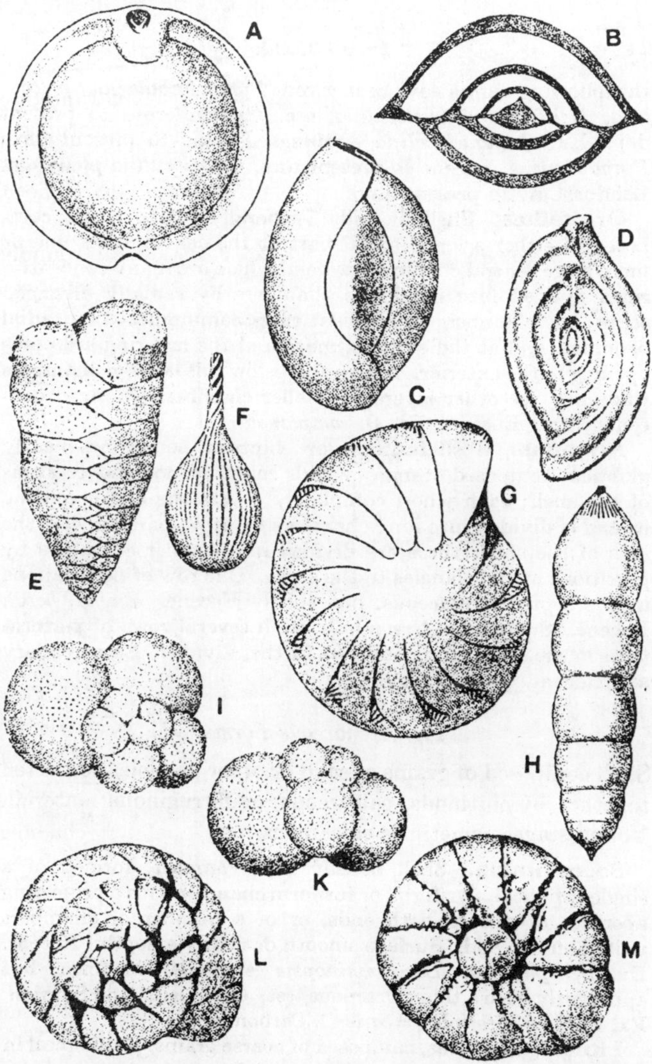

Fig. 3. Foraminifera (recent). A, B, *Pyrgo murrhina*. B, section. C, *Quinqueloculina seminula*. D, *Spiroloculina limbata*. E, *Textularia barretti*. F, *Lagena sulcata*. G, *Cristellaria rotulata*. H, *Nodosaria radicula*. I, K, *Globigerina bulloides*. L, M, *Rotalia beccari*. (After Brady.) All enlarged.

the following names have been given · *Pyrgo, Fabularia, Spiroculina, Miliola, Quinqueloculina*, etc. Carboniferous to present day. Ex. *Quinqueloculina seminula*, Eocene to present day; *Pyrgo ringens*, Eocene to present day; *Spiroloculina planulata*, London Clay to present day.

Orbitolites. Shell discoidal, generally rather large, composed of either a small spiral part at the centre, or of one or more large central chambers, around which are many concentric rings divided into numerous chambers by radially arranged septa; the chambers of adjacent rings communicate by radial openings, and at the external margin of the last ring are pores opening to the exterior. Above and below this layer of chambers there may be other layers of smaller chambers arranged concentrically. Eocene. Ex. *O. complanata.*

Alveolina. Shell fusiform or elliptical, sometimes nearly globular, composed of many whorls coiled around the long axis of the shell; each whorl completely covers the one preceding it, and is divided into long chambers by septa parallel with the axis of the shell; these are divided into smaller chambers by partitions at right angles to the septa. One row of perforations in the septa. Cretaceous, but chiefly Eocene. Ex. *A. bosci*, Eocene. Sub-genus *Alveolinella*, with several rows of perforations in the septa, and chambers further divided. Late Tertiary and Recent. Ex. *A. quoyi.*

II. *Arenaceous Forms*

Shell composed of grains of sand or other particles cemented together by chitinous, calcareous, or ferruginous material. Young stages sometimes calcareous.

Saccammina. Shell usually free, compact, formed of a single spherical, pyriform, or fusiform chamber with a projecting aperture at one or both ends, or of a number of chambers united end to end. Surface smooth or nearly smooth. Recent. Ex. *S. sphærica*. *Saccamminopsis* is similar in form, but apparently with a thin calcareous test. Ordovician and Silurian. Ex. *S. fusuliniformis* (= *carteri*), Carboniferous.

Lituola. Shell free, composed of coarse grains, plani-spiral in the young, later stages uncoiled, straight. Septa labyrinthine.

Aperture single in early stages, later sieve-like. Carboniferous to present day. Ex. *L. nautiloidea*, Chalk.

Orbitolina. Shell partly sandy; conical or flattened, with convex upper, and usually concave lower surface; consisting of central compressed chambers surrounded by concentric rings of subdivided chambers. Cretaceous. Ex. *O. concava*, Upper Greensand.

Endothyra. Shell free, largely calcareous; spiral, nautiloid, or rotaliform; chambers numerous, composed of an outer calcareous, perforated layer, and an inner compact layer formed of small grains cemented together. Aperture simple, at the inner margin of the last chamber. Carboniferous to Trias. Ex. *E. bowmani*, Carboniferous Limestone.

Textularia (fig. 3 E). Shell arenaceous (in the young it is vitreous and perforate); conical, pyriform, or cuneiform; composed of numerous chambers in two alternating parallel series. Aperture slit-like on the inner edge of the last chamber. Carboniferous to present day. Ex. *T. globulosa*, Chalk.

III. *Vitreous Forms*

Shell of calcite, vitreous, perforated by numerous minute canals for the passage of the pseudopodia.

Lagena (fig. 3 F). Shell unilocular, very finely perforated. Form globose, ovate, or flask-shaped. A single terminal aperture, sometimes at the end of a long neck; rarely two apertures. Surface smooth, ribbed, striated, or spinous. Upper Cambrian to present day. Ex. *L. striata*, London Clay to present day; *L. sulcata*, Cretaceous to present day.

Nodosaria (fig. 3 H). Shell composed of a number of chambers which are circular in transverse section, arranged in a straight line, and separated by constrictions. Aperture at the apex of the last chamber. Surface smooth or ornamented with granules, spines, or ribs. Silurian to present day. Ex. *N. zippei*, Gault and Chalk.

Cristellaria (fig. 3 G). Shell compressed, lenticular or elongate, multilocular, coiled in part or entirely in a plane spiral; each coil usually covers the one preceding it. Upper Cambrian to present day. Ex. *C. rotulata*, Chalk to present day.

Globigerina (fig. 3 I, K). Shell perforated by large canals; chambers globular, few, arranged in a helicoid spiral (trochoid), each chamber opening by a large aperture into the central cavity of the spire. No supplemental skeleton. Pelagic forms usually with spines. Cretaceous to present day. Ex. *G. cretacea*, Chalk.

Orbulina. A single spherical chamber, with perforations of two sizes; with smaller chambers (similar to a *Globigerina*) inside the large spherical one. Lias to present day. Ex. *O. universa*, Cretaceous to present day.

Rotalia (fig. 3 L, M). Test very finely perforated, multilocular. The chambers arranged in a helicoid spiral, so that on the upper surface all the whorls are seen, on the lower only the last one. The aperture is in the form of a curved slit on the lower surface of the last chamber. The septa are perforated and usually formed of two layers with canals between the layers. A supplemental skeleton is often present. Lower Cretaceous to present day. Ex. *R. beccari*, Miocene to present day.

Calcarina. Test lenticular, spiral, with only the last whorl visible on the base. Supplemental skeleton greatly developed, traversed by numerous canals, and projecting as long spines from the margin. Chalk to present day. Ex. *C. calcitrapoides*, Chalk.

Fusulina. Shell fusiform, composed of elongated whorls; each whorl completely covers the preceding one, and is divided by septa into a number of chambers, which may be again divided into smaller chambers. Adjoining chambers communicate by a slit at the middle of the base of each septum. Septa folded, each consisting of a single layer. Aperture in the form of a fissure. Carboniferous. Ex. *F. cylindrica*, Carboniferous Limestone.

Amphistegina. Shell lenticular, with sharp edge; the upper and lower surfaces unequally convex; formed of numerous chambers coiled in a plane spiral, each coil completely enclosing the preceding one on one side and partly on the other. Septa formed of a single layer. Supplemental skeleton at the centre of the shell. Aperture similar to that of *Rotalia*. Eocene to present day. Ex. *A. haueri*, Miocene.

Nummulites (figs. 2, 4). Shell lenticular in form, and composed of a large number of whorls coiled in a plane spiral.

Usually each whorl completely covers the preceding one by means of the lateral prolongations of the chambers, so that externally only the last whorl of the shell is visible. The whorls are divided into chambers (*c*) by septa (*b*) which are slightly curved backwards; each chamber communicates with the neighbouring one by means of a median fissure at the inner margin of the septum. Each septum is formed by two imperforate lamellæ between which are irregular spaces. A supplemental skeleton is present, part of it forming what has been termed

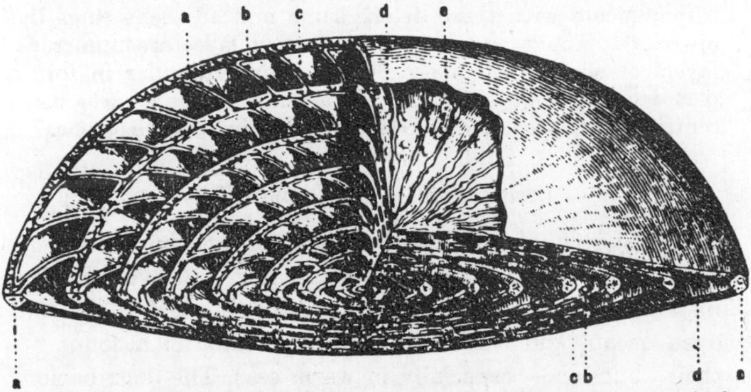

Fig. 4. *Nummulites*, showing vertical and horizontal sections. *a*, marginal cord with canals (supplemental skeleton); *b*, septum, with canals; *c*, chambers; *d*, test; *e*, pillars of the supplemental skeleton. (After Zittel.) Enlarged.

the 'marginal cord' (*a*). The general shell-substance is minutely perforated, and a system of canals traverses the septa and supplemental skeleton. Aperture in the form of a slit at the inner margin of the last chamber. The shell splits readily into two similar parts along the median plane, owing to the relatively large size of the parts of the chambers occurring there. Eocene and Oligocene; maximum development in the Middle Eocene. In the English Eocene the genus is found in the Barton and Bracklesham Beds. Ex. *N. lævigatus*, Bracklesham Beds.

Operculina. Similar to *Nummulites*, but whorls fewer and rapidly enlarging, all visible externally; each of the earlier whorls partly encloses the preceding one. Upper Cretaceous to present day. Ex. *O. complanata*, Miocene.

Lepidocyclina. Test lenticular, circular or stellate, flat to inflated, minutely perforated. In the microspheric form the early chambers show a spiral arrangement; in the megalospheric form the early part consists of chambers which are variable in number and size. The early part is followed by a median layer of chambers arranged in concentric rings, usually alternating with the chambers of adjacent rings, and with rhombic, diamond-shaped, hexagonal or other outline; the chambers communicate with those of the same and adjacent rings by apertures. Above and below the median layer are numerous layers of smaller chambers, flattened and irregular in form, placed one above the other and arranged more or less concentrically. Eocene to Miocene. Ex. *L. mantelli*, Oligocene.

Distribution of the Foraminifera

The majority of the Foraminifera are marine, most of them living on the sea-bottom. A few, however, as for instance *Globigerina*, exist at or near the surface in the open ocean, and these are very important on account of their abundance, especially in warm seas. The distribution of the pelagic Foraminifera in the open ocean, as well as those which live on the sea-floor in shallow water, is influenced largely by temperature; the former are more numerous in warm regions and in warm ocean-currents than in colder water, whilst the species of the latter often have their range determined by temperature and depth.

The Foraminifera found in the Palæozoic deposits are mainly vitreous and arenaceous forms. They appear first in the Upper Cambrian, but are comparatively rare until the Carboniferous, in which some beds are formed largely of their shells, as for instance, the Saccammina limestone of the north of England and Scotland, the *Endothyra*-limestone of North America, and the *Fusulina*-limestone of

Russia, China, Japan and North America. The Foraminifera are mostly of small size in the Permian of England; they are comparatively rare in the Trias, but become abundant in the Jurassic, where, however, rock-building types are generally absent. In the Lias the introduction of numerous vitreous species (*Nodosaria*, *Cristellaria*, etc.), many of which appear to be allied to forms now living in tropical or warm-temperate regions only, is noteworthy; some porcellanous forms belonging to the *Miliola* group are also fairly common. A larger number of genera and species are found in the Middle and Upper Jurassic than in the Lias.

The Order continues to be well represented in the Cretaceous formations, particularly in the Gault and Chalk—*Orbitolina*, *Calcarina*, *Globigerina*, *Rotalia*, etc. being common. Some beds of the Chalk, especially the *Micraster* zones and the Chalk Rock, are largely composed of Foraminifera such as *Globigerina*, *Textularia*, *Bolivina*, *Flabellina*.

The Foraminifera attain their greatest development in Tertiary and recent times. In the Eocene deposits *Nummulites* is often extremely abundant and of large size, forming the greater part of the massive Nummulitic Limestone of Southern Europe, Egypt, Asia Minor, and the Himalayas; *Miliola*, *Orbitolites*, *Alveolina*, *Operculina*, and *Lepidocyclina* are also important rock-building forms in the Eocene period. In the English Eocene, Foraminifera are numerous in the Thanet Sands and the London Clay; in the Barton and Bracklesham Beds *Nummulites*, *Quinqueloculina*, *Alveolina*, etc. occur. In the Oligocene *Nummulites* and *Lepidocyclina* are still present. *Amphistegina* is abundant in the Miocene. A large number of forms occur in the Pliocene deposits of East Anglia and of St Erth in Cornwall.

ORDER II. RADIOLARIA

In the Radiolaria the body consists of a central mass of protoplasm, enclosed in a membrane known as the *central capsule* (fig. 5, 2). The intracapsular protoplasm contains one or more nuclei, and is continuous, through pores in the capsule, with a layer of protoplasm outside the capsule; this layer gives off thread-like pseudopodia, which occasionally unite. A skeleton (fig. 5, 1) is generally present and is usually composed of silica; but in one group of Radiolaria it consists of a substance which was formerly regarded as horny in nature and termed acanthin, but is now believed to consist of strontium sulphate. The skeleton shows great diversity of form and complexity (fig. 6); it may be entirely outside the central capsule or partly within, and consists either of isolated spicules, or of a lattice-like or reticulate structure of varying shape, frequently with projecting spines.

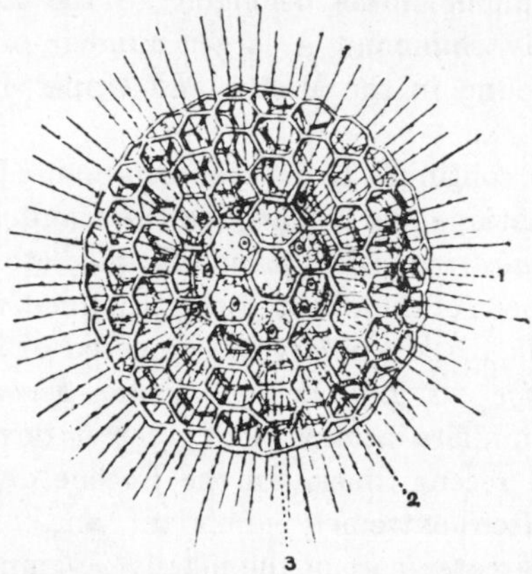

Fig. 5. *Heliosphæra inermis.* ×350. Recent. (After Bütschli.) 1, skeleton; 2, central capsule; 3, nucleus. Pseudopodia project from the surface.

The Radiolaria are all marine and mainly pelagic; the majority live between the surface and a depth of 200 fathoms, but a few forms occur in much deeper water. They have a very wide geographical distribution, being found in all climates, but show the greatest variety of forms in the seas between the tropics; they are also abundant in individuals in the Arctic seas, but the variety of forms is

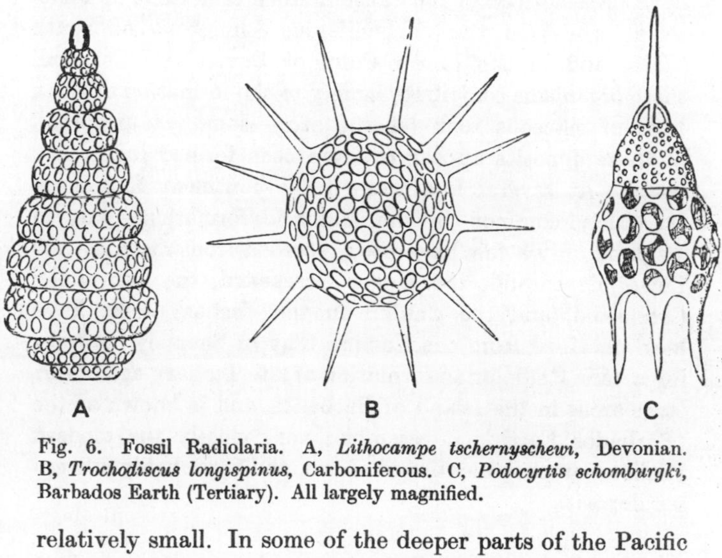

A B C

Fig. 6. Fossil Radiolaria. A, *Lithocampe tschernyschewi*, Devonian. B, *Trochodiscus longispinus*, Carboniferous. C, *Podocyrtis schomburgki*, Barbados Earth (Tertiary). All largely magnified.

relatively small. In some of the deeper parts of the Pacific and Indian Oceans the empty shells of these animals settle and accumulate on the sea-bottom, forming a siliceous deposit known as 'Radiolarian ooze'. Only those Radiolaria in which the shell consists of silica are preserved as fossils.

Cayeux has described as Radiolaria some bodies found in the Pre-Cambrian rocks of Brittany; they are much smaller than later forms of the group, and are thought by some authors to be simply inorganic aggregations. Imperfectly preserved Radiolaria have been recorded from the Cambrian of Thuringia.

In Britain the earliest examples of the Radiolaria occur in the Ordovician rocks of the South of Scotland, where they form beds of chert; others, which are perhaps of nearly the same age, have been found in a chert from Mullion Island (off the west coast of the Lizard). A few specimens have been noticed in the Carboniferous Limestone of Flintshire; whilst in the Carboniferous Limestone of South Wales and in the Lower Culm of Devon and Cornwall these organisms contribute largely to the formation of thick beds of siliceous rock (cherts, etc.)—some, at any rate, of these deposits appear to have been formed in shallow water. At several localities on the continent Radiolaria are fairly common in the Mesozoic formations, but in England only a few have been recorded from the Lias, the Lower Greensand, the Upper Greensand, the Cambridge Greensand, and the Chalk. In the Tertiary some have been obtained from the London Clay of Sheppey. A very important Radiolarian formation of late Tertiary age covers large areas in the Island of Barbados, and is known as the 'Barbados Earth'; it resembles very closely the modern Radiolarian ooze mentioned above, and is probably a deep-sea deposit.

PHYLUM PORIFERA

Classes	*Orders*
1. Hexactinellida	
2. Demospongiæ ...	1. Myxospongida 2. Ceratosa 3. Monaxonida 4. Tetraxonida 5. Lithistida 6. Octactinellida 7. Heteractinellida
3. Calcarea	

Sponges vary greatly in form, size, and complexity of structure. A simple type is similar to a vase or hollow sac, fixed by the lower end, and with an opening or *osculum* at the upper extremity. The wall of such a sponge is thin, and perforated by a large number of pores through which water flows into the central or *gastral cavity* and passes out by the osculum; by this means the sponge is provided with food and oxygen and gets rid of waste matters. The wall of the sponge consists of two layers—an outer or *dermal* and an inner or *gastral*; the dermal (fig. 7, 2) is formed of a surface layer of flattened cells, with a gelatinous layer beneath containing various cells, some of which secrete the elements of the skeleton. The gastral layer (fig. 7, 3) consists of a single layer of cells, each cell being provided with a collar-like projection, in the centre of which is a long flagellum; the circulation of water through the sponge is produced by the movements of these flagella.

A simple form like that just described is found in the young stages of many sponges which afterwards, in their adult condition, are much more complex. Owing to the growth of the sponge-wall being unequal in different parts, either folds

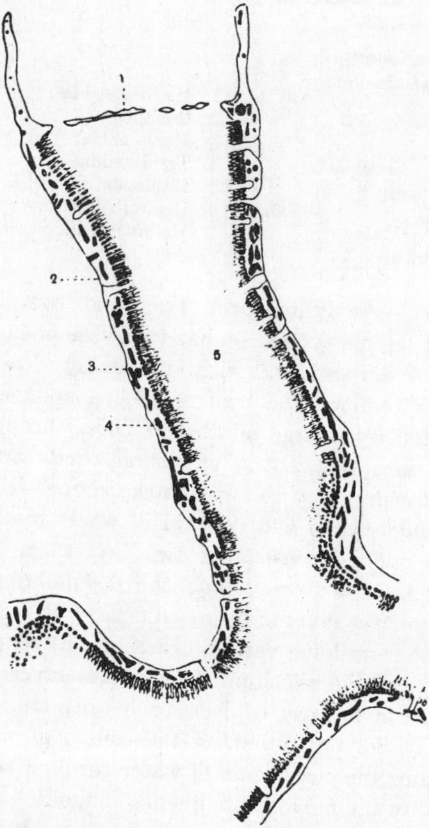

Fig. 7. Vertical section through *Leucosolenia*, a calcareous sponge. Highly magnified. (From Minchin.) 1, sieve-like membrane covering the osculum; 2, outer layer; 3, collar or flagellated cells (the pointer should have been continued to indicate the cells lining 5); 4, spicules: 5, gastral cavity.

or tube-like projections are formed, and these subsequently become more or less completely fused, so that the wall is much thickened (fig. 8) and is traversed by canals which are really spaces enclosed between the folds and outgrowths. In such forms the flagellated cells (choanocytes) are confined

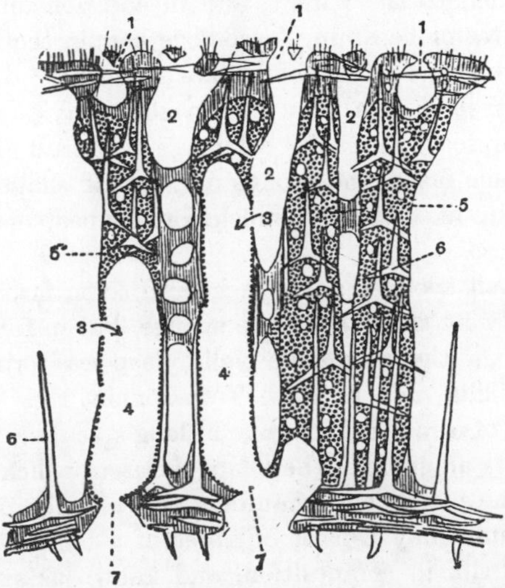

Fig. 8. Section of a portion of *Grantia*, a calcareous sponge. Highly magnified. (From Dendy.) 1, openings of inhalent canals; 2, inhalent canal; 3, openings of inhalent canals into flagellated chamber; 4, flagellated chamber; 5, collar cells; 6, spicules; 7, exhalent opening of flagellated chamber.

to chambers in the sponge-wall (fig. 8, 4). Canals, called *incurrent* or *inhalent* canals (2), pass from the surface of the sponge to these chambers, and others, the *excurrent* or *exhalent* canals, may lead from the chambers and open into the gastral cavity. Further complications, such as branching of the canals, may occur. The thick wall of these more complex sponges is formed mainly of the gelatinous layer.

In a sponge consisting of a single individual, the form depends mainly on the relative rates of growth in different directions, and may be cylindrical, vase-like, globular, discoidal, etc. In a compound sponge the form depends also on the way in which the young individuals of the colony are attached to the parent, and in addition, on their remaining free or becoming fused together; in the latter case the individuals of the colony are frequently distinguishable by their oscula only; when the individuals remain free, arborescent or bushy colonies may result (fig. 12); if they become fused, the sponge may be fan-shaped, funnel-shaped, cup-like, tubular, mushroom-shaped, massive, encrusting, etc.

Nearly all sponges are attached to some foreign object —generally by the base of the sponge, but in forms which are fixed in the mud, especially deep-sea forms of the Hexactinellida, and in some Tetraxonida, this fixation is by means of a root-tuft or rope of long spicules.

In nearly all sponges there is a skeleton, which serves to support the canals and chambers and also for protection. This skeleton may consist of fibres of a horny substance, similar to silk in composition, and known as *spongin*; or of mineral particles, termed *spicules* (fig. 8, 6), composed of carbonate of lime or of colloid silica; or it may consist of both siliceous spicules and spongin. Those forms only which have either a siliceous or calcareous skeleton are definitely known as fossils. Each spicule consists of a number of rays or arms, coming off from a centre, which is the point where the formation of the spicule commenced. In some groups, as for instance in the Monaxonida and Tetraxonida, the spicules are not united or are joined by spongin only; but in others they are fused together or interlocked so as to form a complete scaffolding, and generally it is in these only that the external form of the sponge has been preserved in the fossil state. In most siliceous sponges, two kinds of spicules may be distinguished, the *skeletal-spicules* or *megascleres* which build the main part of the skeleton, and the *flesh-spicules* or *microscleres* which are smaller and isolated and are seldom preserved as fossils. In the axis of each spicule there is a canal known as the *axial canal* (fig. 9, *c*), which in the living sponge is occupied by a thread of organic matter; this is the first part of the spicule to be formed, the mineral matter being subsequently deposited around it.

The spicules of recent siliceous sponges are characterised by the glassy appearance of their surface, and by the silica being colloidal, isotropic, and soluble in heated caustic potash. But in the fossil state the spicules have generally undergone considerable change; occasionally their silica is still colloidal but the surface has no longer the glassy appearance, and the axial canal is frequently filled with secondary silica in a crystalline or crypto-crystalline condition, and is consequently easily distinguished by the aid of polarised light when the spicule itself still remains colloidal. Generally, however, the spicule has become crystalline or crypto-crystalline, and in such cases the axial canal can rarely be detected since it is filled with material in the same condition. Sometimes the silica of the spicules has been entirely removed, a hollow cast only remaining; in other cases it is replaced by another mineral, as for instance by calcite in the sponges from the Lower Chalk of Folkestone, by iron pyrites in *Protospongia* from the Menevian Beds of St David's, by iron peroxide in the sponges of the Upper Chalk of the south of England, and by glauconite in some from the Upper Greensand. Obviously then, the colloidal silica of recent sponges is anything but a stable substance, thus differing widely from crystalline and crypto-crystalline silica.

The spicules of the calcareous sponges are usually smaller than those of the siliceous forms, and their material is not in an isotropic state, but each spicule possesses the optical characters of a crystal of calcite; consequently in polarised light these spicules are readily distinguished from unaltered siliceous forms, which appear dark between crossed Nicol's prisms. Then again the fossil calcareous spicules have undergone much less chemical change than the siliceous ones; generally they are still composed of carbonate of lime, for it is only in rare cases that this is replaced by silica. The external form of the individual calcareous spicules is, however, often less well preserved than in the case of siliceous spicules.

The forms of sponge spicules, both megascleres and microscleres, are very varied, but they can be shown to be modifications of a small number of types or fundamental forms. The spicules, on account of the constancy of their characters, are of great importance in the classification of sponges.

The canal-system is indicated in the skeleton of both recent and fossil forms by spaces in the framework of spicules or spongin, but these spaces represent only the larger canals, the smaller existing in the soft parts alone.

Reproduction in the sponges takes place by budding and by the production of ova and spermatozoa.

The sponges may be divided into three classes: (1) Hexactinellida, (2) Demospongiæ, (3) Calcarea. The first and second are sometimes grouped together as the Non-Calcarea.

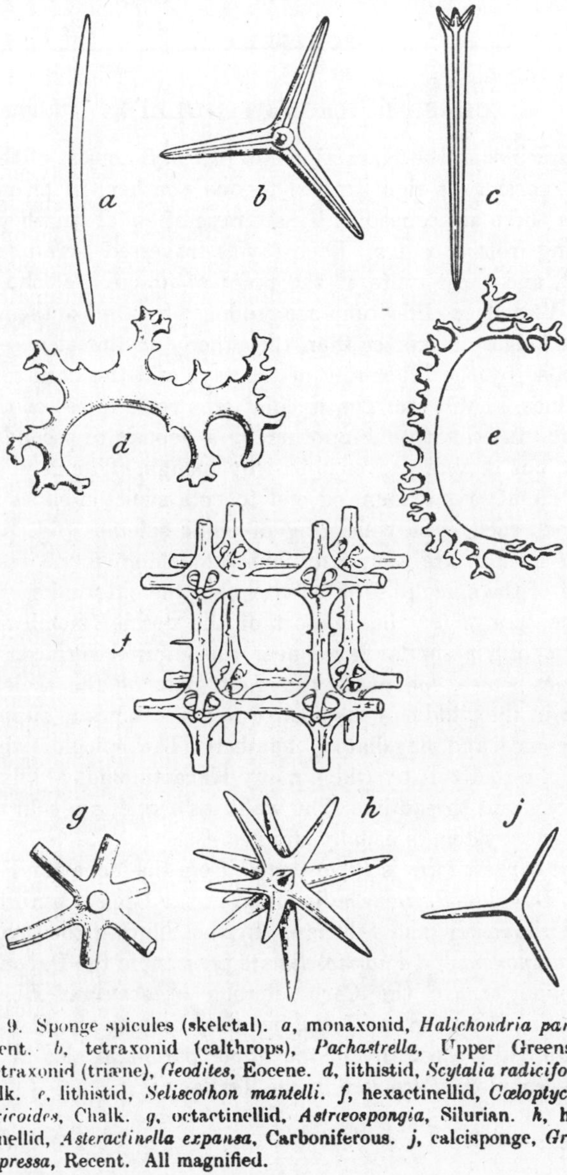

Fig. 9. Sponge spicules (skeletal). *a*, monaxonid, *Halichondria panicea*, Recent. *b*, tetraxonid (calthrops), *Pachastrella*, Upper Greensand. *c*, tetraxonid (triæne), *Geodites*, Eocene. *d*, lithistid, *Scytalia radiciformis*, Chalk. *e*, lithistid, *Seliscothon mantelli*. *f*, hexactinellid, *Cœloptychium agaricoides*, Chalk. *g*, octactinellid, *Astræospongia*, Silurian. *h*, heteractinellid, *Asteractinella expansa*, Carboniferous. *j*, calcisponge, *Grantia compressa*, Recent. All magnified.

CLASS I. HEXACTINELLIDA

The spicules in the Hexactinellida (fig. 9, *f*) consist of three axes crossing at right angles to one another; in primary forms there are consequently six rays of equal length proceeding from a centre. Each ray is traversed by an axial canal, and these unite at the point of junction of the six rays. Various modifications are produced by some of the rays being longer or shorter than the others, or almost absent; and also by the branching of the rays and the occurrence of spines, knobs, etc. The spicules may remain free or they may be fused with one another by a deposit of secondary silica, but they are never united by spongin. When spicules with equal rays are united end to end, skeleton-cubes are formed, each cube consisting of eight spicules (fig. 9, *f*). Flesh-spicules are abundant, but are seldom found fossil. Some of the spicules form a layer near the external surface of the sponge for the support of the dermal membrane; others form a similar layer near the internal surface; the spicules which constitute the main part of the skeleton occur in the middle of the sponge-wall and serve to support the canals and flagellated chambers. The spicules which form the root-tuft by which many Hexactinellids are fixed, are long and thread-like. The walls, as a rule, are thin and the canal-system is usually simple.

The earliest form is *Protospongia* from the Menevian Beds of St David's; *Pyritonema* is found in the Upper Cambrian, the Ordovician, and Silurian. In the Silurian the genera *Astræospongia* and *Phormosella* are present; in the Devonian *Dictyophyton*; in the Carboniferous *Hyalostelia*. Hexactinellids are rare in the Trias, but they become abundant in the Jurassic, especially in the upper part, and also in the Cretaceous; they are rare in the Tertiary.

Protospongia (fig. 10). Form unknown but probably cup-shaped. Spicules cruciform owing to the reduction of one axis and arranged in a quadrate manner, the larger forming a framework, which contains the smaller spicules of two or three sizes, arranged in the same regular way, so that the larger squares enclose four or five series of smaller ones. The spicules were either free or probably partly fused together. Menevian Beds and Lingula Flags. Ex. *P. fenestrata*.

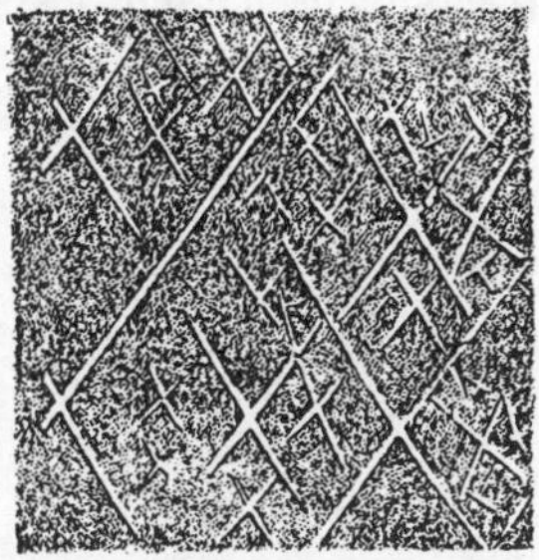

Fig. 10. *Protospongia fenestrata*, Menevian Beds, St David's. ×3¾. (After Hinde.) The original is in the British Museum. Owing to the cleavage of the rock the angles of the spicules are distorted.

Craticularia. Cup-shaped, funnel-shaped, or cylindrical; simple or branching. On both the inner and outer surfaces of the wall are circular or oval canal-openings, which are arranged in vertical and transverse rows crossing each other at right angles. Canals straight, terminating blindly. Inferior Oolite to Upper Chalk (perhaps also Miocene). Ex. *C. fittoni*, Chalk.

Ventriculites. Simple; form variable, but usually cup-shaped, funnel-shaped, or cylindrical. Central cavity large and deep. Walls folded so as to form a series of vertical grooves and ridges; the grooves are divided by transverse extensions of the wall into oval or elongate openings. Canal system well developed; the radial canals are large and start from the central cavity, but end before reaching the outer surface; others start from the outer surface and end before reaching the central cavity. Spicules six-rayed and fused with one another so as to form a mesh-work. The node where the axes cross is hollow, having the form of a negative octahedron, the central part of each face of which is absent; the axial canals cross in the centre of the octahedral space. The sponge was provided with a root consisting of siliceous fibres. Chalk. Ex. *V. radiatus*, *V. impressus*.

Plocoscyphia. Sponge formed of thin-walled tubes and laminæ which anastomose, forming an irregular or rounded mass. Small, close-set openings of canals on the outer surface. Canal system imperfect. Upper Cretaceous. Ex. *P. fenestrata*, Upper Greensand and Chalk Marl.

CLASS II. DEMOSPONGIÆ

The skeleton consists of siliceous spicules, or of spongin, or of both spicules and spongin. In some forms there is little or no spongin, but in others the entire skeleton consists of spongin with no siliceous spicules; between these extremes there is a complete passage. The spicules are never of the hexactinellid type. In some few cases both spicules and spongin are absent.

ORDER 1. MYXOSPONGIDA. Sponges with no skeleton or occasionally with a few isolated spicules. Not known in the fossil state.

ORDER 2. CERATOSA. Sponges with a skeleton composed of a fibrous network of spongin. This Order includes the ordinary bath sponges, etc., and is unknown in the fossil state.

ORDER 3. MONAXONIDA. The skeleton is formed of spongin and spicules in varying proportions. The spicules (fig. 9, *a*) consist of a single rod or axis, which may be straight or curved, and with sharp or blunt ends; each spicule may consist of two rays or of one ray only. In the former the two ends of the spicule are alike and there is a small swelling of the axial canal at the centre of the spicule where growth commenced; in the latter the two ends are dissimilar and the swelling in the axial canal is at one end of the spicule, and growth went on in one direction only. Microscleres or flesh-spicules may also occur but are often absent. Since in this Order the spicules are only united by spongin or other decomposable material, it is extremely rare to find the form of the sponge preserved fossil; usually, detached spicules only occur.

The earliest representatives of the Monaxonida are found in the Middle Cambrian; the Order becomes rather more abundant in the Carboniferous, where the genus *Reniera* occurs. The freshwater form *Spongilla* is found in the Purbeck Beds of the south of England. In the Cretaceous and Tertiary monaxon spicules are sometimes abundant. A large number of Monaxonid sponges are still living.

ORDER 4. TETRAXONIDA. The spicules (fig. 9, *b*, *c*) consist of four rays given off from a common centre, the angle between the rays, when the end of one is taken as a central point, appearing to be 120°. The rays may be equal in length, when the spicule is termed a *calthrops* (fig. 9, *b*), or unequal; frequently one is very much elongated (fig. 9, *c*), and in such forms, known as *triænes*, the three shorter rays are placed near the surface of the sponge-wall and the longer ray is directed inwards. Sometimes the terminations of the rays are bifurcated. Spongin is either absent or occurs in minute quantities only, and since the spicules are not united, the Tetraxonids, like the Monaxonids, are seldom preserved in anything like a perfect condition as fossils. The oldest forms occur in the Carboniferous Limestone, where they are represented by the genera *Geodites* and *Pachastrella*; others are found in the Infra-Lias, the Cretaceous and Tertiary formations.

ORDER 5. LITHISTIDA. The Lithistids have thick stony walls and very variable external form. The spicules (fig. 9, *d*, *e*) are stout and irregular in form, but sometimes show four rays; the extremities branch or expand, and by that means the spicules become firmly interlocked with one another, but do not fuse together. These irregular spicules (sometimes termed *desmas*) are formed by secondary silica being deposited on small spicules of the ordinary kind, which may be four-rayed or consist of a single axis. In addition to these irregular spicules there is generally a surface layer or cortex formed of triæne spicules like those in the Tetraxonids. Flesh-spicules are also present. Several different types of canal system occur. The Lithistids are closely allied to the Tetraxonida, and are sometimes regarded as a division of that Order.

Owing to their solidity the Lithistids are preserved abundantly as fossils. They are rare in the Palæozoic; a few are found in the Upper Cambrian of Canada; in the Ordovician and Silurian *Astylospongia* occurs; in the Carboniferous *Doryderma*, etc. Forms belonging to this Order have been found in the Permian of Timor and Sicily; they are numerous in the Jurassic, attain their maximum in the Upper Cretaceous, and are scarce in the Tertiary.

Verruculina. Irregular, fan- or funnel-shaped, attached by a short stalk. Oscula placed on prominent elevations on the upper, and sometimes also on the under surface. Spicules small, interlacing and forming a fibrous network. Upper Cretaceous. Ex. *V. macromammata* (=*reussi*), Upper Chalk.

Pachinion. Pear-, fig- or club-shaped, sometimes cylindrical, tapering at its lower part to a short stem. Central cavity large and deep, with vertical canals opening into its base. Wall formed of anastomosing fibres, between which are irregular spaces—there are no distinct canals; fibres formed of large spicules, branched and interlaced. There is also a surface layer composed of small spicules. Chalk. Ex. *P. scriptum*, Upper Chalk.

Scytalia. Simple, or formed of two or more individuals growing close together; cylindrical or club-shaped, with a thick wall and a cylindrical stem. Central cavity tube-like, long, continued at its base by several vertical canals; numerous radial canals open into the central cavity and taper towards the external surface. Spicules branching, with root-like prolongations. Chalk. Ex. *S. radiciformis*.

Seliscothon. Mushroom-like, consisting of a flat or concave, circular, plate-like body, and a rounded tapering stem. The circular body has rounded or oblique edges, and numerous, small, rounded oscula on the upper surface; it is formed of fine vertical radiating lamellæ, separated by spaces crossed by fibres—these spaces forming the canal system. Spicules fine, branching irregularly, with bifurcating extremities, and covered with tubercles or spines. Chalk. Ex. *S. planum.*

Doryderma. Cylindrical, pear-shaped, sometimes branching. There are parallel vertical canals opening at the summit of the sponge, and smaller radial canals extending from the surface towards the centre. Spicules large, of various forms; also a surface layer formed of slender trifid spicules. Carboniferous and Cretaceous. Ex. *D. benetti*, Upper Greensand.

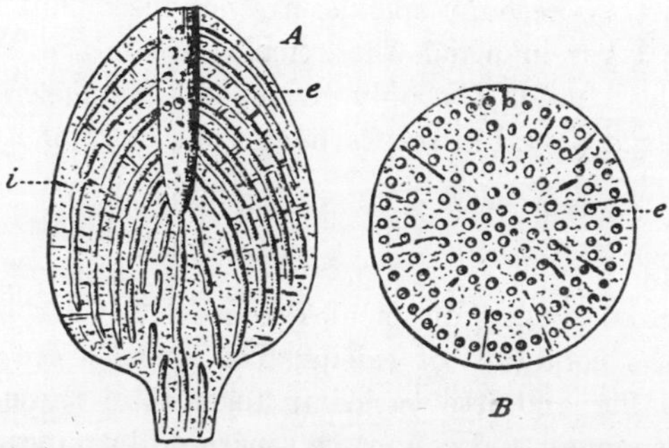

Fig. 11. *Siphonia tulipa*, Upper Greensand, Warminster. A, vertical section. B, horizontal section. *e*, excurrent canals; *i*, incurrent canals. × ⅔.

Siphonia (fig. 11). Pear-, apple- or fig-shaped, provided with a long or short stalk, which is given off from the broad end of the body and terminates in rootlets. The incurrent canals are small, slightly curved, and extend radially from the centre of the sponge to the surface. The excurrent canals are larger, and are arranged parallel with the surface of the sponge, extending from the base to the summit, where they open into the deep central cavity by means of a series of parallel ostia. The skeletal-spicules possess four rays with bifurcated and expanded extremities, by means of which they are interlocked. Upper Greensand to Upper Chalk. Ex. *S. tulipa*, Upper Greensand.

Hallirhoa. Like *Siphonia* but with the sides divided into lobes. Upper Greensand. Ex. *H. costata.*

ORDER 6. OCTACTINELLIDA. The spicules (fig. 9, *g*) consist of eight rays, six of which are in one plane diverging at equal angles, while the other two are at right angles to this plane, forming a vertical axis. Frequently, however, the vertical axis is only slightly developed or altogether absent. The spicules are not united. The only genus is *Astræospongia*, found in the Silurian and Devonian.

ORDER 7. HETERACTINELLIDA. The spicules are unusually large (fig. 9, *h*), the number of rays varying from six to thirty. The body spicules are not fused, but there is a surface layer in which the spicules are interwoven and more or less fused. The only genera are *Tholiasterella* and *Asteractinella*, found in the Carboniferous rocks of Ayrshire.

CLASS III. CALCAREA

The skeleton consists of spicules composed of carbonate of lime in the condition of calcite. The spicules are usually much smaller and less varied in form than those of the siliceous sponges, and cannot be separated into megascleres and microscleres. There are three kinds, the simple uniaxial, the three-rayed—with the rays in one plane (fig. 9, *j*), and the four-rayed; they are sometimes fused with one another, but often are either arranged close together so as to form fibres, or are loosely distributed. Spongin is never present. There are three grades of structure in the Calcarea. In simple forms, known as the Ascon type, the flagellate cells (choanocytes) line the gastral cavity (fig. 7). In others, in which the sponge wall is folded, the flagellate cells are limited to the flagellate chambers; in the Sycon type these chambers open directly to the gastral cavity (fig. 8), but in others (the Leucon type) excurrent canals connect the chambers with the gastral cavity. The Ascon type of sponge is not definitely known to occur in the fossil state. A group of sponges known as the Pharetronids appears to belong to the Leuconids. In this group the spicules are arranged to form solid anastomosing fibres, and some of the spicules have a characteristic tuning-fork shape. The Pharetronids are found mainly in the Mesozoic, and the genera described below belong to this group with the exception of *Barroisia* which is of the Sycon type.

A few examples of the Calcarea have been recorded from the Ordovician, Silurian and Carboniferous, and several forms occur in the Permian of Sicily. The group becomes important in the Trias, and is well represented in the Jurassic and Cretaceous.

Peronidella. Cylindrical, simple or branched; central cavity tubular and extending from the summit to the base of the sponge. Walls thick and with no definite canals, but having irregular spaces between the spicular fibres. Spicules three- or four-rayed, forming anastomosing fibres. Carboniferous (possibly also Devonian) to Cretaceous; most abundant in the Jurassic and Cretaceous. Ex. *P. pistilliformis*, Great Oolite and Cornbrash.

Corynella. Form similar to *Peronidella.* Radial excurrent canals open into the central cavity, which often does not extend to the base of the sponge, but is continued downwards by vertical canals. Incurrent canals fine, directed obliquely downward. Osculum usually with radial furrows. Jurassic and Cretaceous (? Trias). Ex. *C. foraminosa*, Lower Greensand.

Holcospongia. Simple or compound; individuals usually spherical, hemispherical, or club-shaped; their summits rounded, with a central area in which a number of excurrent canals open, and from which furrows extend down the sides of the sponge. Spicules large and three-rayed, and some also filiform; and a surface layer of three-rayed spicules, of various sizes, felted together. Inferior Oolite to Cretaceous. Ex. *H. polita*, Corallian.

Rhaphidonema Cup- or funnel-shaped or leaf-like, usually with definite canals. Oscula on either the inner or the outer surface. Spicules of three rays, one of which is but slightly developed. On one (or sometimes both) surfaces is a thin, compact or finely porous layer of spicules. Trias to Cretaceous. Ex. *R. macropora*, Lower Greensand.

Porosphæra. Small simple sponges, commonly more or less spherical, but sometimes pear-, thimble-, or melon-shaped; often free, but sometimes attached to foreign bodies. Numerous, simple, straight, radiating canals open at the surface by minute apertures. Spicules with four rays, of which three are short and blunt and fused to the rays of adjoining spicules, whilst the fourth ray is longer and tapering. A surface layer (not often preserved) consists of a mixture of minute three- and four-rayed spicules and simple rods. Upper Cretaceous. Ex. *P. globularis*, Chalk.

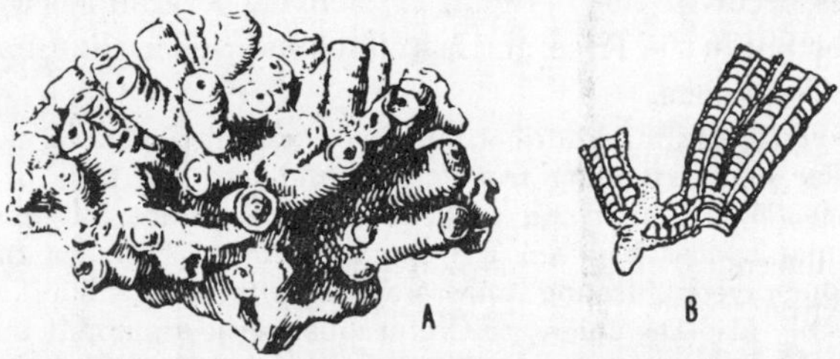

Fig. 12. *Barroisia anastomans*, Lower Greensand, Faringdon. A, colony. B, vertical section of three individuals of a colony. $\times \frac{3}{4}$.

Barroisia (fig. 12). Usually compound and bushy. Individuals cylindrical, each divided into a series of chambers by transverse partitions, which have a central circular opening, through which a tube usually passes. Canals simple, numerous, minute. Spicules slender, three-rayed; also a surface layer of larger spicules. Lower Greensand to Chalk. Ex. *B. anastomans*, Lower Greensand.

Distribution of the Porifera

The Sponges are all aquatic, and with the exception of the Monaxonid genus *Spongilla* and its allies, all marine. They are found in the seas of all parts of the world and are more numerous between the shore-line and 200 fathoms than at greater depths; many of the genera have a very wide distribution. All the Orders except the Octactinellida and the Heteractinellida have living representatives. The Monaxonids are abundant between the shore-line and 200 fathoms, and gradually decrease in numbers beyond that limit. The Tetraxonids are also common in water of less depth than 200 fathoms, but extend down to 2000 fathoms. The Lithistids range from 7½ to 1075 fathoms, and are most abundant between 100 and 150 fathoms. The Hexactinellids occur in deeper water than the Lithistids, being found down to a depth of 2900 fathoms; but they are abundant between 100 and 200 fathoms, and again between 300 and 700 fathoms. The Calcarea are mainly shallow water forms.

The fossil forms are comparatively rare in the Palæozoic rocks until we reach the Carboniferous; and throughout the geological formations they are much less abundant in argillaceous than in calcareous and arenaceous rocks. Sponges are first found in the Lower Cambrian rocks: the earliest British form is *Protospongia* from the Menevian Beds and Lingula Flags; in the Lingula Flags and Tremadoc the Hexactinellid genus *Pyritonema* occurs. In the Ordovician we have in the Llandeilo Beds the first appearance of *Ischadites*,[1] associated with *Pyritonema*: in the Bala Beds we meet with *Astylospongia*. The most abundant Silurian form is *Ischadites*; *Astræospongia*, *Phormosella*, and *Pyritonema* also occur. Sponges are rare in the Devonian, but *Astræospongia*, *Sphærospongia*,[1] and *Receptaculites*[1] have been recorded. In the Carboniferous rocks sponges become much more common, the siliceous spicules often forming

[1] The sponge-character of the Silurian and Devonian genera *Ischadites*, *Receptaculites*, and *Sphærospongia*, which have been placed by some authors in the Hexactinellida, is now disputed; if they are sponges it is probable that they belong to the Calcarea. *Archæocyathus*, Lower Cambrian, is probably a sponge.

thick beds of chert: the Monaxonids are represented by *Reniera*, the Tetraxonids by *Geodites*, the Lithistids by *Doryderma*, the Hexactinellids by *Hyalostelia* and *Erythrospongia*, and the Heteractinellids by *Tholiasterella* and *Asteractinella*. Sponges are usually absent in the Permian but Lithistids have been found in Timor, and also Calcarea in Sicily. They are rare in the Trias, except in the St Cassian Beds of the Tyrol, where the Calcarea are, for the first time, numerous.

In the Jurassic, sponges are extremely abundant; the only Monaxonid is *Spongilla* from the Purbeck Beds; Lithistids and Hexactinellids, although common in Germany and Switzerland, are comparatively rare in England; the first group is represented by *Platychonia*, the second by *Craticularia*, *Verrucocœlia*, etc.; the Calcarea are numerous in this country as well as in France and Germany, common genera being *Peronidella*, *Corynella*, and *Holcospongia*. The occurrence of Hexactinellids in the Inferior Oolite is noteworthy, since other evidence shows that that deposit was laid down in shallow water, but at the present day Hexactinellids are characteristic of deep water.

Sponges are more abundant in the Cretaceous than in any other system. In England they are found chiefly at four horizons: (1) in the Lower Greensand of Faringdon, Upware, Kent, and Surrey, where the Calcarea are much better represented than the other groups, *Peronidella*, *Barroisia*, and *Rhaphidonema* being common forms: (2) in the Upper Greensand and Chloritic Marl of Warminster, Blackdown, Haldon, and the Isle of Wight, where the Lithistids (*e.g. Doryderma*, *Siphonia*, *Hallirhoa*) are very abundant, exceeding the Hexactinellids (*e.g. Craticularia*, *Plocoscyphia*, *Stauronema*); the Calcarea are also common in places: (3) in the Lower Chalk of the south of England, where we find *Siphonia*, *Craticularia*, *Stauronema*, *Plocoscyphia*, etc.; the Calcarea are rare: (4) in the Upper Chalk, where the siliceous sponges are very common; amongst the Lithistids the following occur: *Seliscothon*, *Verruculina*, *Scytalia*, *Doryderma*, and *Siphonia*; the Hexactinellids are represented by *Craticularia*, *Verrucocœlia*, *Guettardia*, *Ventriculites*, *Cephalites*, *Plocoscyphia*, and *Camerospongia*; the Calcarea are represented by *Porosphæra*. Lithistids and Hexactinellids are present in the Miocene of Algeria, but generally in the Tertiary formations, although detached spicules are sometimes abundant, few perfect sponges are found.

PHYLUM CŒLENTERA

Classes	*Orders*
1. Hydrozoa...	1. Gymnoblastea 2. Calyptoblastea 3. Graptolithina 4. Hydrocorallina 5. Stromatoporoidea 6. Trachomedusæ (not fossil) 7. Narcomedusæ (not fossil) 8. Siphonophora (not fossil)
2. Scyphozoa	1. Stauromedusæ (not fossil) 2. Peromedusæ (not fossil) 3. Cubomedusæ (not fossil) 4. Discomedusæ
3. Anthozoa or Actinozoa ...	1. Zoantharia 2. Alcyonaria
4. Ctenophora (not fossil)	

The Cœlentera include hydroids, jelly-fishes, sea-anemones, corals, and allied forms. The individuals are radially symmetrical, and have only one internal cavity, the *cœlenteron*, which opens to the exterior by the mouth. The body-wall consists of an outer layer of cells, the *ectoderm*, and an inner layer, the *endoderm*; between these is a gelatinous layer (*mesoglœa*)—usually quite thin, but in the jelly-fishes of considerable thickness. Stinging cells known as nematocysts or thread-cells are generally present in the ectoderm.

This Phylum is divided into four classes, (1) Hydrozoa, (2) Scyphozoa, (3) Anthozoa or Actinozoa, (4) Ctenophora.

CLASS I. HYDROZOA

The simplest type of the Hydrozoa is the common freshwater *Hydra*. In this the body has the form of an elongated sac, about a quarter of an inch in length, and is attached by one end, whilst at the other is the mouth surrounded by a row of long processes, called *tentacles*. The large undivided cavity in this sac, which opens into the hollow tentacles above, is the cœlenteron. The whole body is very contractile and constantly changing its shape. Reproduction may take place in three ways, (1) by the growth of buds, which ultimately separate from the parent, (2) sexually, by the production of ova and spermatozoa in the ectoderm, and (3), in rare cases, by fission.

Other Hydrozoa consist of a number of individuals (polyps or hydranths) similar to *Hydra*, but growing together as a colony (fig. 13); in such cases the cœlentera of all the individuals are placed in living communication with one another by means of a tube-like extension from the base of each polyp; this common connecting portion of the colony is called the *cœnosarc* (fig. 13, 5). Frequently the cœnosarc is much branched, giving rise to tree-like forms; it is usually attached to some foreign object by a horizontal branching portion.

In such hydroid colonies the polyps are asexual, and the reproductive elements are produced in another individual of a somewhat different character, known as a *medusa* or *gonophore*: this arises by budding from the hydroid (fig. 13, 9), and is often more or less bell-shaped, and may become detached from the colony, or it may be less perfectly developed and remain attached; at the inner edge of the bell is a shelf-like fold, the *velum*. The generative cells are of ectodermal origin, and from them the hydroid develops.

Hydra possesses no hard parts, but in other forms an external skeleton composed of chitin or of carbonate of lime is secreted; it commonly forms a tube-like sheath around the cœnosarc and is called the *perisarc* (fig. 13, 6). In one group the perisarc is produced at the base of each polyp into a cup-like structure or *hydrotheca* (fig. 13, 7), into which the polyp can retract. The gonophores may also be protected by a chitinous capsule called the *gonotheca* or *gonangium* (fig. 13, 10) The vertical branching part of the cœnosarc, together with the perisarc around it, is called the *hydrocaulus*; the horizontal root-like portion and its perisarc form the *hydrorhiza*.

The principal characters which distinguish the Hydrozoa from the other Cœlenterates are: the cœlenteron being undivided by radial partitions or ridges; the absence of a digestive tract projecting into the cœlenteron; the usual occurrence of an asexual (hydroid) generation alternating with a sexual (medusoid) generation; the medusa having a velum; the ova and spermatozoa being derived from the ectoderm.

Nearly all the Hydrozoa are marine. They are divided into eight Orders, of which five occur fossil: (1) Gymnoblastea, (2) Calyptoblastea, (3) Graptolithina, (4) Hydrocorallina, (5) Stromatoporoidea.

ORDER I. GYMNOBLASTEA

The Gymnoblastea have no hydrothecæ into which the polyps can retract; gonangia (gonothecæ) are also absent.

Well-known living forms are *Tubularia*, *Bougainvillea*, and *Hydractinia*. The last has been found fossil in Eocene and later deposits; it forms a crust over the shells of gasteropods, especially those tenanted by Hermit-crabs. The hard part of this crust is chitinous, or rarely calcareous, and consists of laminæ separated by irregular or cubical spaces and crossed by vertical pillars; on the surface are projecting spines. The soft parts form a layer over this skeleton, and consist of a sheet of ectoderm on the surface, and another sheet next the skeleton: between these are branching and anastomosing cœnosarcal tubes. The skeleton is secreted by the lower ectoderm. From the cœnosarc arise the polyps, which are placed on long vertical stalks and are of four kinds, (1) gastrozooids—the ordinary nutritive individuals, (2) blastostyles, which are individuals specially modified for bearing medusæ, (3) dactylozooids—individuals modified for catching prey and having short knob-like tentacles crowded with nematocysts, (4) tentacular polyps, which are very slender, without a mouth, and occur near the edge of the colony.

Parkeria, which is found in the Cambridge Greensand, probably belongs to this Order. A few other forms have been described from the Alpine Trias and the Jurassic of southern Europe.

ORDER II. CALYPTOBLASTEA

This Order is distinguished by the presence of hydrothecæ into which the polyps can completely retract, and by the possession of gonangia (gonothecæ). (Fig. 13, 7, 10.)

The arrangement of the polyps and hydrothecæ on the hydrocaulus varies considerably in different genera. Sometimes they are placed on stalks as in *Obelia* (fig. 13) and *Campanularia*; in many others they are sessile. They may be in rows or placed in various positions on the hydrocaulus. In *Plumularia*, *Aglaophenia*, etc., they form a single row; in *Sertularia*, etc., there are two rows placed on opposite sides of the branches. Sometimes the hydrothecæ are close together, but more usually they are separated.

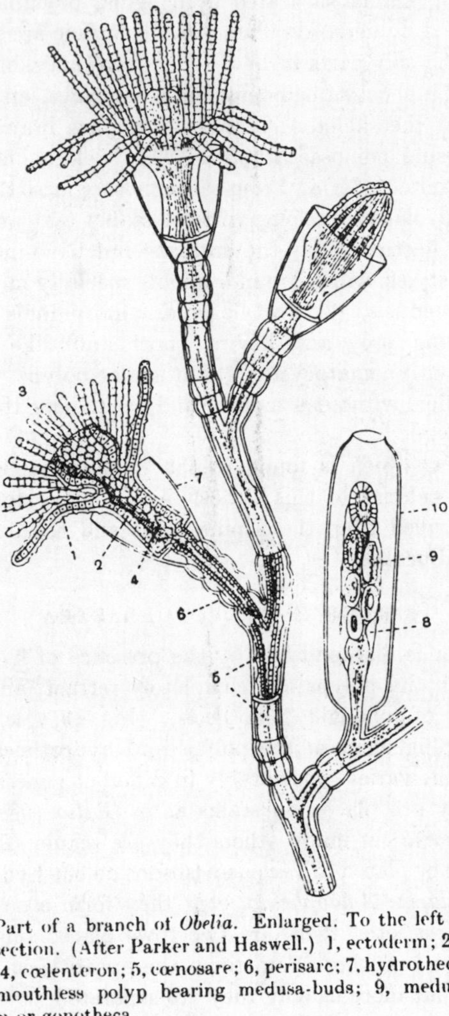

Fig. 13. Part of a branch of *Obelia*. Enlarged. To the left a portion is shown in section. (After Parker and Haswell.) 1, ectoderm; 2, endoderm; 3, mouth; 4, cœlenteron; 5, cœnosarc; 6, perisarc; 7, hydrotheca; 8, blastostyle, a mouthless polyp bearing medusa-buds; 9, medusa-bud; 10, gonangium or gonotheca.

In the Plumulariidæ there are, in addition to the ordinary polyps, others which are solid and tentacle-like; they are usually provided with nematocysts and are called *nematophores*; each one is placed in a hydrotheca.

Although the Calyptoblastea possess a well-developed chitinous skeleton they are rarely found fossil. *Archæolafœa* and *Archæocryptolaria* from the Cambrian of Victoria, and *Mastigograptus* from the Ordovician of New York appear to belong to this group. One form has been found in the Pleistocene.

ORDER III. GRAPTOLITHINA

SUB-ORDER 1. *GRAPTOLITOIDEA*

The graptolites are found only in the Lower Palæozoic rocks, where, owing to their abundance and to the limited range in time of both genera and species, combined with their wide geographical distribution, they are of great importance to the stratigraphical geologist. They occur most commonly in argillaceous rocks, especially in black carbonaceous shales, whilst they are relatively rare in sandstones and limestones. The graptolites were compound animals, and the soft parts were protected by a skeleton of chitin which shows a general resemblance to that found in the Calyptoblastea, *e.g. Sertularia* and *Plumularia.* But the original material of the skeleton is seldom preserved unaltered; in some cases it has been replaced by iron pyrites, but usually it has become carbonised.

The entire skeleton of the graptolite is termed the *rhabdosome* or *polypary*; this in an unbranched form like *Monograptus* consists of a row of small cells known as *hydrothecæ* (fig. 14, *hy*). Each hydrotheca is hollow and opens on the one hand into an internal space (the *common canal*, fig. 14 *cc*) and on the other to the exterior (*ap*); the latter aperture known as the *mouth* of the hydrotheca, is frequently circular, but sometimes quadrangular, or more or less constricted by apertural outgrowths. Embedded in the wall or *periderm* on the side opposite to the row of hydrothecæ is a chitinous thread or rod, termed the *virgula* (fig. 14, *v*). In some species of *Monograptus* the virgula projects beyond the distal[1] end

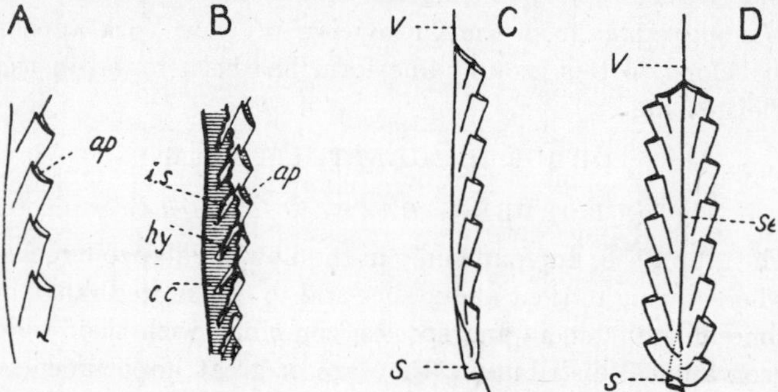

Fig. 14. Enlarged figures to show the structure of graptolites. A, B, C, *Monograptus*. D, *Diplograptus*. *ap*, aperture of hydrotheca; *cc*, common canal; *hy*, hydrotheca; *is*, interthecal septum; *s*, sicula; *se*, median septum; *v*, virgula.

of the common canal. At the proximal end of the rhabdosome there is a small conical body, termed the *sicula* (fig. 14, *s*), which will be described more fully below (p. 64).

The soft parts of the graptolites are unknown, but from comparison with living hydroids, which have a similar skeleton, we may consider it probable that each hydrotheca lodged an individual polyp, and that these were connected by means of the cœnosarc in the common canal.

[1] The *proximal* end is that next the sicula and is the part which is formed first; the distal end is furthest from the sicula and is formed last. The side of the graptolite on which the hydrothecæ occur is spoken of as the *ventral*, and the opposite side as the *dorsal*.

In the form just described (*Monograptus*) the rhabdosome is always simple, but in many genera it consists of two or more branches or stipes. When there are several radiating branches their proximal parts are sometimes enclosed in a horny sheath, termed the *central disc*, as in some species of *Tetragraptus* (fig. 15) and *Dichograptus*. In those genera

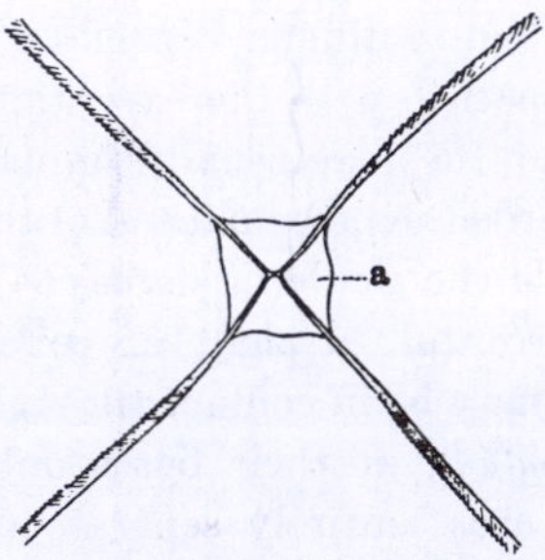

Fig. 15.

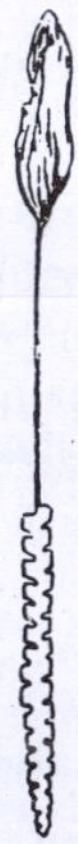

Fig. 16.

Fig. 15. *Tetragraptus headi*, Arenig Rocks. *a*, central disc. × $\frac{1}{5}$.

Fig. 16. *Climacograptus parvus*, Ordovician. With vesicle at end of virgula. (After Ruedemann.) × $\frac{1}{4}$.

which have two branches (fig. 19), the angle between the two is termed the *angle of divergence*; it is measured from the hydrothecal side of each branch. In some graptolites (*e.g. Monograptus*, fig. 14 A—C) there is only a single row of hydrothecæ, such forms are said to be *uniserial*; others (*e.g. Diplograptus*, fig. 14 D) possess two rows on opposite sides of the rhabdosome—these are the *biserial* forms, and they may have a single common canal as in *Retiolites*, or there may be two canals separated by a septum (fig. 14 D, *se*), as in *Climacograptus*: in many forms of *Diplograptus* there is only one common canal, but others possess an incomplete septum which, to some extent, divides the canal into two parts. In *Dicranograptus* the proximal part of the rhabdosome is biserial, whilst the distal part consists of two uniserial branches. In *Dimorphograptus*, on the other hand, the proximal part is uniserial and the distal part biserial; this genus therefore serves to connect the biserial *Diplograptus* with the uniserial *Monograptus*.

The hydrothecæ vary considerably in form in different genera, and sometimes even in different species of the same genus; but in any one species, with the exception of a few of the earlier hydrothecæ, they are usually similar in form, but diminish in size towards the proximal end of the rhabdosome; they may resemble the sicula in shape, or they may be tubular, prismatic, conical, straight or coiled (*Monograptus lobiferus*). They may be in contact throughout their entire length (*Phyllograptus*), at their bases only (*Nemagraptus*), or, in a few cases, entirely separate (*Rastrites*). Frequently they are provided with one or more spines near the mouth. In most graptolites the hydrothecæ communicate freely with the common canal, and in this respect differ from living hydroids, in which there is a constriction or an imperfect diaphragm at the base of each hydrotheca, separating it from the common canal (fig. 13).

A microscopic examination of thin sections of *Monograptus* shows that the periderm consists of three layers, the external and internal layers being much thinner than the middle layer. In a few graptolites the middle layer of the periderm is more or less extensively perforated and may become reticulate: this modification is especially noteworthy in *Retiolites*, in which the middle layer is reduced to a network of fibres (fig. 17), whilst the inner and outer layers are very thin and often not preserved.

In *Monograptus* the virgula, when present, is found in the periderm opposite the row of hydrothecæ, but in the biserial forms it is central (fig. 14 D), being situated either in the middle of the common canal, as in some forms of *Diplograptus* and *Climacograptus*, or in the septum separating the two canals, as in other forms of these genera; in such biserial forms the virgula is commonly enclosed in a tube-like covering. In the earlier uniserial genera (*Didymograptus*, *Tetragraptus*, *Dichograptus*, *Leptograptus*, *Dicellograptus*, etc.) the virgula is not found in the wall of the common canal, but is represented by a thread or *nema* which projects from the pointed end of the sicula and commonly ends in a chitinous disc by which the graptolite was attached (as in fig. 24 A).

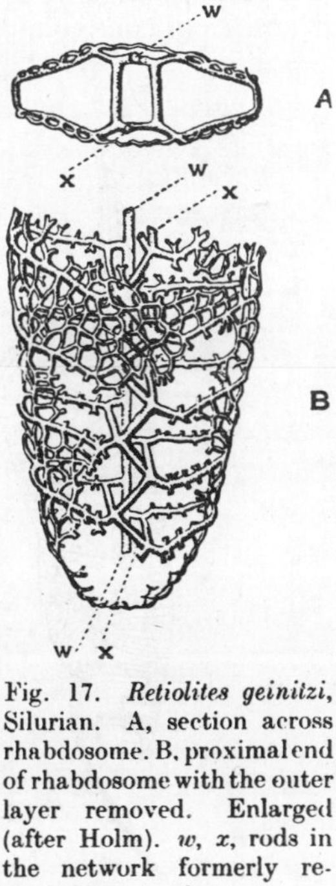

Fig. 17. *Retiolites geinitzi*, Silurian. A, section across rhabdosome. B, proximal end of rhabdosome with the outer layer removed. Enlarged (after Holm). *w*, *x*, rods in the network formerly regarded as virgulæ.

The position of the sicula, in relation to the rest of the rhabdosome, varies in different genera, and depends on the directions in which the hydrothecæ grow from it. In *Monograptus* the sicula is situated on the dorsal surface of the rhabdosome, the pointed end being directed distally (fig. 14 C, *s*). In *Diplograptus* it has a similar position but is more or less completely enclosed between the hydrothecæ (fig. 14

D, *s*). In *Didymograptus* the two branches of the rhabdosome diverge from the broad end, the pointed end being directed proximally (fig. 19 A—C). In *Dicellograptus* the

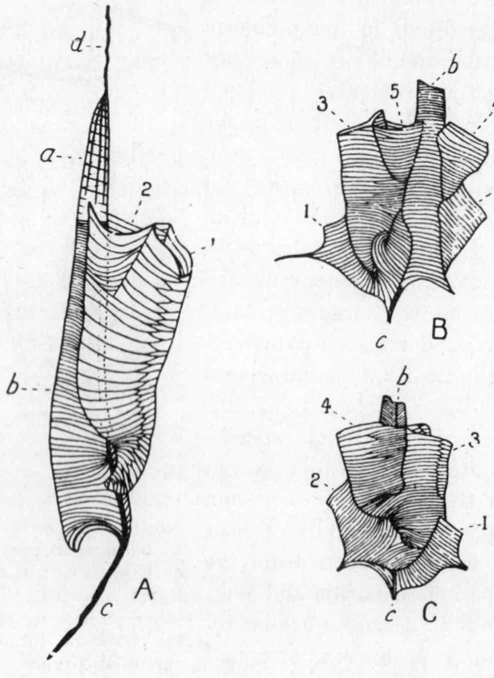

Fig. 18. Early stages of (A) *Monograptus* (after Kraft); B, C, opposite sides of *Diplograptus* (after Wiman). Enlarged. *a*, the early part, and *b* the later formed part of the sicula; *c*, spine; *d*, virgula; 1–5, hydrothecæ in order of their development.

sicula has a similar position, but the apex projects like a spine between the two branches (fig. 19 E).

The appearance of even the same species of graptolite varies considerably according to its mode of preservation.

Frequently it is flattened to a film, and when this is the case we may get a side view, a front view showing the mouths of the hydrothecæ, or a back view; in the two latter cases the margins will be parallel. But when the

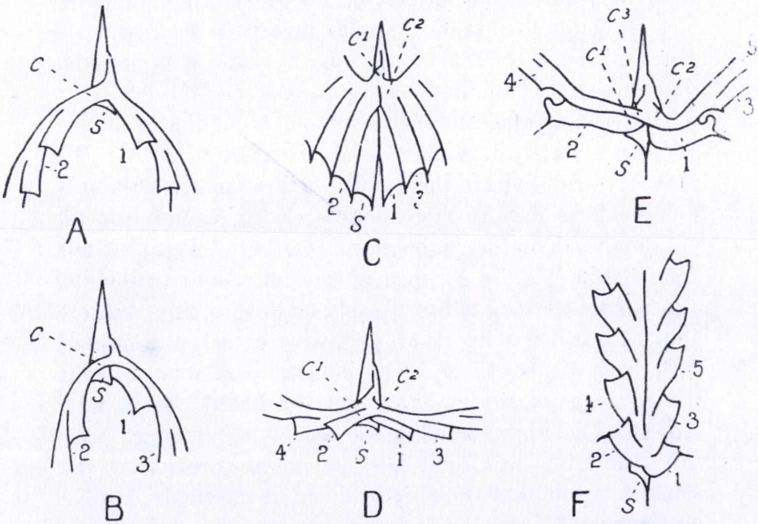

Fig. 19. A, *Didymograptus bifidus*; B, *D. minutus*; C, *D. gibberulus*; D, *Leptograptus flaccidus*; E, *Dicellograptus geniculatus*; F, *Diplograptus teretiusculus*. A—C, Lower Ordovician; D—F, Middle Ordovician. All enlarged. Proximal ends showing the order and mode of development of the early hydrothecæ. c 1, c 2, c 3, first, second and third crossing canals; *s*, sicula. 1–5, hydrothecæ in the order of their development.

interior has been filled in with iron pyrites, or when the graptolite is preserved in a limestone, the natural form of the polypary is often retained.

No medusoid form is known in the graptolites; but sac-like bodies, which may be gonangia, are sometimes present. In a few biserial graptolites such bodies have been found attached to the polypary; they are not joined to the hydrothecæ, but come off at right angles to them along the middle line of the sides of the polypary, but since no siculæ have been found in or near these sacs their nature and function must be regarded as uncertain. In other cases, however, sacs or vesicles containing siculæ have been seen near the distal extremity of the virgula (fig. 21), and it is probable that these are of the nature of gonangia (fig. 13, 10).

The earliest stage in the development of the graptolite at present known is the *sicula* (figs. 18 A, 20 A); this probably arises within the sac-like bodies mentioned above. The sicula is usually more or less clearly exposed at the proximal end of the rhabdosome (figs. 14 C, 19, *s*); it is a hollow cone (fig. 18 A) open at the base, and consists of two parts—the pointed or apical end with a thin wall (*a*), and the broader or apertural part with a thick wall marked with lines of growth (*b*). The pointed part was probably the covering of the embryo, and the broad part a later growth. The apical end of the sicula is prolonged as a thread forming the virgula (*d*). A spine-like projection (*c*) is sometimes found at the apertural end of the sicula, but has no connection with the virgula. In the development of *Didymograptus* (fig. 19 A—C) a bud is formed on one side of the sicula and from this arise (1) the first hydrotheca (fig. 19 A, 1) and (2) a tubular body known as the crossing-canal (*c*); the latter grows across the sicula and gives rise to the second hydrotheca (2) which is on the side opposite to the first hydrotheca. From each of these two hydrothecæ a stipe or branch is developed, owing to the fact that each hydrotheca gives rise by budding to another hydrotheca and in this way two continuous linear series are formed. More complex branching (as in *Tetragraptus*, *Dichograptus*) is produced when one hydrotheca buds off two hydrothecæ instead of one.

In later genera, the first-formed hydrothecæ gradually acquire an alternating mode of origin. Thus, for example, in *Didymograptus gibberulus* the first hydrotheca of the second branch (2) (fig. 19 C) is derived, by the crossing canal (*c* 1), from the first hydrotheca of the first branch (1), but the second hydrotheca (3) of the first branch is derived by a second crossing canal (*c* 2) from the first hydrotheca of the second branch, and only after these three hydrothecæ have been produced do the branches become independent. In *Dicellograptus* (fig. 19 E) there are formed in this manner three crossing canals (*c* 1, *c* 2, *c* 3), and the hydrothecæ (1, 2, 3, 4) are alternating in their origin. In *Diplograptus* the development may be similar to that of *Dicellograptus*, but the crossing canals are reduced in size and the hydrothecæ grow up the virgula. In other species of *Diplograptus* the alternate development of the hydrothecæ may continue throughout (as in figs. 19 F, 20), and in such forms there is no median septum. In *Monograptus* (fig. 18 A, 1, 2) the hydrothecæ grow directly upwards and produce only a single series.

Whilst each rhabdosome consists of a colony of individual polyps there is evidence which seems to indicate that in some biserial graptolites a number of rhabdosomes were grouped together to form larger colonies. Thus Ruedemann has described specimens of *Diplograptus foliaceus* (fig. 21), from the Utica Slate (Ordovician) of New York, which consist of a number of individuals radiating from a centre where they unite by the distal prolongations of their virgulæ; at the point of union there is a small, nearly square, chitinous sheath which is similar in appearance to the central disc of *Tetragraptus*; below this is a larger quadrate body, apparently vesicular, which may have been either a float (pneumatocyst) or an organ of fixation. Around the small disc are from four to eight globular vesicles, which Ruedemann considers to be gonangia, since they contain siculæ; the siculæ sometimes pass out and develop into fresh colonies, but in other cases they remain attached to the parent, and, by the growth of the virgula, extend outwards, and sub-

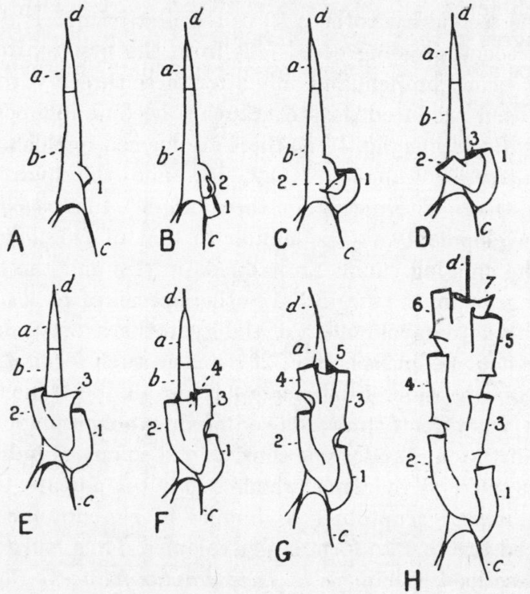

Fig. 20. *Climacograptus typicalis*, Upper Ordovician, showing the successive stages of growth (A to H). ×7. (After Bulman.) *a*, the early part and *b*, the later part of the sicula; *c*, spine from sicula; *d*, virgula. 1–7, hydrothecæ in the order of their formation.

sequently hydrothecæ arise in the usual way. In some other species of *Diplograptus* (*D. vesiculosus*, etc.) and in *Climacograptus* (fig. 16) a single vesicle is sometimes found at the distal end of the virgula. A vesicle may also occur at the proximal end of the polypary, but its function is unknown.

Owing to the fact that the soft parts of the graptolites are entirely unknown it is difficult to speak of their affinities with any degree of certainty. It seems probable, however, that they belong to the Hydrozoa; Allman and others consider them to be related to the Calyptoblastea, especially to such forms as *Sertularia* and *Plumularia*, with which they agree in the general characters of the hydrothecæ and perhaps also in the possession of gonangia. But they differ

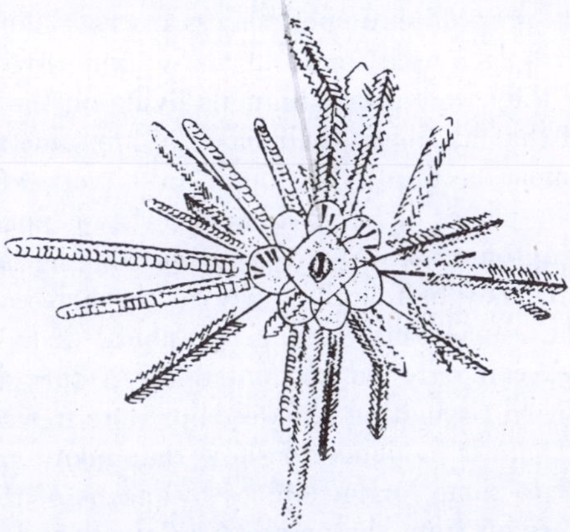

Fig. 21. *Diplograptus foliaceus*, from the Utica Slate, New York. × ⅔. (After Ruedemann.)

in some important respects from the Calyptoblastea, *e.g.* in possessing a virgula and sicula, in the diminution in size of the hydrothecæ towards the proximal end of the rhabdosome, in the hydrothecæ being nearly always in contact, and in the free communication which exists in most cases between the hydrothecæ at their bases; their development is also different—in the graptolites each hydrotheca is budded off from another hydrotheca, but in the Calyptoblastea the new polyps are budded off from the cœnosarc.

Further, the graptolites never form the much-branched tree-like colonies which occur so commonly in recent hydroids, and the graptolites are never firmly fixed by any root-like structure corresponding to the hydrorhiza. On the other hand the dendroid graptolites (p. 74), to which the true graptolites appear to be closely related, do form much-branched colonies with, in some cases, a root-like hydrorhiza.

Since the graptolites do not possess any root-like structure (hydrorhiza) such as is found in many living hydroids it is not likely they were sessile animals living on the sea-floor; further, if that had been their mode of life some specimens would almost certainly be found in a vertical position crossing the planes of lamination, but that is not the case, for the graptolites are found lying flat on the lamination planes as if they had sunk slowly to the bottom in quiet water. The remarkably wide geographical distribution of the species can only be accounted for if the graptolites lived attached to floating sea-weeds or were free-swimming animals. There is evidence to show that many graptolites were fixed to some foreign object by means of the thread which comes off from the point of the sicula and ends in a chitinous disc; it is possible that some of the later biserial graptolites were free-swimming animals, since the vesicle found at the centre of the radiating colonies (fig. 21) and at the distal end of the single stipes (fig. 16) may have served as a float; also the perforate or reticulate structure of the wall of some biserial forms (fig. 17) appears to be an adaptation for a floating mode of life. The much greater abundance of graptolites in thin, fine-grained, carbonaceous shales than in thicker and coarser deposits, suggests that the graptolites lived mainly at some distance from the shore where sediment was deposited slowly in tranquil water; and the carbonaceous matter in those shales may have been derived from the decomposition of sea-weeds.

The genera of graptolites at present accepted are based, to a large extent, on the number of branches of the rhabdosome; but it is now considered that this feature is of less importance than was formerly supposed, and that a classification which shows the genealogical relationships of the forms should be founded chiefly on the characters of the hydrothecæ and, to some extent, on the angle of divergence of the branches. The early graptolites, such as *Bryograptus*, appear, at first sight, to be more advanced than the later types (*e.g. Monograptus*), on account of their more complex branching; but in the early forms the hydrothecæ are very simple, differing but little from the sicula, whereas in the later ones they exhibit considerable modification. In some genera the hydrothecæ of different species show great variety of form, those of one species being often much more like those of a species belonging to another genus than to other species of the same genus: thus we get the same type of hydrotheca in the three forms *Bryograptus callavei*, *Tetragraptus hicksi*, and *Didymograptus affinis*, and another type in *Bryograptus retroflexus*, *Tetragraptus denticulatus*, and *Didymograptus fasciculatus*. It is contended that each of these groups is a genealogical series and should be regarded as a genus—that *T. hicksi* has descended from *B. callavei*, and *D. affinis* from *T. hicksi* According to the old view all the species of *Didymograptus* were thought to have descended from one common ancestor, but this will not account for the close resemblance which the hydrothecæ of certain species of *Didymograptus* bear to those of certain species of *Tetragraptus*; on the other hand, this is readily explained if we consider that the species of *Didymograptus* have descended from various species of *Tetragraptus*. Then again, the remarkable diversity in the hydrothecæ of *Monograptus* can be easily understood if we grant that the forms included under this term are the descendants of different species of one or more genera. But since species which have a different ancestry cannot be placed in the same genus, we must regard *Monograptus* as an assemblage of forms which agree merely in consisting of a single uniserial branch or stipe, and have descended, through *Dimorphograptus*, from various groups of *Diplograptus* and perhaps of *Climacograptus*.

Didymograptus (fig. 19 A—C). Rhabdosome bilaterally symmetrical, consisting of two uniserial stipes diverging at an angle which varies, in different species, from 0° to 180° (or occasionally more). Hydrothecæ sub-cylindrical, in contact for a considerable part of their length. Lower Arenig to Upper Llandeilo. Ex. *D. murchisoni*, Lower Llandeilo; *D. patulus*, *D. extensus*, Arenig.

Phyllograptus (fig. 22). Rhabdosome leaf-like, consisting of four uniserial stipes united along the whole of their length. Hydrothecæ cylindrical or sub-cylindrical, in contact throughout their entire length. Sicula pointing distally. Arenig. Ex. *P. typus*.

Tetragraptus (figs. 15, 23). Rhabdosome bilaterally symmetrical, uniserial, consisting of four simple radiating branches which arise from the bifurcation of two short branches coming off from opposite sides of the sicula (constituting a *Didymograptus* stage). Hydrothecæ cylindrical or sub-cylindrical, in contact for a considerable part of their length. A central disc may or may not be present. Arenig. Ex. *T. quadribrachiatus*.

Dichograptus. Rhabdosome typically bilaterally symmetrical, consisting of eight uniserial main stipes produced by bifurcation through *Didymograptus* and *Tetragraptus* stages. Hydrothecæ cylindrical or sub-cylindrical. A central disc is frequently present. Lower Arenig. Ex. *D. octobrachiatus*.

Loganograptus (Arenig) and *Clonograptus* (Tremadoc and Arenig) are forms in which bifurcation has proceeded further than in *Dichograptus*.

Bryograptus. Bilaterally sub-symmetrical, uniserial, consisting of two main stipes diverging at a small angle from the sicula, which has its point directed distally. From the inner margins of the main stipes similar secondary stipes (which may bear other stipes) arise. Hydrothecæ like those of *Dichograptus*. Thecæ of smaller size than hydrothecæ, known as *bithecæ*, are present in some species, and are also found in *Clonograptus*. Tremadoc. Ex. *B. kjerulfi*.

Leptograptus (fig. 19 D). Rhabdosome consisting of two simple, slender, flexuous, uniserial stipes given off in opposite directions from the sicula at angles greater than 1[illegible]. Hydrothecæ are long tubes with slight sigmoid curvature, in contact for about half their length. Upp[illegible]andeilo to Lower Bala. Ex. *L. flaccidus*, Lower Bala.

Fig. 22.

Fig. 23.

Fig. 22. *Phyllograptus*, Arenig Rocks. The graptolite has been cut in two, and the upper part raised so as to show the four branches. Natural size.

Fig. 23. *Tetragraptus Bigsbyi*, Lower Ordovician. Early part of the rhabdosome showing four branches; *c*, crossing canal; *s*, sicula. (After Holm.) ×4.

Pleurograptus. Two principal branches as in *Leptograptus*; these bear secondary branches on both sides, often arising alternately, and sometimes bearing smaller branches. Lower Bala. Ex. *P. linearis*.

Nemagraptus (=*Cœnograptus*). Bilaterally symmetrical, uniserial, consisting of two slender, more or less flexuous main stipes coming off from the middle of a well-defined sicula; from each of these stipes secondary branches may be given off in a symmetrical or nearly symmetrical manner. Hydrothecæ as in *Leptograptus*. Llandeilo. Ex. *N. gracilis*.

Dicranograptus. Bilaterally symmetrical, biserial in the proximal portion, dividing distally into two uniserial branches.

Hydrothecæ tubular, with sigmoid curvature and inturned apertures. Upper Llandeilo to Lower Bala. Ex. *D. clingani*, Bala.

Dicellograptus (fig. 19 E). Like *Dicranograptus*, uniserial, the two branches united at the sicula only, which points distally. Angle of divergence greater than 180°. Hydrothecæ usually with strongly sigmoid curvature. Upper Arenig to Upper Bala. Ex. *D. anceps*, Upper Bala.

Diplograptus (figs. 14 D, 18, 19, 21). Rhabdosome biserial. Hydrothecæ subprismatic or sub-cylindrical tubes, overlapping and placed obliquely. Virgula prolonged beyond the distal extremity of the rhabdosome. Sicula more or less completely concealed. Arenig to Tarannon. Ex. *D. foliaceus*, Llandeilo to Lower Bala. *Petalograptus* and *Cephalograptus* are sub-genera; Llandovery and Tarannon.

Climacograptus (figs. 16, 20). Biserial. Hydrothecæ tubular, with sharp sigmoid curvature, apertures placed in depressions. Sicula often concealed. Upper Arenig to Tarannon. Ex. *C. normalis*, Llandovery and Tarannon.

Retiolites (fig. 17). Biserial, straight. Hydrothecæ like those of *Diplograptus*. Periderm consists mainly of a network of threads and rods. Lower Bala to Lower Ludlow. Ex. *R. geinitzianus*, Upper Tarannon and Wenlock.

Dimorphograptus (see p. 60). Llandovery.

Monograptus (figs. 14, A—C, 18 A). Rhabdosome unbranched, uniserial; straight, curved, or spiral. Hydrothecæ very variable in form in different species. Sicula situated at the proximal end of the rhabdosome, and its pointed end directed distally. Lower Llandovery to Lower Ludlow. Ex. *M. nilssoni*, Lower Ludlow; *M. leptotheca*, Llandovery; *M. priodon*, Wenlock; *M. spinigerus*, Llandovery.

Rastrites. Closely allied to *Monograptus*, but the hydrothecæ are long, tubular, and widely separated. Llandovery to Tarannon. Ex. *R. peregrinus*, Llandovery to Tarannon.

Cyrtograptus. Similar to *Monograptus*, but coiled into a plane spiral with branches given off from the external (hydrothecal) margin. Upper Tarannon to Wenlock. Ex. *C. murchisoni*, Wenlock Shale.

Distribution of the Graptolitoidea

Cambrian. In Britain the earliest graptolites occur in the Tremadoc Beds, where we find the branching forms *Bryograptus* and *Clonograptus*.

Ordovician. The graptolites in the Arenig division are mainly uniserial forms without a virgula in the wall of the common canal and are often branched. The most characteristic genera are *Didymograptus*, *Tetragraptus*, *Dichograptus*, and *Phyllograptus*; *Clonograptus* survives from the Cambrian into the lower part of the Arenig. In the Llandeilo *Didymograptus* is still found and is fairly common in the lower part of the formation; other important genera in the Llandeilo are *Dicellograptus*, the biserial *Diplograptus* and *Climacograptus*, and *Nemagraptus* and *Dicranograptus* which now appear for the first time. In the Bala Beds, the biserial genera *Diplograptus*, *Climacograptus* and *Dicranograptus* become much more abundant, and with them occur *Leptograptus* and *Pleurograptus*.

Silurian. The only genera which pass up from the Ordovician to the Silurian are *Climacograptus*, *Diplograptus*, and *Retiolites*, and nearly all the species in the two systems are different, so that between the Ordovician and Silurian there is a great break in the graptolitic succession. As a whole, the Silurian formations are characterised by the uniserial genera *Monograptus*, *Rastrites* and *Cyrtograptus*, which appear first in the Lower Silurian. In the lower part of the Llandovery the genera *Diplograptus* and *Climacograptus* are fairly abundant, but they become extinct in the Tarannon, and in the Wenlock and Ludlow Beds the only forms are *Monograptus*, *Cyrtograptus*, and *Retiolites*. The last traces of graptolites occur in the Downtonian Beds, but they are too imperfect for determination.

SUB-ORDER 2. *DENDROIDEA*

In the Palæozoic formations there are organisms, usually termed 'dendroid graptolites', which present considerable resemblance to the Calyptoblastic hydroids, but they are closely related to the Graptolitoidea, which were probably

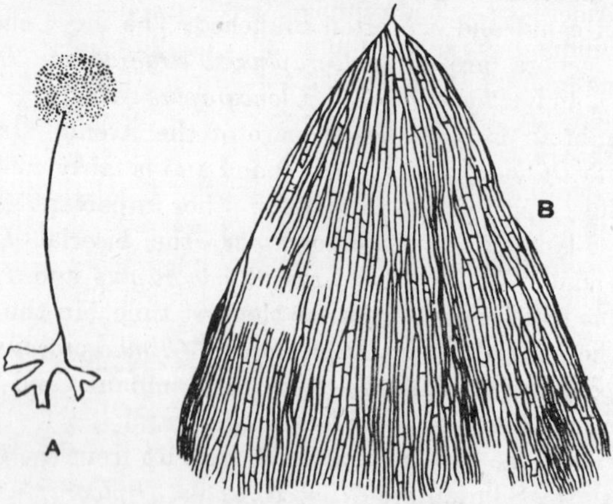

Fig. 24. *Dictyonema flabelliforme*, Upper Cambrian. (After Ruedemann.) A, young form with thread and disc for attachment. $\times 3$. B, complete colony. $\times \frac{2}{3}$.

derived from them through the genera *Clonograptus* and *Bryograptus*. The best known forms are *Dendrograptus*, *Ptilograptus*, *Dictyonema* (fig. 24), and *Callograptus*. These are usually much branched and tree-like, and some are fixed by a root-like structure, others by a thread or *nema* coming off from the point of the sicula (fig. 24 A). Transverse sections of *Dictyonema* and *Dendrograptus* show that the branches are composed of two or three different types of individuals, somewhat resembling in this respect the recent

Calyptoblastean forms *Clathrozoon* and *Grammaria*. Those genera which possess a sicula and were suspended by a thread coming off from its pointed end appear to be closely related to the true graptolites; this relationship is more marked in the early species of *Dictyonema* than in the later.

Dictyonema (=*Dictyograptus*) (fig. 24) is found in the Cambrian, Ordovician and Silurian, and in the Devonian of North America, and has a funnel-shaped skeleton which consists of numerous radiating branches, placed nearly parallel with one another, and united by transverse bars. A sicula is present, but in later forms the nema is replaced by a root-like structure and the colony probably grew erect from the substratum. Of other dendroid graptolites *Dendrograptus* occurs in the Upper Cambrian and Ordovician, *Callograptus* ranges from Arenig to Carboniferous, and *Ptilograptus* from the Arenig to the Ludlow Beds.

ORDER IV. HYDROCORALLINA

The skeleton in the Hydrocorallina is calcareous and has the form of encrusting or branching masses. It consists of a network of rods, in which there are tubes of two sizes opening on the surface; the larger are called *gastropores*, and have horizontal partitions or tabulæ; the smaller are named *dactylopores*. The skeleton is of ectodermal origin, and is perforated by a network of cœnosarcal tubes, above which is a superficial layer of ectoderm. The polyps project above this layer, and are of two kinds: nutritive individuals or gastrozooids, which are placed in the gastropores, and dactylozooids placed in the dactylopores. The soft parts in the branching forms may extend throughout the skeleton, but in the massive forms they are limited to the superficial layers.

Millepora is an important rock-building organism at the present day, often contributing largely to the formation of coral reefs; it has been recorded from Cainozoic deposits, but whether these examples really belong to that genus appears to be somewhat doubtful. *Stylaster* is a living form, and is stated to occur in the Miocene. *Milleporidium* from the Upper Jurassic of Stramberg and *Millestroma* from the Upper Cretaceous of Egypt may belong to this Order.

ORDER V. STROMATOPOROIDEA

In the Stromatoporoids the skeleton is calcareous, and very variable in form; it may be hemispherical, spheroidal, dendroid, encrusting, or altogether irregular, and frequently forms large masses. It consists of a series of concentric laminæ separated by spaces; these are crossed at right angles by rods or pillars, which give off horizontal processes at definite intervals; these processes join together and really form the laminæ which are perforated by openings of various sizes. The pillars may pass only from one lamina to the next, or may be continued through a considerable number of laminæ. The under surface of the skeleton is often covered by a thin imperforate layer, with concentric furrows, similar to the epitheca of many compound corals. On the upper surfaces of the laminæ there are, in many forms, shallow grooves, having a stellate arrangement, and known as *astrorhizæ*; these have been compared with the cœnosarcal grooves of *Hydractinia* and *Millepora*. In some genera, as for example *Actinostroma* (fig. 25), the two elements of the skeleton, the laminæ and pillars, remain quite distinct, but in others, like *Stromatopora*, they become to a great extent blended together so as to form a more or less netted structure; between these two types, however, there are intermediate forms. The first type (*Actinostroma*) has been compared with *Hydractinia* (see p. 54), but is always calcareous and forms larger masses; the second (*Stromatopora*) shows some resemblance to *Millepora* (see p. 75), and, like that genus, possesses vertical tubes with horizontal partitions (or 'tabulæ'), but the tubes seem to be of one size only, and consequently there is nothing to indicate that this type was dimorphic; it differs also in possessing radial pillars.

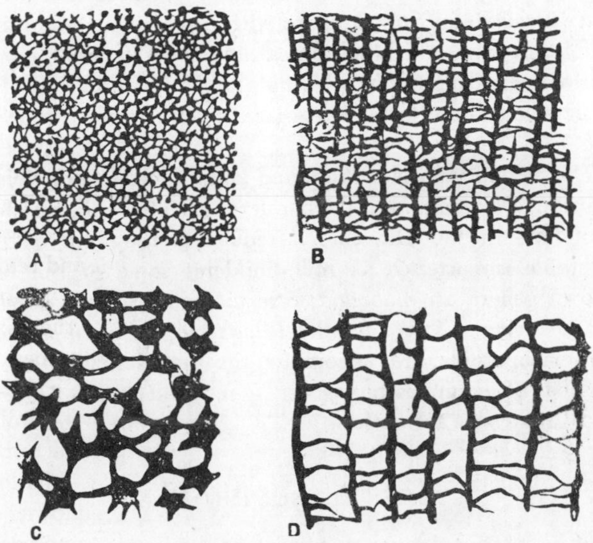

Fig. 25. A, tangential section of *Actinostroma intertextum* showing the radial pillars. B, vertical section showing the radial pillars and the formation of the concentric laminæ by processes given off from these. ×12. C and D, parts of A and B further enlarged. From the Silurian Rocks. (After Nicholson.)

From the structure of the skeleton it has been inferred that the Stromatoporoids are Hydrozoa connected with both *Hydractinia* and the Hydrocorallinæ. According to this view the soft parts in the Stromatoporoids formed a continuous layer on the surface of the skeleton, and it is believed by some authors that the polyps in some cases (*e.g. Stromatopora*) were placed in definite tubes, but in others tubes are absent and there are pores only in the external lamina. Another view of the systematic position of some of the Stromatoporoidea is that they are Foraminifera. This has been supported by a comparison of the structure of *Actinostroma* and *Stromatopora* with *Gypsina*, a foraminifer which is common in the tropical and sub-tropical regions; it is of encrusting habit, growing attached to corals or other objects, and often forms irregular spreading masses of large size.

The Stromatoporoids are found mainly in the Ordovician, Silurian, Devonian, and Carboniferous Systems, being most abundant in the Devonian; frequently they are of considerable importance as rock-builders; some of the best known genera are *Labechia*, *Stromatopora*, *Stromatoporella*, *Actinostroma*, *Clathrodictyon*, *Idiostroma*, and *Amphipora*. Stromatoporoids are uncommon in the Mesozoic, but a number of examples have been found in Triassic, Jurassic and Cretaceous formations.

CLASS II. SCYPHOZOA

The Scyphozoa (Scyphomedusæ) or Acalephæ include the larger and more conspicuous jelly-fishes, such as *Aurelia*, *Rhizostoma*, and *Pelagia*. They possess no hard parts; nevertheless their impressions have been found in various formations. One of the best preserved is *Rhizostomites*, belonging to the Order Discomedusæ, from the Lithographic Limestone (Upper Jurassic) of Solenhofen in Bavaria.

Even in the oldest fossiliferous formations traces of supposed Scyphozoa have been found; the most satisfactory of these is *Medusina* from the Lower Cambrian of Sweden, referred to the Discomedusæ, and from the Silurian of Thuringia. *Palæonectris* is found in the Devonian. Others, but of which the nature is doubtful, have been described by Walcott from the Middle Cambrian of Alabama and British Columbia.

CLASS III. ANTHOZOA (ACTINOZOA)

This Class includes the corals and sea-anemones. They differ from the Hydrozoa (1) in possessing an œsophageal tube of *stomodæum*, which is distinct from the cœlenteron, though opening into it; (2) in having the cœlenteron divided up into chambers by vertical radiating partitions known as mesenteries; (3) in the reproductive elements being developed in the endoderm of the mesenteries and never on a medusa.

The Anthozoa possess an apparent radial symmetry, but closer examination reveals a bilateral arrangement of their parts. In a typical form, such as the common sea-anemone or a simple coral (fig. 26), the body has a more or less cylindrical shape, and is attached by one end, the other having an opening, the mouth (fig. 26, 2), surrounded by tentacles (1). The mouth leads into the *stomodæum* (3), which opens at its lower end into the cœlenteron. The latter is divided into chambers by radiating partitions, the *mesenteries* (fig. 26, 4, and fig. 27, *a*—*c*), each of which consists of a thin gelatinous layer in the middle and a layer of endoderm on each side. In the upper part of the polyp the inner edges of the principal mesenteries join the stomodæum, but in the lower part they remain free, and a section in the former region (fig. 27) will show the body wall and also the stomodæum, but in the latter the body wall only. The tentacles (fig. 26, 1) are placed immediately above the intermesenteric chambers, and the space in each tentacle is continuous with that of the chamber below. A bilateral symmetry is indicated by the oval or slit-like mouth, and the similarly compressed stomodæum; also by

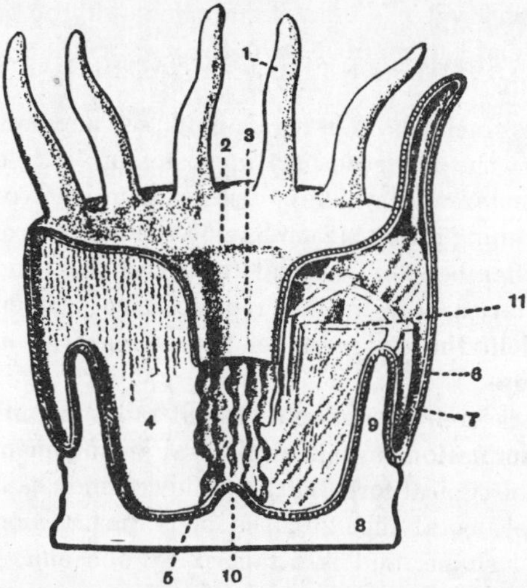

Fig. 26. Semi-diagrammatic view of half a simple Coral. (Partly after Bourne.) On the right side the tissues are represented as transparent to show the arrangement of the theca and septa; on the left a mesentery is seen. 1, tentacle; 2, mouth; 3, stomodæum; 4, mesentery; 5, mesenteric filaments, free edge of mesentery; 6, ectoderm; 7, endoderm; 8, basal plate of skeleton; 9, outer wall ('theca'); 10, columella; 11, septum.

the arrangement of the longitudinal muscles which occur on one face of each mesentery, extending from the base of the polyp upwards (fig. 27). The sea-anemones have no hard parts, but the majority of Anthozoa possess a skeleton, which in many cases is quite external to the body, and is formed of carbonate of lime (fig. 26, 8, 9); in others it is internal and may consist of calcareous spicules, or of an axial rod of horny or calcareous material. The Anthozoa are divided into two Orders, (1) the Zoantharia, (2) the Alcyonaria.

ORDER I. ZOANTHARIA

In the Zoantharia the tentacles are generally numerous and are never eight in number, as is the case in the Alcyonaria; occasionally there are six only, but frequently a multiple of six, and they usually form several circles around the mouth. The tentacles are nearly always simple (fig. 26, 1). The mesenteries (fig. 27, *a*, *b*, *c*) are usually numerous also and form several cycles; those belonging to the primary cycle are formed first and reach to the stomodæum; the other cycles (secondary, tertiary, etc.) are successively smaller. The mesenteries are arranged in couples (fig. 27) with the longitudinal muscles of each couple facing one another, except in the case of the couples situated at the grooved ends of the stomodæum, where the muscles are turned away from each other (*d*, *e*). Commonly there are six couples of primary mesenteries, six of secondary, twelve of tertiary, and so on. The narrow space between the two mesenteries of a couple is known as an *entocœle*; the wider space between two mesenteries of adjacent couples is known as an *exocœle*. A skeleton is often present in the Zoantharia and may be calcareous or horny; when calcareous it is never composed of spicules but consists of aragonite fibres.

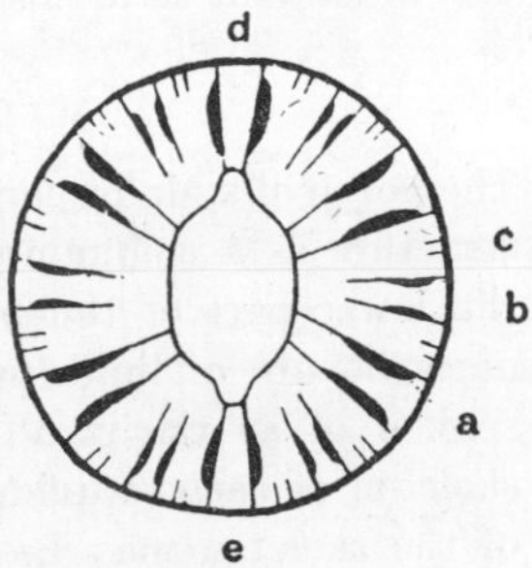

Fig. 27. Diagrammatic section of a Zoantharian polyp passing through the stomodæum. *a*, primary mesenteries; *b*, secondary mesenteries; *c* tertiary mesenteries; *d*, *e*, primary mesenteries at the ends of the compressed stomodæum. The muscles are indicated by the thickenings on the mesenteries.

The Zoantharia comprise, (1) the sea-anemones, which have usually been grouped together as the Actinaria, and are unknown in the fossil state, since they possess no hard parts; (2) the Antipatharia—colonial forms in which the skeleton consists of an internal horny rod secreted by the ectoderm; these also are not found fossil; (3) the Madreporaria, including the well-known stony corals, which are very abundant as fossils.

MADREPORARIA

The polyp of a Madreporarian coral has essentially the same structure as a common sea-anemone, but the ectoderm of the lower part of the body secretes a skeleton consisting of carbonate of lime (fig. 26, 8, 9). The entire skeleton is spoken of as the *corallum*, and in compound corals the skeleton of each individual is termed a *corallite*. The parts of the skeleton may be solid, or they may be perforated, or formed of a network of rods.

In a typical simple coral (fig. 29) the skeleton has a more or less conical form; the base of the cone, on which the polyp is placed, is usually depressed, and is termed the *calyx*. The corallum is bounded by an outer wall (fig. 26, 9; fig. 28, *d*), and sometimes there is, outside this, another calcareous layer, the *epitheca* (p. 85). The whole space enclosed by the outer wall is termed the *visceral chamber*; it is divided up by various partitions, the most important of which are the *septa* (fig. 26, 11; fig. 28, *b*). These are vertical plates extending from the margin towards the centre, and alternating in position with the mesenteries. The septa are of different sizes, some reaching the centre, others being shorter; they frequently occur in series or cycles, of which three or more may often be distinguished, the largest being the primary (*b*), the others the *secondary*, *tertiary*, etc. Each cycle of septa agrees in position with the corresponding cycle of mesenteries, *e.g.* the primary septa are within the entocœles of the first cycle of mesenteries. In many corals found in the Palæozoic formations one of the primary septa (the cardinal septum) is much smaller than those formed after it, and consequently appears, at the surface of the calyx, to lie in a pit or cavity, which is called a *fossula* (figs. 45, *a*; 40 A, *a*). Usually only one fossula is present—the *cardinal fossula*, but sometimes others, known as the *counter* and

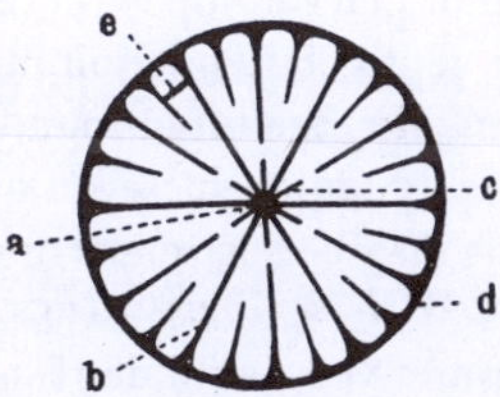

Fig. 28.

Fig. 29.

Fig. 28. Diagrammatic section (horizontal) of a simple coral. *a*, columella; *b*, primary septa; *c*, pali; *d*, outer wall ('theca'); *e*, dissepiments.
Fig. 29. *Montlivaltia trochoides*, Inferior Oolite, showing exsert septa. × ½.

alar fossulæ, are found (fig. 40 A, *dd*, *e*) (see p. 92). The tabulæ are depressed where they cross the fossulæ.

When the septa project upwards above the edge of the wall they are said to be *exsert* (fig. 29). The faces of the septa are sometimes plane, but usually have ridges, granules, or spines. In *Heliophyllum* and some other Palæozoic corals the septa have narrow shelf-like ridges (*carinæ*) which are inclined downwards from the centre of the coral to the margin (fig. 39). In some corals the septa are poorly developed, and may be represented by ridges only or by rows of spines.

In the centre of the coral, where the larger septa meet, there is often a vertical rod, which extends from the base of the chamber to the bottom of the calyx; this is the *columella* (figs. 28, *a*; 42, *c*). Its structure varies considerably; when it is solid and ends in a knob or point in the calyx, it is said to be *styliform*; sometimes the top is porous or *spongy*. When the columella is formed by the twisting together of processes given off from the inner edges of the septa, it is *false*. In Rugose corals the columella is produced by the dilation of either the cardinal or the counter septum. Other vertical partitions, somewhat similar to the septa, are the *pali* (fig. 28, *c*); these are radiating plates attached to the columella and placed opposite the inner edges of some of the shorter septa, but not joining them. Bars or rods, known as *synapticulæ*, are often found joining one septum to another. Similarly, adjacent septa are often connected by thin plates, which may be horizontal or oblique, usually curved, and are called *dissepiments* (figs. 28, *e*; 42, *d*); in some genera they are very abundant near the margin of the visceral chamber and form a spongy or *vesicular* tissue (fig. 41, *d*). *Tabulæ* are more or less horizontal plates which cross the septa, and occupy the central part of the visceral chamber, or, when well developed, extend quite across it (figs. 41, *t*; 40 B); they may be flat, concave or convex, and are arranged one above another, so that the visceral chamber is divided into horizontal compartments. In some genera (*e.g. Turbinolia*) the septa project outside the wall of the coral forming vertical ridges known as *costæ*.

In one family of Rugose Corals found in the Carboniferous there is a large cylindrical column in the centre of the coral which projects up into the calyx; it is formed of vertical radiating plates, representing the axial parts of the longer septa, crossed by transverse plates, representing the central parts of the tabulæ. Sometimes (fig. 44) the two elements are distinct, in other cases they are less regular and form a vesicular tissue (fig. 45).

The young coral polyp is a free-swimming animal; when it becomes fixed the first part of the skeleton to appear is a circular plate between the base of the polyp and the surface to which it is attached; on the basal plate radial ridges—the first traces of the septa—are secreted in folds formed in the base of the polyp between each couple of primary mesenteries. The wall next appears at the edge of the septa, and is formed either by the union of the thickened ends of the septa, or as an independent secretion between the ends of the septa. At the edge of the basal plate an upgrowth may occur forming the epitheca outside the wall. For some time the polyp extends down to the base of the cup-like skeleton (fig. 26) and a fold hangs over the outside (fig. 26, 6, 7); but as the septa and wall increase in height the lower part of the visceral chamber becomes (in most cases) more or less completely cut off by the development of dissepiments or tabulæ which are secreted by the basal part of the polyp, and below which the soft parts do not extend. As growth proceeds more of these partitions are formed, and eventually a large part of the coral ceases to have any direct connexion with the polyp. On account of the septa and columella the basal wall of a coral polyp, unlike that of a sea-anemone which remains flat, becomes greatly infolded; the infolds occur between every two mesenteries. The parts of the coral skeleton described above are entirely *external* to the polyp; but the synapticulæ, on the other hand, perforate the mesenteries and the basal wall of the polyp, and are formed by the growth and ultimate fusion of two opposite granules on the faces of adjacent septa.

Some corals remain simple (*i.e.* consist of a single individual) throughout life. Others, which are simple in the young state, afterwards become compound and form colonies by giving off buds. In budding, new individuals may arise from the part of the polyp below the circle of tentacles (fig. 26, 6, 7), in which case a branching coral like *Dendrophyllia* (fig. 30 A) is frequently formed; this mode of increase is termed *lateral budding*. In other cases buds arise on the upper surface of the polyp, and then the young corallites are found inside the calyx of the parent—hence this is known as *calicular* budding (fig. 30 B). In *basal budding*

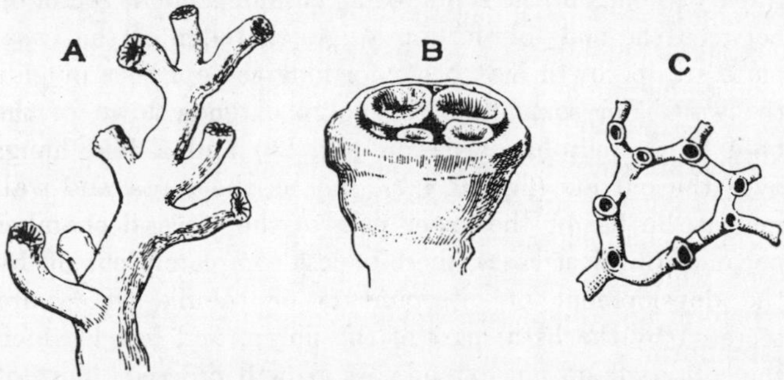

Fig. 30. A. *Dendrophyllia nigrescens*, showing corallites which have been produced by lateral budding, Recent. × ½. B. *Kodonophyllum truncatum*, showing calicular budding, Wenlock Limestone. Natural size. C. *Cladochonus crassus* (seen from above), showing basal budding, Carboniferous Limestone. Natural size.

(fig. 30 C), which is common in the Alcyonaria, the buds spring from creeping prolongations or stolons, which are given off from the base of the coral. In different corallites of some branching corals all stages may be seen in the division of a single corallite into two separate corallites. The calyx becomes oval, and then slightly constricted in the middle; the constriction increases until two separate corallites are formed. This division appears to be due to the development of a bud on the oral surface of the polyp inside the circle of tentacles.

When the individual corallites in a compound coral are free and diverge from one another, the corallum is termed *dendroid* (fig. 30 A): when they are nearly parallel to one another it is *fasciculate* (fig. 31). If the corallites are in contact and, owing to growth-pressure, polygonal in outline, the corallum is *massive*. When in a massive coral the corallites are nearly parallel to one another the corallum becomes *basaltiform*. In some massive corals the boundaries of the corallites tend to be indefinite; such corals are termed *astræiform*. In the evolution of various groups of corals it is commonly found that the earliest form of a series is simple, and gives rise to a dendroid type, which later develops into a massive form, and this in turn may become astræiform.

Fig. 31. *Lithostrotion junceum*, Carboniferous Limestone. Portion of a colony showing fasciculate form. × $\frac{2}{3}$.

If the corallites are not in contact the spaces between the individual corallites are sometimes filled up with calcareous material formed by the cœnosarc, and known as *cœnenchyma*. In massive corals (*e.g. Acervularia*) the base of the corallum is sometimes covered by a thin epithecal plate—the *basal epitheca*.

In dendroid corals the polyps on the different corallites may be quite separate from one another; but in massive corals, whilst the upper parts of the polyps are more or less separate, the lower parts are united and the cœlentera of adjacent individuals communicate with one another. When cœnenchyma is present the polyps are united by an extension of the part which ordinarily occurs outside the theca, and now forms a sheet called the cœnosarc.

Microscopic examination of thin sections shows that each part of a coral is formed of thin layers or growth-lamellæ which consist of fine needle-like crystals placed more or less perpendicularly to the surfaces of the lamellæ. In a dissepiment the upper surface only is covered by the soft parts, and a section shows (fig. 32) a series of lines parallel to the surface and other finer lines crossing at right angles are seen—the former mark the growth-lamellæ, the latter the crystalline fibres. In a septum both sides are covered by the soft parts, and a transverse section shows (fig. 33) a

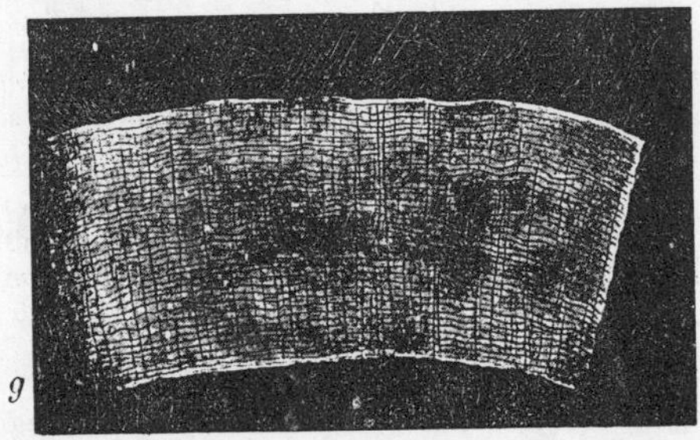

Fig. 32. Section of a dissepiment of *Galaxea*. Magnified. *g*, growth-lamellæ. (From M. M. Ogilvie.)

median dark line or row of dark spots, on each side of which the structure is symmetrical. When the surface of the septum is plane, the lamellæ are straight, or nearly straight, and parallel with the surface, and the fibres are perpendicular; but when the surface is ridged the lamellæ are curved so as to be parallel with the ridges, and the fibres radiate out from the dark median spots toward the curved surface of the ridge (fig. 33). When the septa bear striæ, granules, or spines, in addition to ridges, the folding of the lamellæ and the radiating arrangement of the fibres become more complex; but in all cases the structure is directly related to the form of the surface. The dark lines and spots represent the part of the septum which was first secreted; their dark appearance may be due either to the less regular arrangement of the fibres or to the imperfect

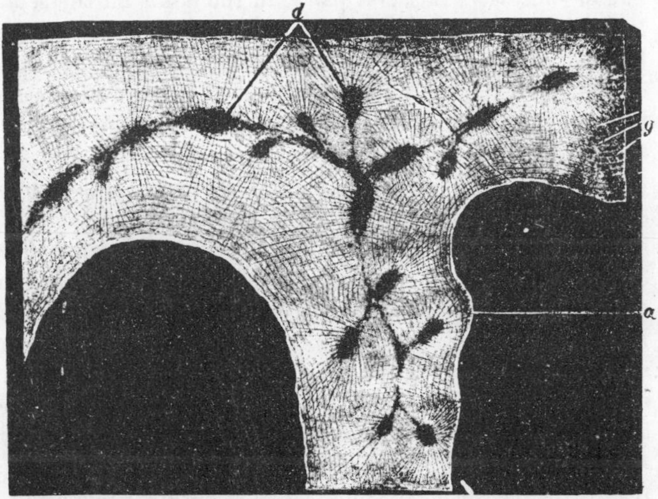

Fig. 33. Transverse section of part of a septum and theca of *Galaxea*. Highly magnified. *d*, dark spots; *g*, growth-lamellæ; *a*, granule on septum. (From M. M. Ogilvie.)

calcification of the material of that part. In fossil corals the dark part has often undergone secondary changes which give it a more distinct appearance.

In the development of a living Zoantharian coral six primary septa are first formed and appear simultaneously,[1] one septum between each couple of primary mesenteries;

[1] In some corals twelve septa are first developed simultaneously, of which six grow more rapidly than the others and become the primary septa.

next six secondary septa are introduced between the primary septa, either simultaneously or in bilateral pairs from the dorsal to the ventral border; other cycles may subsequently be added in a somewhat similar manner, not simultaneously, but in successive bilateral pairs; in the adult all the septa have generally a completely radial arrangement. In the Rugose corals of the Palæozoic period the development[1] of the septa follows a different course. Instead of the six primary septa appearing simultaneously, two septa (fig. 34 A), one on each side, are first formed and meet in the centre of the coral—representing the cardinal (1) and counter septa (1′) of the adult, on the ventral and dorsal sides respectively (fig. 40 A, *a*, *b*); next, two more septa (fig 34 B, 2) appear, one on each side of the cardinal septum, and as growth proceeds these become more widely separated from the cardinal septum, and eventually form the alar septa of the adult (fig. 40, *c*); afterwards, two septa (3) are added, one on each side of the counter septum, and these also spread outwards as growth proceeds (as indicated by the arrows in fig. 34 D). The six septa now present are regarded as the primary septa (1, 2, 3). The later septa (sometimes termed *metasepta*) are introduced in pairs; these appear at four points—one septum on each side of the cardinal septum (1), and one between each alar septum (2) and the primary septum (3). The two pairs which are first added (fig. 34 E, *a*) are attached to the cardinal sides of the primary septa 2 and 3; similarly later pairs (fig. 34 F, G, *b*, *c*) are introduced and are joined to the cardinal sides of the previously formed septa. As growth proceeds all the later septa (*a*—*c*), unlike the primary septa, gradually move

[1] This can be studied by gradually grinding down the tip of a perfect specimen. The arrangement of the septa in Rugose corals can also be seen either on the surface of the wall or by removing the theca.

towards the counter septum, as indicated by the arrows in fig. 34 G.

In the adults of some Rugose corals (fig. 40 A) the arrangement of the septa is similar to that just described, so that on each side of the cardinal fossula and on the counter side of each alar septum the later septa (metasepta) have a

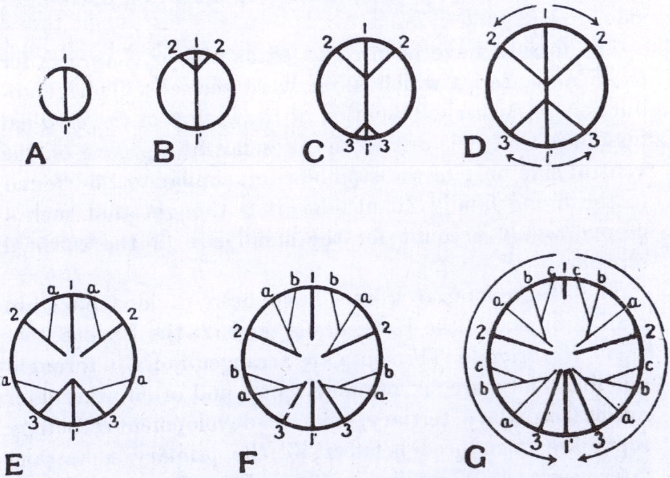

Fig. 34. Development of the septa in a simple Rugose Coral, *Zaphrentis*. 1—3, primary septa; 1, cardinal septum; 1 , counter septum; 2, alar septa; 3, counter-lateral septa; *a*, *b*, *c*, later septa (metasepta). (After Carruthers.)

pinnate arrangement. In other genera, however, the pinnate plan is not seen in the adult (figs. 44, 45), since in the later stages of growth all the septa either become free at their inner edges or unite only at the centre of the coral; and in such cases, unless a fossula is present, the symmetry of the coral is nearly or quite radial.

From the description of the septal development given above, it will be seen that the fossulæ are breaks in the sequence of the septa. The cardinal fossula (fig. 40 A, *a*) is limited by the later septa added on each side of the small cardinal septum. The counter fossula, on the opposite side, where no new septa are introduced, is bounded by the two primary septa (*d*, *d*) which enclose the counter septum (*b*). The alar fossulæ are the spaces between each alar septum (*c*) and the newer septa which have been added on its counter side.

The fossulæ have been regarded as pits or chambers for those mesenteries which alone were specialised for reproduction. Another explanation of the nature of the cardinal fossula is that it is due to the presence of a groove on the ventral side only of the stomodæum, similar to that found in the living family Zoanthidæ; it is thought that such a groove would account for the small size of the cardinal septum.

The Madreporarian corals have been divided into two main groups: (1) the Hexacoralla, and (2) the Rugosa.

(1) *Hexacoralla*. The septa are arranged radially throughout life. There are six primary septa, and often other later cycles (secondary, tertiary, etc.). In development secondary septa are introduced between all the primary septa, and later septa in a similar manner. The Hexacoralla were formerly divided into (1) Aporosa, in which the septa and wall are solid, and (2) the Perforata in which they are perforated. When the perforations are numerous the skeleton is light and porous and appears to consist of a network of rods. In living corals the perforations are traversed by canals of the soft parts. The separation of the Aporosa from the Perforata cannot be maintained since it has been shown that corals with a perforate skeleton have arisen independently from more than one group of Aporose corals. Moreover, in a few Rugose corals (*e.g. Calostylis*) and a few

Tabulate corals (*e.g. Aræopora*) the same perforate character is developed.

(2) ***Rugosa***. Septa and theca usually solid; tabulæ and dissepiments generally well developed and clearly differentiated. The coral is usually bilaterally symmetrical owing to the pinnate arrangement of the septa and to the presence of one or more fossulæ (fig. 40). New septa are introduced along four lines only. In the adult coral the bilateral symmetry may be more or less completely lost. The septa are usually of two sizes; long (or *major*) septa alternately with short (or *minor*) septa. Increase takes place by budding. The Rugosa are limited to the Palæozoic formations; the name of the group is taken from the vertical ridges often seen on the wall of the coral.

Formerly it was maintained that the Rugosa possess only four primary septa—the cardinal, the counter, and two alar septa, which divide the coral into quadrants; on account of this the name Tetracoralla has sometimes been used for this group. The study of the development of the septa has shown that there are really six primary septa, and the Rugose corals consequently agree in that respect with the Hexacoralla, so that it is possible that both may have descended from the same ancestors; the difference in the mode of development of the later septa, however, seems to indicate that the two groups soon diverged. A difficulty in accepting this view of their common ancestry is due to the fact that Rugose corals are found as early as the Ordovician, whereas the Hexacoralla are not known to occur in the Palæozoic formations. Consequently it would be necessary to assume that the Palæozoic ancestors of the Hexacoralla did not secrete any hard parts capable of being preserved. Another view is that various families of Rugose corals are the ancestors from which a number of families of the Hexacoralla have sprung independently. If this were proved it would follow that the separation of the Hexacoralla as a group distinct from Rugosa would be unnatural. But for the present it is convenient to retain these two divisions.

1. RUGOSA

Columnaria. Compound, typically massive, with small corallites. Septa long or short, with expanded bases. Tabulæ generally complete. Dissepiments in a single series, vertically elongated, large, but sometimes absent. Ordovician to Devonian; abundant in Upper Ordovician of North America. Ex. *C. alveolata*, Ordovician.

Streptelasma. Simple, conical or turbinate, bilateral. Septa numerous, alternately long and short, dilated at the periphery. The major septa twist together at the axis. Cardinal fossula is present, but is sometimes indistinct. Tabulæ irregular, sloping towards the periphery. Dissepiments poorly developed. Ordovician and Silurian. Ex. *S. corniculum*, Ordovician.

Palæocyclus. Small, simple, discoidal, with flat base. Septa strong, spinose, almost touching, exsert, the longer reaching the centre of the coral. Fossula distinct. No tabulæ or dissepiments. Silurian. Ex. *P. porpita*.

Cetophyllum (fig. 35). Turbinate or conical. Septa numerous, alternately long and short, but extending only a short distance into the visceral chamber, the central part being occupied by tabulæ. Four shallow fossulæ are present. No columella. Peripheral zone is formed of large dissepiments and is relatively narrow. The epitheca often gives off root-like processes. Bala to L. Ludlow. Ex. *C. subturbinatum*, Wenlock Limestone.

Cystiphyllum (fig. 36). Nearly always simple, conical or cylindrical. Septa absent or rudimentary; visceral chamber filled with vesicular tissue, the outer part consisting of dissepiments, the central part representing tabulæ. Fossula sometimes present. Columella absent. Calyx often a shallow basin, commonly with ridges representing septa. Silurian. Ex. *C. cylindricum*, Silurian. A similar form, *Mesophyllum*, is common in the Devonian. Ex. *M. vesiculosum*.

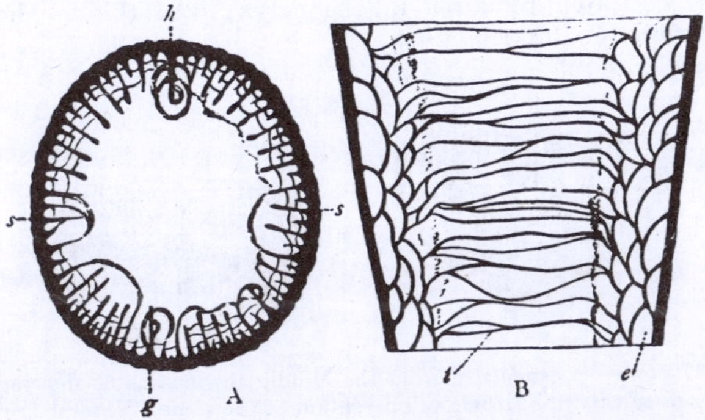

Fig. 35. *Cetophyllum subturbinatum*, Silurian. A. Horizontal section. B. Vertical section. *h*, cardinal septum; *ss*, alar septa; *g*, counter septum; *e*, dissepiments; *t*, tabulæ. (From Nicholson.) The name *Omphyma* is not valid. *O. subturbinata* belongs to the genus *Cetophyllum*.

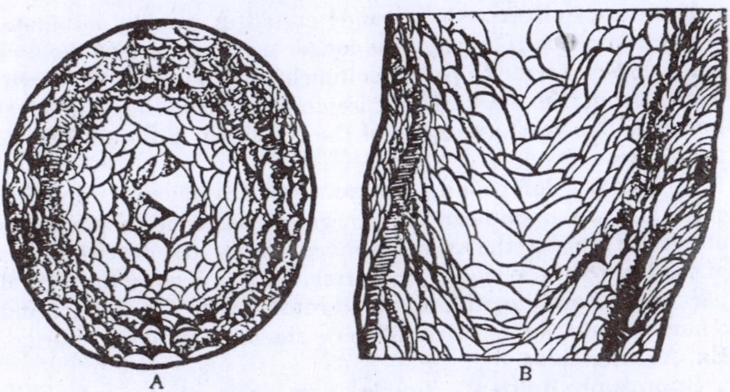

Fig. 36. *Cystiphyllum cylindricum*, Wenlock Limestone. A, horizontal; B, vertical section. (From Nicholson.) ×2.

Calceola (fig. 37). Simple, conical or slipper-shaped, one side is flat, the other convex; calyx very deep and closed by a semilunar operculum, which has on its inner surface a strongly-marked median ridge and several less prominent lateral ridges; septa indicated by striæ in the calyx; wall thick. Middle Devonian. Ex. *C. sandalina.*

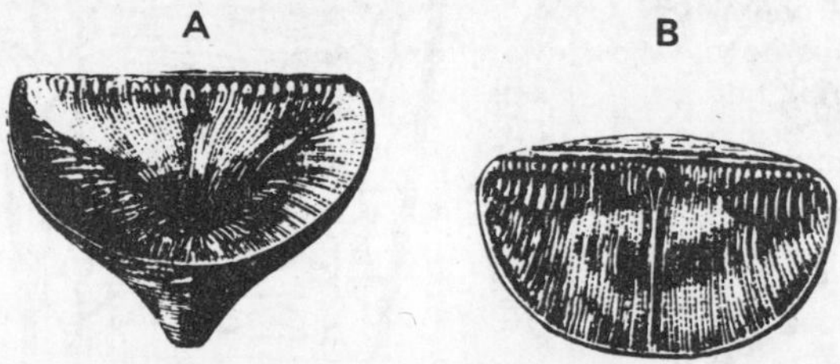

Fig. 37. *Calceola sandalina,* from the Middle Devonian. A, showing interior of calyx; B, inside of operculum of the same. Natural size.

Goniophyllum. Similar to *Calceola,* but quadrangular; operculum consists of four plates forming a pyramid over calyx. Visceral chamber filled with vesicular tissue. Silurian. Ex. *G. fletcheri,* Wenlock Limestone. An operculum also occurs in the genus *Rhizophyllum,* Silurian.

Kodonophyllum. Compound; corallites usually turbinate. Septa dilated peripherally; the longer usually meet at the axis where they form a pseudocolumella. Tabulæ slope downwards from the axis. No dissepiments. Budding calicular; buds small, near the margins of the corallites. Silurian. Ex. *K. truncatum.*

Xylodes. Compound, dendroid or fasciculate. Budding calicular, the buds near the margin. The longer septa reach almost or quite to the axis; the shorter septa may be two-thirds of the length of the longer, or shorter. Tabulæ occupy the central part of the corallite, differentiated into a broad inner series and a narrow outer series. Dissepiments small, numerous. Silurian. Ex. *X. articulatus.*

Spongophylloides. Simple, sub-turbinate to sub-cylindrical. Septa wavy or zigzag, reaching or nearly reaching the axis. Tabulæ small, close together. Dissepiments form a broad zone of vesicular tissue, into which the septa do not penetrate. Silurian. Ex. *S. grayi*.

Acervularia (fig. 38). Compound, massive; corallites with an outer polygonal (frequently hexagonal) wall, and an inner circular wall formed by the thickening of the septa. Septa well developed, the longer reaching the centre where they may twist together. Tabulæ extend across the greater part of the visceral chamber; those inside the circular wall are irregular and slope downwards towards the axis; those outside the wall horizontal and flat. Dissepiments form a narrow peripheral zone of vesicles. Silurian. Ex. *A. ananas*.

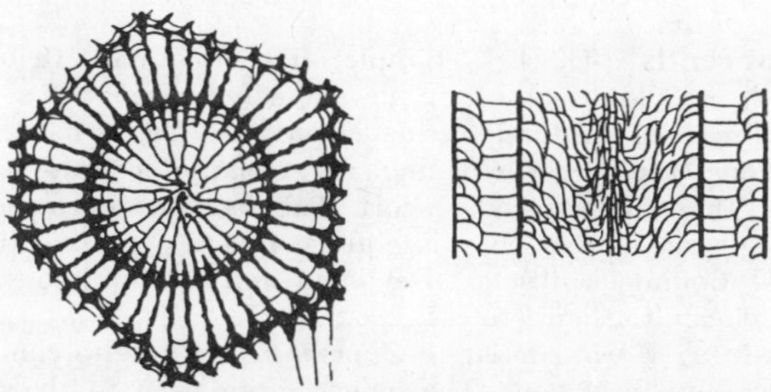

Fig. 38. *Acervularia luxurians*, Wenlock Limestone. Horizontal and vertical sections of one corallite. (After Lang and Smith.) ×2.

Phillipsastræa (= *Smithia*). Compound, massive. Septa numerous, becoming thickened between the margin and centre of the corallite; only the longer septa extend inside this thickening towards the centre of the corallite. Septa of adjacent corallites often confluent. Septa usually with carinæ. Outer wall of corallites thin or absent. No columella. Tabulæ and dissepiments well developed. Devonian. Ex. *P. hennahi* and *P. pengellyi*.

Prismatophyllum. Compound, massive, with thin wall. Septa may or may not reach the axis; typically with carinæ. Main part of tabulæ horizontal, but the peripheral parts sloping downwards. Typically numerous, small, globose dissepiments. Devonian. Ex. *P. davidsoni*.

Heliophyllum (fig. 39). Usually simple, turbinate. Septa bear carinæ. Devonian. Very common in North America. Ex. *H. Halli.*

Petraia. Simple, conical or sub-cylindrical, slightly curved; calyx very deep, sometimes reaching almost to the base of the coral. Septa close together, of two sizes, little developed except in the lower third of the coral where they reach the centre. Tabulæ and dissepiments usually absent; no columella. Silurian. Ex. *P. radiata.*

Cyathaxonia. Simple, turbinate or elongate-conical. Septa reach the columella, which is large and solid. Fossula present. Tabulæ sometimes present. No dissepiments. Carboniferous. Ex. *C. cornu.*

Zaphrentis (fig. 40). Simple, free, bilateral; turbinate, conical, or cylindrical, often curved; calyx deep; theca thick. A well-marked cardinal fossula is present. Septa moderately numerous, the larger reaching very nearly or quite to the centre, the smaller usually short. Tabulæ well developed, extending quite across the visceral chamber. No true dissepiments. Columella absent. Devonian and Carboniferous. Ex. *Z. delanouei*, Carboniferous Limestone.

Caninia. Form similar to *Zaphrentis* but often cylindrical, slender, and very long. The longer septa meet in the centre (as in *Zaphrentis*) in the young stages, but in the later stages the septa are short and the central part of the coral is occupied solely by tabulæ. No columella. In the simplest form, *C. cornucopiæ*, there are no dissepiments, but in all other species a peripheral ring of more or less vertical dissepiments is present in the adult part. Carboniferous. Ex. *C. cornucopiæ*, *C. cylindrica.*

Amplexus. Similar to *Zaphrentis*, but generally cylindrical and with the septa only fully developed on the upper surface of the tabulæ; in later stages the septa become progressively shorter. Devonian and Carboniferous. Ex. *A. coralloides*, Carboniferous.

Palæosmilia (fig. 41). Simple or compound: often massive. Septa numerous, of two sizes, alternating, the longer reaching the centre. Fossula often absent. Tabulæ rather small, occupying the central part only of the visceral chamber. Dissepiments form an extensive peripheral zone of vesicular tissue. Carboniferous. Ex. *P. murchisoni*, *P. regium*, Carboniferous.

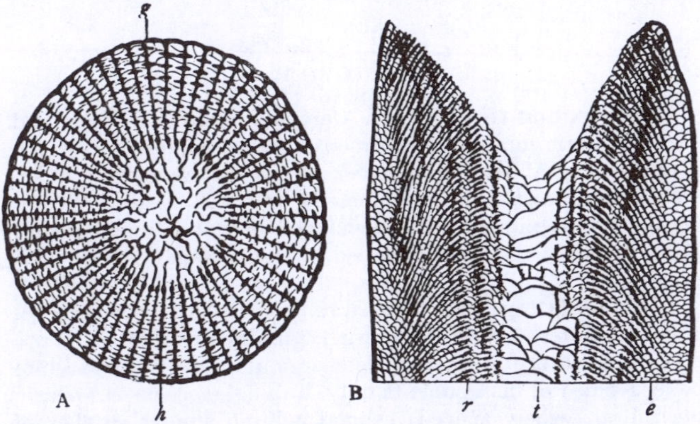

Fig. 39. *Heliophyllum elegantulum*, Devonian. A. Horizontal section. B. Vertical section. *h*, cardinal septum; *g*, counter septum; *e*, vesicular dissepiments; *t*, tabulæ; *r*, carinæ on septa. (From Nicholson.) ×2.

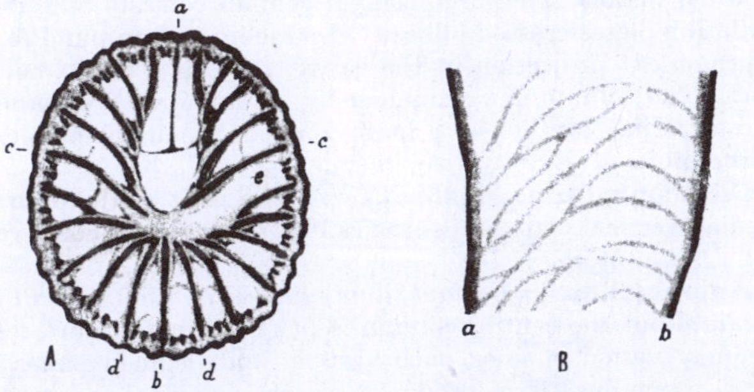

Fig. 40. *Zaphrentis delanouei*, Carboniferous Limestone. A. Horizontal section; *a*, cardinal septum in fossula; *b*, counter septum; *c*, alar septa; *d*, counter-lateral septa bounding the counter fossula; *e*, alar fossula. B. Vertical section showing tabulæ bending down into the cardinal fossula (*a*); (*b*), counter side. ×5. (Drawn by R. G. Carruthers.)

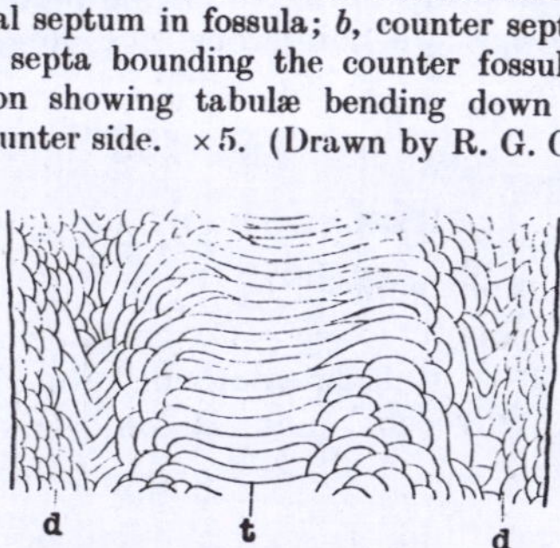

Fig. 41. *Palæosmilia murchisoni*, Carboniferous Limestone. Portion of a vertical section. *d* dissepiments; *t* tabulæ.

Lithostrotion (figs. 31, 42). Compound, either massive and with prismatic corallites, or formed of separated, nearly parallel, cylindrical corallites (fasciculate). Septa well developed, alternately long and short. Columella rod-like, laterally compressed. Dissepiments well developed except in small forms. Tabulæ wide, usually tent-shaped. Fossula often distinct. Carboniferous. Ex. *L. basaltiforme.*

Lonsdaleia (fig. 43). Compound, either massive with polygonal corallites, or fasciculate with cylindrical corallites. Septa do not reach the epitheca, the marginal part of the corallite being formed of dissepiments only. Tabulæ more or less nearly horizontal, widely spaced; central column similar to that of *Dibunophyllum.* Carboniferous. Ex. *L. duplicata.*

Clisiophyllum (fig. 44). Simple, turbinate or sub-cylindrical. Septa numerous, alternately long and short; a well-marked cardinal fossula. The large central column consists of vertical radiating plates crossed obliquely by inclined plates, and forms a prominent projection in the calyx; there is a short median plate. The column is surrounded by a zone formed of tabulæ, and external to this is a large zone of small dissepiments. Carboniferous. Ex. *C. bipartitum.*

Dibunophyllum. Like *Clisiophyllum* but with a strong median vertical plate across the central column. Carboniferous. Ex. *D. muirheadi.*

Aulophyllum (=*Cyclophyllum*) (fig. 45). Similar to *Clisiophyllum* but the central column is more distinctly limited and is finely vesicular since both vertical and inclined plates are more numerous; it is produced on the fossular side into an angular or ridge-like projection. No medial plate. Tabulæ slope towards the periphery. Carboniferous. Ex. *A. fungites.*

2. HEXACORALLA

Turbinolia. Simple, conical, free; calyx circular, with projecting columella. Septa exsert. Costæ lamellar, projecting, with pits in the grooves between them. No dissepiments or tabulæ. Eocene, Oligocene, and Recent. Ex. *T. humilis,* Barton Beds.

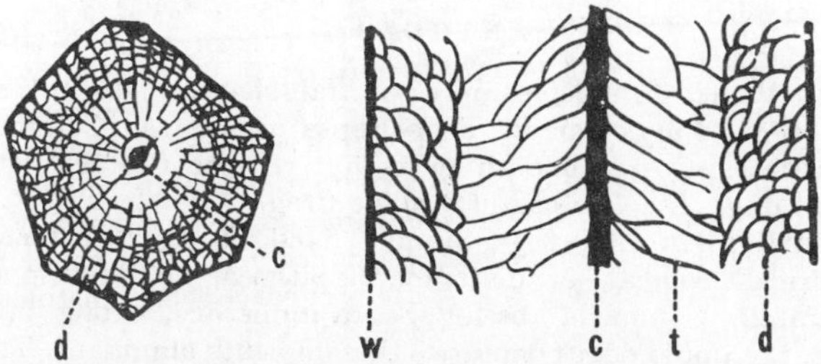

Fig. 42. *Lithostrotion basaltiforme*, Carboniferous Limestone. A. Horizontal section of a single corallite, ×2½. B. Vertical section, ×5. *c*, columella; *t*, tabulæ; *d*, dissepiments; *w*, theca.

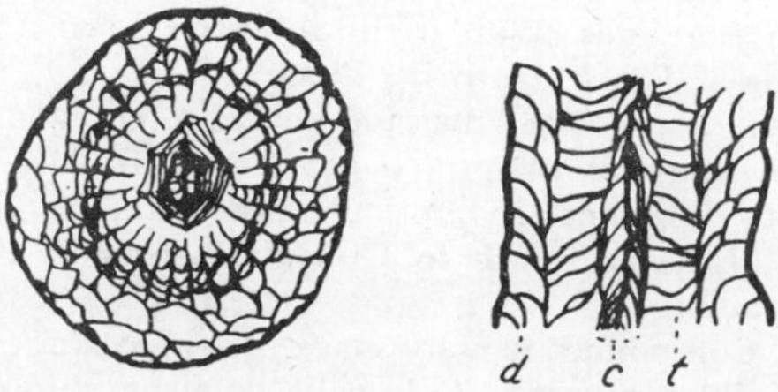

Fig. 43. *Lonsdaleia duplicata*, Carboniferous Limestone. *c*, central column; *d*, dissepiments; *t*, tabulæ. ×1½.

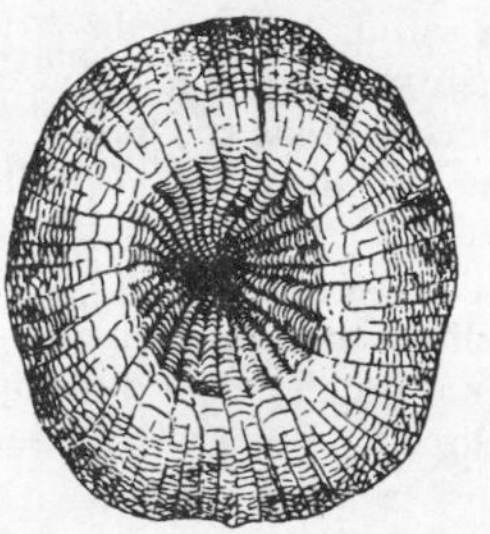

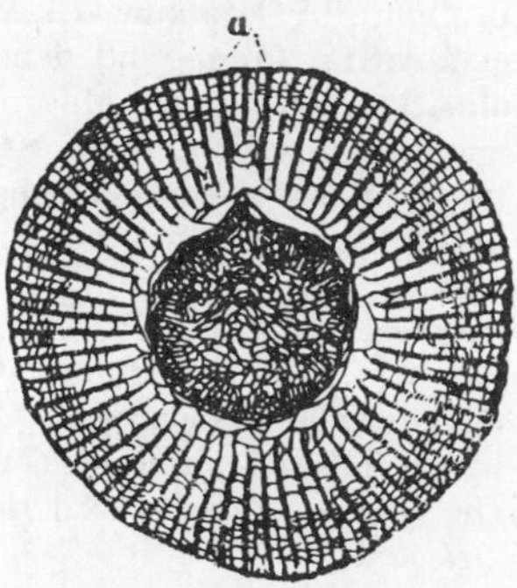

Fig. 44. Fig. 45.

Fig. 44. *Clisiophyllum bipartitum*, Carboniferous Limestone. Horizontal section showing the large central column. Natural size.

Fig. 45. *Aulophyllum* [*Cyclophyllum*] *fungites*, Carboniferous Limestone. Horizontal section. *a*. cardinal fossula. ×1½.

Flabellum. Simple, compressed, fan-shaped, free or fixed by rootlets. Calyx narrow, deep; septa numerous. Columella trabeculate. Costæ smooth or spiny. Upper Cretaceous to present day. Ex. *F. woodi*, Coralline Crag.

Montlivaltia (fig. 29). Simple, fixed or free; turbinate, cylindrical, conical, or discoidal. Epitheca well developed, corrugated. Columella absent. Septa numerous, strong, often exsert, the upper edges dentate. Dissepiments abundant. Trias to Cretaceous; in England, Inferior Oolite to Corallian. Ex. *M. trochoides*, Inferior and Great Oolite.

Parasmilia. Simple, fixed, turbinate or elongate. Calyx circular. Columella spongy. Septa well developed, exsert, granular on the sides. Wall with vertical ridges. Cretaceous to present day. Ex. *P. centralis*, Chalk.

Isastræa. Compound, massive; calyces polygonal. Columella rudimentary or absent. Septa thin and close together. Dissepiments abundant. Synapticulæ present. Trias to Eocene; in England, Inferior Oolite to Upper Greensand. Ex. *I. explanata*, Corallian.

Stylina. Compound, usually massive; calyces circular, projecting, usually separated. Columella small, styliform. Septa exsert. Dissepiments fairly abundant, flat. Corallites united by costæ. Basal epitheca with folds. Trias to Cretaceous; in England, Inferior Oolite to Corallian. Ex. *S. tubulifera*, Corallian.

Thecosmilia. Compound, dendroid or rarely almost massive. Multiplication by fission. Margins of calyces irregular. Columella rudimentary or absent. Septa strong, upper edges dentate, more or less exsert. Dissepiments abundant. Epitheca thick and folded, but often not preserved. Trias to Cretaceous; in England, Lias to Corallian. Ex. *T. annularis*, Corallian and Kimeridgian.

Holocystis. Compound, massive, convex; calyces polygonal. Columella very small or absent. Corallites united by their walls or by costæ. The four principal septa are much better developed than the others. Tabulæ well developed. Lower Greensand. Ex. *H. elegans*.

Thamnasteria. Compound, massive; convex or laminar. Walls of the corallites indistinct. Calyces shallow. Septa formed of fan-shaped rows of rods; the septa of adjoining corallites confluent; faces of septa with granulations. Columella small, trabeculate. Dissepiments present, synapticulæ numerous. Usually a basal epitheca. Trias to Miocene; in England, Inferior Oolite to Upper Greensand. Ex. *T. lyelli*, Great Oolite.

Micrabacia. Simple, free, discoidal, base concave. Columella false. Septa numerous, with their outer edges perpendicular. Synapticulæ present. Wall on the base only, thin; costæ granular. Upper Cretaceous. Ex. *M. coronula*.

Goniopora (=*Litharæa*). Compound, massive, perforate. Calyces more or less polygonal. Septa well developed, the faces spiny, the upper edges dentate. Walls of the corallites reticulate. Columella formed by the ends of the septa. Cretaceous to present day, common in Eocene. Ex. *G. websteri*, Bracklesham Beds.

ORDER II. ALCYONARIA

The Alcyonaria are nearly all colonial organisms; the polyps possess eight mesenteries and eight tentacles, the latter being provided with pinnules (fig. 46, 4). In the stomodæum there is only one groove with cilia, and the longitudinal muscles (fig. 47, 6) on the mesenteries are all directed toward the groove. All the mesenteries reach the stomodæum (1). The nature of the skeleton varies considerably; in *Alcyonium* it consists of isolated spicules of carbonate of lime embedded in the common gelatinous base from which the polyps arise. In some cases it has the form of an axial rod surrounded by the cœnosarc; this rod may consist of horny material (*e.g. Gorgonia*) or of carbonate of lime (*e.g. Corallium*, the red coral), or it may be formed of alternating segments of horny and of calcareous material as in *Isis*. In the 'organ-pipe coral' (*Tubipora musica*, fig. 48) the skeleton consists of numerous parallel tubes or corallites (*a*) which are not in contact but are held together by horizontal calcareous plates or 'platforms' (*b*). The walls of the corallites, although apparently quite compact, are really composed of spicules which have serrated edges and are firmly fitted together. A single polyp lives at the summit of each corallite;

spicules occur in the middle gelatinous layer (mesoglœa) of the polyp, and in the lower part become interlocked to form the solid wall of the corallite. The interior of each corallite is divided up by tabulæ which are often funnel-shaped (fig. 48, *c*).

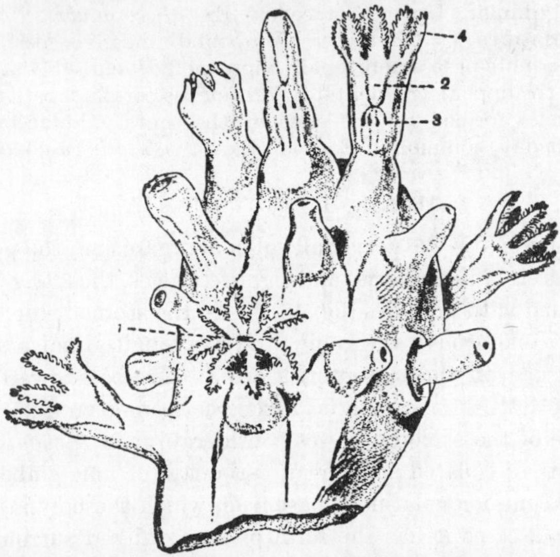

Fig. 46. Part of a colony of *Alcyonium digitatum* showing thirteen polyps in various stages of retraction and expansion. (From Shipley and MacBride.) 1, mouth; 2, stomodæum; 3, mesenteries; 4, tentacles. × 8.

In some of the Alcyonaria, as for example *Pennatula*, there are in addition to the ordinary polyps (or *autozooids*) others of a more rudimentary character, known as *siphonozooids*, in which tentacles are absent.

The blue coral (*Heliopora*), which is abundant in the Indian and Pacific Oceans, differs from other living Alcyonaria in that the skeleton consists of calcareous fibres

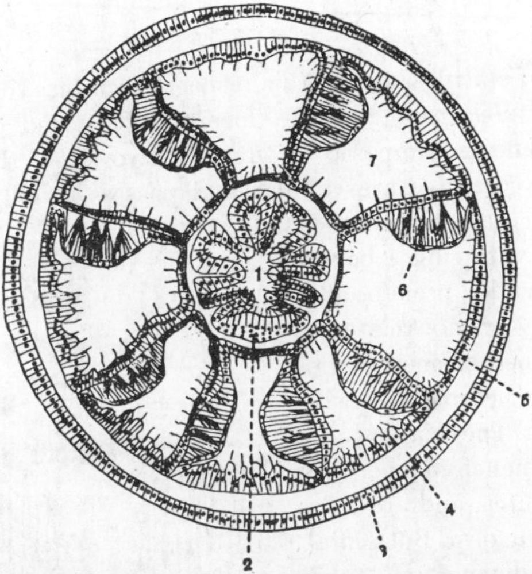

Fig. 47. Transverse section through a polyp of *Alcyonium digitatum* in the region of the stomodæum. ×about 120. 1, cavity of stomodæum; 2, ventral groove with cilia (siphonoglyph); 3, ectoderm; 4, gelatinous layer; 5, endoderm; 6, muscles of mesenteries; 7, cavity between mesenteries. (After Hickson.)

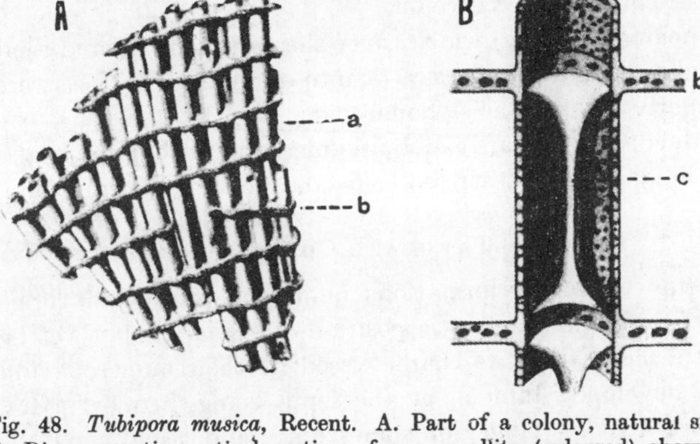

Fig. 48. *Tubipora musica*, Recent. A. Part of a colony, natural size. B. Diagrammatic vertical section of one corallite (enlarged) showing canals in the wall and platform. *a*, corallite; *b*, platform; *c*, tabula.

instead of spicules, and in this respect resembles the Madreporaria. *Heliopora* has the form of branched or lobed masses, and is composed of tubes of two sizes; the larger tubes or corallites are circular and possess usually fifteen spine-like projections at their summits with ridges below; these are called pseudosepta, since they are not related to the number of mesenteries and do not correspond with true septa. The smaller tubes form a cœnenchyma between the corallites, and are more irregular in form. Both corallites and cœnenchymal tubes are divided by horizontal plates or tabulæ. The soft parts form a thin sheet over the surface of the skeleton; polyps (fig. 49, *a*, *b*) are placed in the corallites and give off branching tubes (*d*) which cover the cœnenchyma and send blind prolongations or cæca (*e*) into its tubes. The cæca were formerly regarded as siphonozooids.

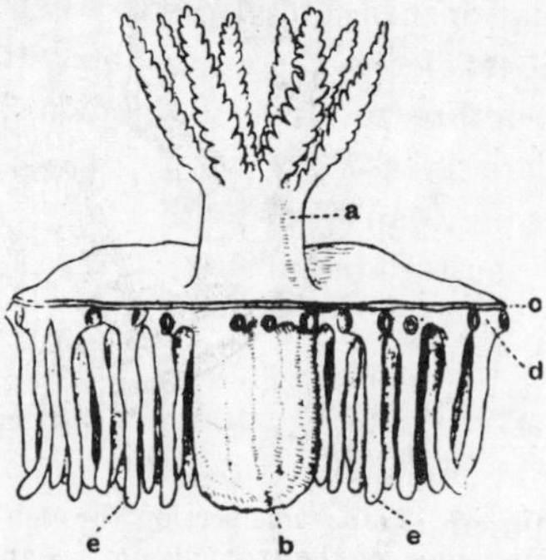

Fig. 49. *Heliopora cœrulea*. A single polyp and the adjacent soft parts. *a*, the projecting part of the polyp with eight pinnate tentacles; *b*, lower part of the polyp; *c*, ectoderm; *d*, sheet of canals; *e*, cæca. (After Bourne.)

Alcyonaria are rare as fossils, unless the 'Tabulate Corals' of the Palæozoic, described below, be included in that group.

TABULATE CORALS

In the Palæozoic formations numerous compound corals are found, the systematic position of which cannot yet be established; they are characterised by their numerous and well-developed tabulæ, by the septa being, in most cases, represented by ridges or spines only, and usually by the long, slender, tube-like corallites. Some of these corals present considerable resemblance to living Alcyonaria; for example, *Syringopora* is similar to *Tubipora*, and *Heliolites* to *Heliopora*: on account of this, many authors maintain that these fossil forms belong to the Alcyonaria, but this relationship is denied by other writers who point out that the skeleton is not formed of spicules but is similar in structure to that of Zoantharian corals, and further that there is a close resemblance between *Favosites* and the living Zoantharian *Alveopora*. Other views of the affinities of these Palæozoic corals are (1) that they do not belong to either the Zoantharia or the Alcyonaria, but constitute an isolated group of the Anthozoa, (2) that they have been derived from early forms of the Rugose corals, of which they form a specialised offshoot; the evidence for this view appears to be furnished chiefly by the Heliolitidæ.

A few species which appear to be allied to the Palæozoic forms have been found in deposits of Mesozoic age.

Syringopora (fig. 50). Compound; corallites tubular, for the most part not in contact, more or less parallel to one another. The interiors of the different corallites communicate by means of horizontal connecting tubes. Septa feebly developed, generally represented by spines. Tabulæ numerous, more or less funnel-shaped. Budding basal. Llandovery to Carboniferous Limestone. Ex. *S. reticulata*, Carboniferous.

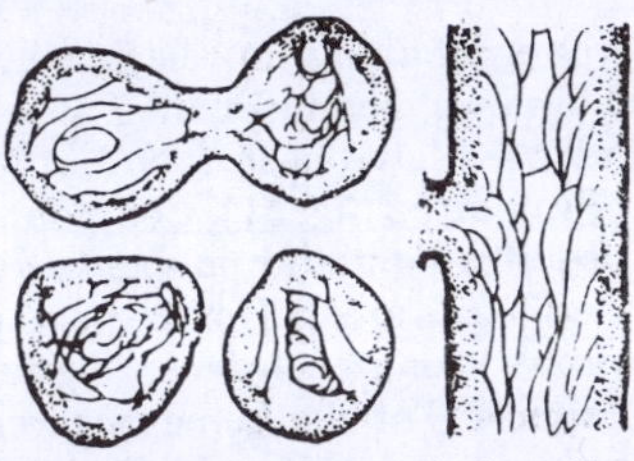

Fig. 50. *Syringopora reticulata*, Carboniferous Limestone. Horizontal and vertical sections of corallites. × 5.

Syringopora agrees with *Tubipora* (fig. 48) in consisting of parallel, cylindrical corallites, which have funnel-shaped tabulæ, and in its basal budding; it differs from *Tubipora* in having much thicker walls which are not composed of spicules, and are not perforated by minute canals; also in the tabulæ being much less regular in form and position, and in possessing septa in the form of spines. The platforms of *Tubipora* (which are traversed by canals opening into the corallites) are represented by the connecting tubes of *Syringopora*; in one species of *Syringopora* (*S. tabulata*) the resemblance is particularly close, since the connecting tubes are given off from the corallites at definite levels in a radiating manner. On the other hand it must be noted that *Heterocœnia provincialis*, an Aporose coral from the Chalk, closely resembles *Tubipora* in its general build,

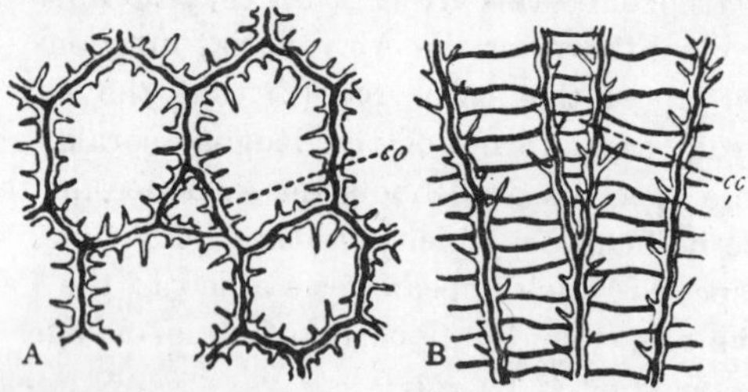

Fig. 51. *Favosites*, Silurian. A, horizontal; B, vertical section. *co*, young corallite. (From Nicholson.) ×5.

although having no relationship to the latter. No fossil forms which would connect the Palæozoic *Syringopora* with the recent *Tubipora* have been found in Mesozoic or Tertiary formations.

Favosites (fig. 51). Compound, massive, sometimes branched. Corallites long and polygonal; the walls are in contact but not fused, and are perforated by pores ('mural pores') arranged in rows along each face. Septa absent or represented by rows of spines. Tabulæ numerous, regular, generally extending quite across the corallite. Basal epitheca present. Bala to Carboniferous Limestone. Ex. *F. gothlandica*, Silurian.

Favosites is related to *Syringopora*, but the corallites are in contact, and consequently connecting tubes are absent, though probably represented by the mural pores. The living Madreporarian *Alveopora* agrees in general structure with *Favosites*, but its walls are less compact, and its basal epitheca is quite small; the development of the colony also differs, and it is probable that the resemblances between the two genera are due to homœomorphy rather than relationship. Some corals (*e.g. Koninckia, Ubaghsia*) which resemble *Favosites* are found in the Upper Cretaceous. *Alveopora* is found living in the Pacific and Indian Oceans and the Red Sea, and has been recorded from the Upper Cretaceous of Portugal, and from the Oligocene of Styria.

Pachypora. Similar to *Favosites,* but the walls of the corallites are greatly thickened, especially near the surface of the coral, by a secondary deposit of carbonate of lime. Silurian to Carboniferous. Ex. *P. cervicornis*, Devonian.

Alveolites. Allied to *Favosites.* Massive, encrusting, or branching. Corallites small, laterally compressed, and more or less triangular in section; walls moderately thick, with a few large mural pores. Usually a single septum. Silurian and Devonian. Ex. *A. labechei*, Silurian.

Pleurodictyum. Compound, discoidal, attached by part of the base, upper surface slightly convex. Corallites diverge from the centre of the base; walls thick, with irregular pores. Septa rudimentary. Tabulæ not numerous, more or less united. A basal epitheca. Devonian and Carboniferous. Ex. *P. problematicum*, Devonian.

Michelinia. Similar to *Pleurodictyum*, but the tabulæ are more numerous and form a vesicular tissue, and root-like processes are usually given off from the epitheca on the base of the coral. Devonian and Carboniferous. Ex. *M. favosa*, Carboniferous.

Heliolites (fig. 52). Corallum compound, massive or branching, formed of tubes of two sizes; the larger circular ones are the corallites, between which are the smaller polygonal tubes forming the cœnenchyma. Tabulæ occur in both, and are complete and horizontal; in the corallites septa may be absent, but, when present, they are short, lamellar or spinose, and generally twelve in number. Columella sometimes found in the corallites. Bala to Devonian. Ex. *H. porosus*, Devonian.

In general structure *Heliopora* is similar to *Heliolites*, but is more branching, whilst *Heliolites* forms rounded or encrusting masses; further, the smaller tubes which form the cœnenchyma branch dichotomously in *Heliolites*, but in *Heliopora* new tubes are introduced between the older ones. By many writers these two genera are considered to be closely allied, but the relationship is denied by others, who state that important differences are found in the structure of the corallite walls and septa. According to Lindström and others, the corallites of *Heliolites* possess a distinct and independent wall (theca) and also have true septa, whilst in *Heliopora* the corallites are simply bounded by the walls of the cœnenchymal tubes, and possess pseudosepta instead of septa and these have the form of ridges except at the opening of the corallites. Bourne, on the other hand,

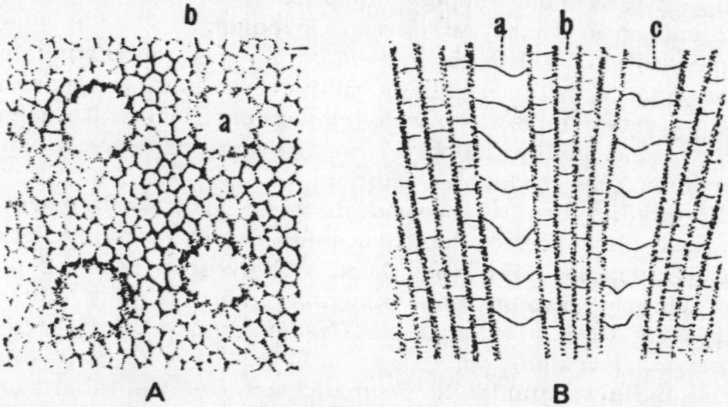

Fig. 52. *Heliolites porosus*, Devonian. A. Horizontal section. B. Vertical section. *a*, corallites; *b*, tubes forming the cœnenchyma; *c*, tabulæ. × 5.

considers that the corallites of *Heliolites* possess no independent wall, and agree in this respect with *Heliopora*. Although the cœnenchyma of *Heliolites* resembles closely that of *Heliopora*, yet Lindström and Kiär maintain that it has originated independently in the two genera, and cannot be taken as evidence of relationship; this view is based on a study of the development and phylogeny of *Heliolites*, and leads to the conclusion that that genus and its allies constitute a specialised offshoot from the early Rugosa; it is claimed that *Heliolites* has descended from an earlier Heliolitid in which the cœnenchyma is vesicular instead of tubular, and that the circular corallite wall of the Heliolitids is equivalent to the inner circular wall of *Acervularia* and *Endophyllum*, whilst the cœnenchyma represents the vesicular dissepiments of those genera. The great interval of time between the last appearance of *Heliolites* and first appearance of *Heliopora* lends support to the view that these genera are not closely allied; the former and its allies are not known in rocks of later age than the Devonian, while the latter has been recorded in rocks of Cretaceous and later date only.

Plasmopora. Allied to *Heliolites.* Usually discoidal or hemispherical. Walls of smaller tubes incomplete or absent, and their tabulæ forming a vesicular tissue. Septa in corallites lamellar, and prolonged outside each calyx, so as to enclose large spaces of uniform size. Basal epitheca with concentric ridges. Ordovician to Devonian. Ex. *C. petaliformis*, Silurian.

Propora. Allied to *Plasmopora.* Edges of calyces projecting; septa represented by spines, and not prolonged outside the calyx to enclose large spaces. Ordovician to Silurian. Ex. *P. tabulata*, Wenlock Limestone.

Halysites. Compound; corallites long and tubular, arranged in a single row and united at their sides so as to form laminæ, which intersect; in some species the corallites are of two sizes—the smaller perhaps represent the cœnenchymal tubes of *Heliolites.* Epitheca thick. Septa absent or represented by spines. Tabulæ well developed, horizontal or concave. Llandeilo Beds to Wenlock Limestone. Ex. *H. catenularia*, Wenlock Limestone.

Chætetes. Massive, often laminar, consisting of slender, tube-like polygonal corallites which are contiguous; walls often incompletely formed and may give the appearance of a solitary septum. Tabulæ thin, complete, widely separated. No septa. Probably no mural pores. Chiefly Carboniferous. Ex. *C. radians.*

Distribution of the Anthozoa

Zoantharia. From the point of view of their distribution at the present day, the Madreporaria may be divided into two groups, the solitary and the reef-building.

The solitary corals (*i.e.* the corals which do not form reefs) are found in almost all latitudes, but live mainly in rather deep water, the larger number occurring between depths of 50 and 1000 fathoms; some few (*e.g. Caryophyllia*) live in quite shallow water, whilst others inhabit the depths between 1000 and 2900 fathoms. Those which live in the deep sea, where the temperature is low and the light weak or absent, are mainly simple cup corals or delicately branching forms, and many have thin, fragile skeletons. The species of solitary corals have a wide geographical distribution, but they extend only a short way back into the geological record; thus not a single living species is found fossil in the English Cainozoic formations; about a third of the living genera, however, are represented in Cainozoic rocks, and a few (*e.g. Caryophyllia, Parasmilia, Trochocyathus*) occur in Mesozoic formations.

The distribution of reef-building corals is limited by both depth and temperature, and is also influenced by light and salinity. Thus they are found mainly between the shore-line and 14 fathoms, but some extend down to 26 fathoms and a few even lower. The maximum depth at which a true reef will form is 25 fathoms. Reef-building corals thrive only where the average temperature of the coldest month of the year is not less than 70° F. and where the usual temperature is from 77° to 86° F. Since, owing to currents, the waters along the western shores of continents are colder than those along the eastern shores the great coral reefs to-day are found in the tropical parts of the mid- and west-Pacific, the Indian Ocean and the Red Sea, and the tropical and sub-tropical parts of the west Atlantic. Similarly reef corals extend outside the tropics under the influence of warm currents, as in the Florida reefs and the Bahamas, and southern Japan. At the present day the reef-coral faunas of the Atlantic and the Indo-Pacific regions differ considerably. The latter are more numerous in species and more luxuriant in growth than those in the Atlantic, and include several genera not present in the Atlantic. Similarly there are some genera in the Atlantic not known in the Indo-Pacific region. The two faunas, however, were not always so distinct, since several genera which are now confined to the Indo-Pacific region have been found fossil in the Oligocene and Miocene of the West Indies and the south-eastern part of the United States.

The growth-form of reef corals, even those of the same species, is influenced largely by external conditions. In the quiet water of the lagoon, and outside the reef below the depth at which wave action is felt (18–25 fathoms), the corals are mainly forms which are only weakly attached to the bottom or have fragile skeletons consisting of slender branches or laminæ. While on the exposed part of the reef the corals have a massive growth-form or are composed of stout branches. The rate of growth of corals differs in different species, and in each species varies according to local conditions. In the reefs of Florida *Orbicella annularis* is the predominating coral, and its upward growth is from 5 to 7 mm. per annum. In *Acropora palmata* the upward growth is from 25 to 40 mm. per annum. But a comparison based on the increase in weight shows that *Acropora* grows nearly four times as fast as *Orbicella*, and it is estimated that a reef composed of it would grow upward at a rate of one inch in a year. Branching corals increase in dimensions more rapidly than massive forms. In the former the growth rate of those with perforate skeletons is more rapid than those with dense skeletons. In the Pacific the growth of massive corals appears to be more rapid than that of similar forms in the Atlantic Ocean. Under favourable conditions some colonies attain a diameter of from 6 to 10 feet.

Although corals are sedentary animals, some of the species and genera have a wide distribution. This is due to the long duration of the free-swimming larval stage, during which distribution is effected by ocean currents.

Corals, with possibly one or two exceptions, can only exist in salt water; but *Madrepora cribripora* is said to inhabit nearly fresh water. Clear water is likewise generally necessary, but one species, *Porites limosa*, thrives in muddy situations. In geological times, and especially in the Palæozoic and Mesozoic periods, the reef-building corals had a much wider geographical range than they have at the present day, and their remains occur abundantly in various formations in temperate and even polar regions; but in the course of the later Cainozoic period the range of the reef-builders became more and more restricted until the present limits were reached.

The Zoantharia found in the Palæozoic formations belong to the Rugose group. The other common corals of the Palæozoic are the Tabulates, the systematic position of which is uncertain. In the Mesozoic and later formations the Hexacoralla are abundantly represented.

Alcyonaria. The Alcyonaria occur in all parts of the world, and are found at all depths from the shore-line down to 2300 fathoms, but they are most abundant at depths of less than 100 fathoms; beyond this limit the number of species gradually diminishes as the depth of the water increases.

Very few of the modern Alcyonarian families occur fossil, but the Pennatulidæ are represented in the Trias by *Prographularia*, in the Lower Lias by *Mesosceptron*, in the Cretaceous by *Pavonaria*, and in the Cainozoic by *Graphularia*. The red coral, *Corallium*, is found in the Cretaceous and Cainozoic (perhaps also in the Jurassic): forms allied to *Gorgonia* occur in the Cretaceous and Tertiary rocks: *Isis* is found in the Cainozoic, and perhaps also in Cretaceous formations. Spicules, similar to those of *Alcyonium*, have been detected in the Upper Cretaceous. *Heliopora* is first recorded from the Cretaceous. The organ-pipe coral, *Tubipora*, which now lives on coral reefs in the Tropics, has not been found fossil.

Fossil corals are comparatively rare in argillaceous and arenaceous beds but often abundant in calcareous rocks, many limestones being formed largely of coral remains. This is indeed what might be expected, since existing forms can, as a general rule, live only in clear water. The chief features in the geological distribution of the Anthozoa are given in the following table.

Cambrian. *Archæocyathus*, found in the Cambrian in North America, Sardinia, Spain, and Australia has sometimes been regarded as a coral, but is probably a sponge.

Ordovician. In North America corals (especially *Streptelasma* and *Columnaria*) are common in this system, but in England only a few forms have been found, the most important being *Favosites, Heliolites, Halysites.*

Silurian. Corals are very abundant, especially in the Wenlock Limestone. Rugosa: *Xylodes, Acervularia, Omphyma, Cystiphyllum, Kodonophyllum, Chonophyllum, Spongophylloides, Tryplasma, Calostylis, Palæocyclus, Petraia, Goniophyllum.* Tabulate corals: *Syringopora, Favosites, Heliolites, Plasmopora, Propora, Halysites.*

Devonian. Rugosa: *Cyathophyllum, Heliophyllum, Phillipsastræa, Endophyllum, Spongophyllum, Mesophylloides, Zaphrentis, Prismatophyllum, Mesophyllum, Calceola.* Tabulate corals: *Favosites, Alveolites, Pachypora, Pleurodictyum, Heliolites.*

Carboniferous. Rugosa: *Palæosmilia, Lithostrotion, Orionastræa, Clisiophyllum, Dibunophyllum, Aulophyllum, Lonsdaleia, Zaphrentis, Cyathaxonia, Caninia, Amplexus.* Tabulate corals: *Michelinia. Syringopora, Chætetes.*

Permian. Rugose and Tabulate corals, generally similar to those of the Carboniferous, have been found in Russia, China, Timor, etc. *Waagenophyllum, Lonsdaleia, Corwenia, Zaphrentis, Caninia.*

Trias. Corals are absent in England, but occur in the Alpine Trias; the Palæozoic forms have become extinct and in place of them are *Rhabdophyllia, Montlivaltia, Thecosmilia, Stylophyllum, Styllophyllopsis, Isastræa, Phyllocœnia, Astrocœnia, Stylina, Omphalophyllia.*

Jurassic. *Styllophyllopsis, Heterastræa, Astrocœnia* and *Thecosmilia* are found in the Lias but are not common. In the Oolites corals become very abundant, *e.g. Montlivaltia, Isastræa, Thamnasteria, Thecosmilia, Stylina, Cyathophora, Cladophyllia, Calamophyllia, Chomatoseris.*

Cretaceous. Corals are not abundant in England; the chief forms are *Parasmilia, Trochocyathus, Micrabacia, Holocystis.* In some parts of Europe, especially in the Gosau beds (of Chalk age) of the Austrian Alps, corals are very numerous and include *Astrocœnia, Montlivaltia, Isastræa, Cyclolites, Synastræa,* etc.

Cainozoic. Corals are rare in English Cainozoic formations: *Turbinolia, Dendrophyllia, Oculina* and *Goniopora* (*Litharæa*) occur in the Eocene; *Madrepora* in the Oligocene; *Flabellum* in the Pliocene. In the middle and south of Europe, and in the south-eastern part of the United States. corals are found abundantly in various Cainozoic deposits.

PHYLUM ECHINODERMA

Sub-Phyla		*Classes*
1. Eleutherozoa	...	1. Asterozoa 2. Echinoidea 3. Holothuroidea
2. Pelmatozoa ...	...	1. Crinoidea 2. Cystidea 3. Blastoidea 4. Edrioasteroidea

The Echinoderms are all marine and comprise the star-fishes, brittle-stars, sea-urchins, sea-lilies, sea-cucumbers, and the extinct blastoids and cystideans. The body is very often radially symmetrical, the symmetry being generally pentamerous. But in many cases there is also a more or less well-marked bilateral arrangement of parts. In the majority of cases the alimentary canal terminates in an anus. A body-cavity or cœlom is present and surrounds the alimentary canal. The water-vascular system (fig. 56) is one of the distinguishing features of the group: it consists of a set of vessels containing a watery fluid and generally placed in communication with the sea-water by means of a canal; one vessel forms a ring round the œsophagus from which radiating trunks are given off. The water-vascular system functions in respiration and as a sensory organ, and often also in locomotion. A nervous system is present; one part of it has a distribution similar to that of the water-vascular system. Reproduction is mainly sexual; as a rule the sexes are separate, but do not differ externally.

In nearly all echinoderms there is a dermal skeleton. This is calcareous and consists sometimes of isolated pieces, but more usually of rods or plates united by fibres of connective tissue and forming a complete shell or test, which may be either flexible or rigid; spines and other processes are often attached to the plates. When examined microscopically each part of the skeleton is found to be formed of a network of calcareous rods (fig. 53), with a jelly-like substance in the spaces of the network. The details of the structure vary in different forms, depending on the size and shape of the spaces between the rods. In the spines of sea-urchins the network of rods has usually a radial arrange-

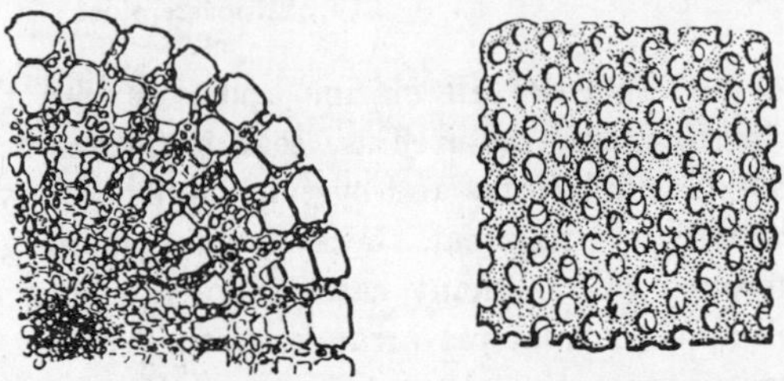

A B

Fig. 53. A. Portion of transverse section of a spine of a sea-urchin, *Echinometra*, Recent. Magnified. B. Section of interambulacral plate of recent *Cidaris* cut parallel to the surface. Magnified.

ment, with polygonal or rectangular spaces (fig. 53 A), except at the centre, where the structure is more irregular. Another characteristic feature of the skeleton is that each component part shows the optical characters of a crystal of calcite, and differs only from an ordinary crystal in not having crystal contours and in the possession of the netted structure. In a plate the principal crystallographic axis is at right angles to the surface, in a spine it is parallel with the length. In fossil specimens the spaces in the network of rods usually become filled with calcite, which is deposited in crystalline continuity with that forming the plate or spine. In such cases the characteristic cleavage of calcite becomes well marked, so that when the plate or spine is broken, the fracture passes along the cleavage planes, instead of being irregular as in the recent forms. By the infiltration of calcite and the development of cleavage, the organic structure in fossil echinoderms is sometimes partly or almost completely destroyed.

The Echinoderma are divided into two main groups, (1) the Eleutherozoa, (2) the Pelmatozoa.

I. ELEUTHEROZOA

The Eleutherozoa possess no fixing organ and are able to move about freely. This group is divided into three classes (1) Asterozoa, (2) Echinoidea, (3) Holothuroidea.

CLASS I. ASTEROZOA

The Asterozoa are represented in the older Palæozoic rocks by a great diversity of forms, but these, in the main, can be arranged in two groups which have survived to the present day—the Asteroidea (or starfish) and the Ophiuroidea (or brittle-stars). The Asteroidea are the simpler of these two groups, and have undergone less modification from the parent Asterozoan stock. All the forms of the Asterozoa are built round the water-vascular system (fig. 56) in a more or less similar way; there is a central mouth inside the water-vascular ring, and a disc of varying extent around the mouth; five arms (occasionally secondarily multiplied) come off from the disc. The main variations in the structure of the skeleton appear to be connected with the manner of life of the forms, and can be best illustrated by an account of the structure of the two groups.

SUB-CLASS I. ASTEROIDEA

In the Asteroidea the arms are usually short and merge gradually into the disc. Occasionally, as in the recent genus *Pentagonaster* and in the Chalk genus *Metopaster*, the arms are so short that the whole body is almost a pentagon. Other genera, such as *Astropecten*, have longer arms, but no Asteroidea, except a few deep sea forms, have the long thin arms which usually characterise the Ophiuroidea.

The two surfaces of the Asteroid are readily distinguishable. The under surface (known as the *oral*, *ambulacral*, or *actinal*, fig. 54) is marked by the mouth, and the five deep ambulacral grooves (*Amb. gr.*) along the arms. In each of these grooves one of the five radial water vessels (fig. 56, *b*) is placed and from it arise the tubular offshoots known as the *tube-feet* (*f*). The upper surface (known as the *apical*, *aboral*, *anti-ambulacral*, or *abactinal*) is completely covered over; a distinct ossicle on this surface is the *madreporite*, the porous plate through which water is admitted into the water-vascular system.

The ambulacral grooves extend from the mouth to the extreme tip of the arms. Each groove is formed by two rows of ossicles (the *ambulacral ossicles*, fig. 55, *a*) which meet at an angle making an arch, and is bordered on each side by another row of ossicles, the *adambulacrals* (fig. 55, *b*). Between the ambulacrals are pores for the passage of the *ampullæ* or reservoirs (*f*) attached to the tube-feet (*g*). The tube-feet themselves are used for pulling open Lamellibranchs on which star-fish feed, and for climbing and walking. The ambulacral groove can be closed for the protection of the tube-feet by muscles placed ventrally to the radial water vessel or opened by muscles dorsal to the same structure. Longitudinal muscles occur between the adambulacrals and at the dorsal tips of the ambulacral ossicles, by

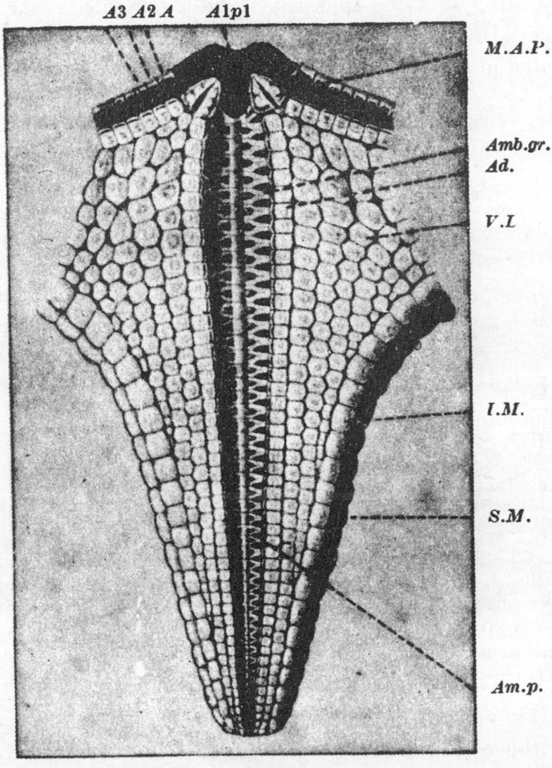

Fig. 54. Oral surface of a fifth part of the skeleton of *Pentaceros reticulatus*. *A* 1 *p* 1, anterior process of first ambulacral; *A*, *A* 2, *A* 3, the first three ambulacral ossicles; *M.A.P.*, mouth-angle plates; *Amb.gr.*, ambulacral groove; *Ad.*, adambulacrals; *V.L.*, ventro-lateral plates; *I.M.*, infero-marginal ossicles; *S.M.*, supero-marginal ossicles; *Am.p.*, ambulacral pore. (From Spencer after Agassiz.)

means of which each side of the arm can be contracted. The ossicles nearest the mouth, in series with the adambulacrals, are called *mouth-angle plates* (fig. 54, *M.A.P.*); they are often so stout that they give the mouth a star-shaped form. In the inter-radial angles supporting the mouth-angle plates is a stout plate, the *odontophor*; this is not usually

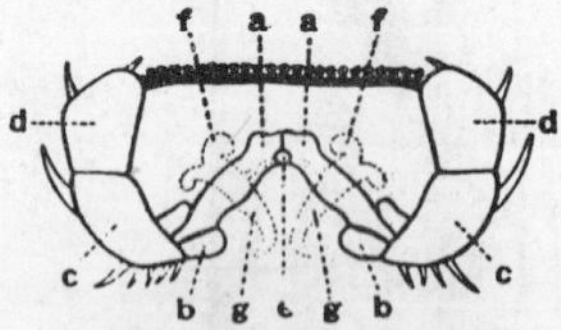

Fig. 55.

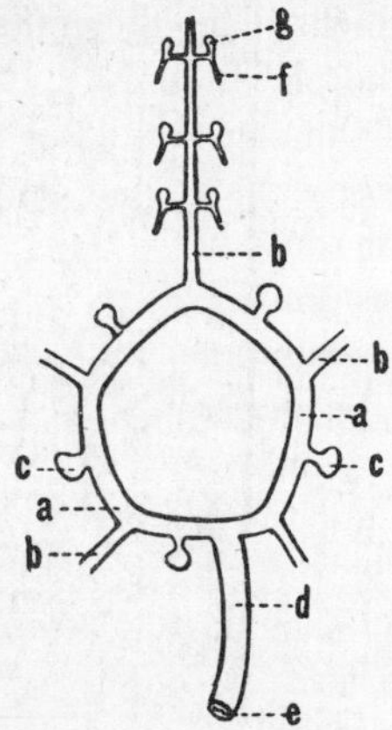

Fig. 56

Fig. 55. Section of the arm of a star-fish (*Astropecten*). *a*, ambulacral ossicles; *b*, adambulacral plates; *c*, infero-marginal plates with spines; *d*, supero-marginals; *e*, radial water vessel; *f*, ampulla; *g*, tube-feet. Enlarged.

Fig. 56. Diagram of the water-vascular system of a star-fish. *a*, circular vessel round the mouth; *b*, radial vessels; *c*, Polian vesicles; *d*, stone-canal; *e*, madreporic plate; *f*, tube-feet (only a few shown); *g*, ampulla.

visible on the oral surface in recent forms as it is covered by the ventro-lateral plates, but in many Palæozoic genera which do not possess ventro laterals it is seen distinctly.

In the remaining parts of the skeleton, which are known collectively as the *interambulacral skeleton*, the following parts are usually clearly differentiated: (1) a double series of plates, the *supero-* and *infero-marginals* which form the sides of the arms and disc (figs. 55, *c*, *d*; 54, *S.M.*, *I.M.*): (2) small plates, the *ventro-laterals* placed on the oral surface of the disc between the marginal plates and the adambulacral ossicles (fig. 54, *V.L.*): (3) a central primary circlet of radial and inter-radial plates on the aboral surface, usually more distinct in the young than in the adult form: (4) plates which fill in the remaining portions of the aboral surface. The plates running down the middle of the aboral surface of the arm are known as radials. The terminal member of this series is notched for the reception of the most distal tube-foot which possesses an eye-spot, and the plate is therefore known as an *ocular* Some or all of the plates of the interambulacral skeleton may be partly cut away to allow of tube-like projections of the skin which form simple respiratory organs known as *dermal branchiæ* or *papulæ*.

All the plates except the ambulacrals may carry spines. The disposition of the spines is of importance in classification. In the genera found in the Chalk the ornament formed by the pits in which the spines are sunk may be used to distinguish genera and even species. Frequently some of the spines are modified into pincer-like organs (*pedicellariæ*) which serve for protection and as a means of clearing the surface of the body.

The soft parts follow the general radiate symmetry already noticed in the water-vascular system. The mouth leads into a short œsophagus which opens into a globular stomach; above the stomach is the pentagonal pyloric sac, from the angles of which are given off branches which soon divide into two and extend down the arms near the aboral surface. From the pyloric sac a short narrow intestine leads to the anus at the centre of the aboral surface. The distribution of the main part of the nervous system is similar to that of the water-vascular system: it consists of a ring round the mouth and of a branch which extends down the ambulacral groove of each arm; there is also a layer of fine nerve fibres under the ectoderm. The genital glands occur in pairs at the base of each arm and open to the exterior between the rays. The water-vascular system communicates with the exterior by means of a canal (fig. 56, *d*) which passes from the circular vessel to the madreporite on the aboral surface of the disc; this is known as the *stone canal* on account of the deposit of carbonate of lime in its walls.

Metopaster. Body flattened, pentagonal in outline, the rays only slightly produced. Marginal plates thick, with rabbet edge which bears shallow spine pits. Supero-marginal plates few in number, forming a broad border to the disc; the terminal pair of plates the largest. Aboral surface covered with small polygonal (usually hexagonal) plates. Infero-marginal plates more numerous than the supero-marginals. Plates on the oral surface small, polygonal. Cretaceous. Ex. *M. parkinsoni*, Upper Chalk.

Mitraster. Similar to *Metopaster*, but rounded (or slightly pentagonal) in form, with supero-marginal plates few and of more nearly equal size. Chalk. Ex. *M. hunteri*.

Crateraster. Body almost pentagonal. Lateral faces of marginal plates with crater-like pits. Apical faces of marginals usually with rugosities. Chalk. Ex. *C. quinqueloba*.

Pycinaster. General shape of the body similar to *Calliderma*. Marginals high and almost smooth. Supero-marginals wedge-shaped. Upper Greensand and Chalk. Ex. *P. angustatus*, Upper Chalk.

Calliderma. Body flattened, pentagonal-stellate, with the rays moderately long. Marginal plates large, forming a broad border to the disc, covered with shallow spine pits. Aboral surface of disc with small plates arranged regularly. Cretaceous to present day. Ex. *C. smithiæ*, Chalk.

Stauranderaster. Body high; arms produced. Plates with a rabbet edge. Ornament on plates, when present, confined to the central raised area. Proximal marginals breast-plate shaped. A distinct central circlet of plates is often present on the aboral surface. Chalk. Ex. *S. bulbiferus*.

SUB-CLASS II. OPHIUROIDEA

The Ophiuroidea are a highly modified group. The arrangement of the nervous and water-vascular systems is similar to that found in the Asteroidea, but the tube-feet no longer have any locomotory function, being merely sensory or respiratory organs. The arms are long and thin, and are capable of wriggling and writhing movements. The disc is round and sharply marked off from the arms. Many Ophiuroids live on mud from the sea bottom which they push into their mouths by means of the tube-feet nearest the mouth.

The structure of the arm is shown in fig. 57. The ambulacral ossicles are no longer pairs of rod-like bodies, but consist of a single series of stout *vertebræ* (fig. 57, *d*) which articulate upon each other. The derivation of these vertebræ from pairs of ambulacral ossicles can be followed in the young forms and in the older Palæozoic fossils. The adambulacrals are represented by thin plates, known as lateral plates or side shields (*b*), which usually possess a ridge carrying a comb of long spines. The aboral surface is protected by a series of dorsal plates (*a*) analogous to the radials of the Asteroidea. The groove is covered by ventral plates (*c*) not represented in the Asteroidea. Neither generative organs nor diverticula from the alimentary canal enter the arms as they do in the Asteroidea.

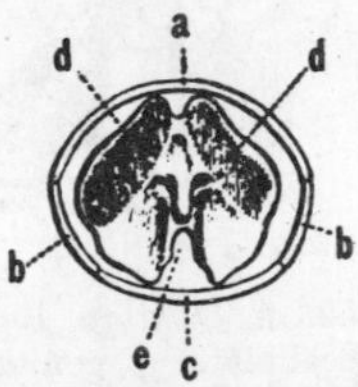

Fig. 57. Section of the arm of an Ophiuroid (*Ophioglypha*). *a*, dorsal plate; *b*, lateral plate; *c*, ventral plate; *d*, ambulacral ossicles fused along the median vertical line; *e*, ambulacral groove. Enlarged.

The oral surface of the disc (fig. 58 A) is formed by interradial pouches covered with scaly plates and granules. The slits (g) between the pouches and the arms serve as genital openings and for the entrance of water for respiratory purposes. In the inter-radial angles between the mouth plates are five large *buccal plates* (b), one of which serves as a madreporite. The aboral surface of the disc (fig. 58 B) is in most cases covered with numerous small plates, but usually there is at the bases of the arms on each side a

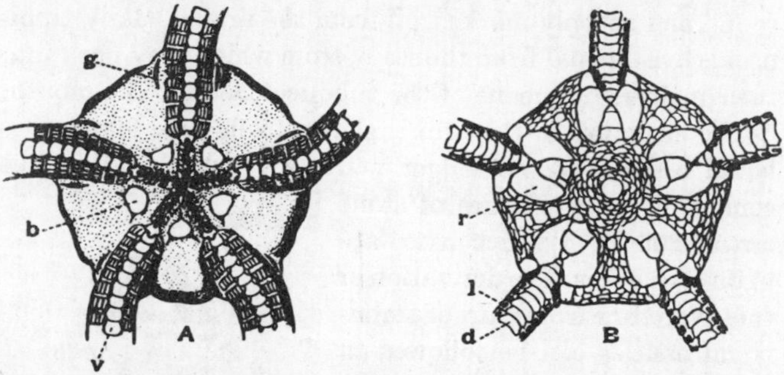

Fig. 58. A. *Ophiura*, Recent. Oral surface of disc and part of the arms. *b*, buccal plates; *g*, genital slits; *v*, ventral plates of arms. B. *Ophioglypha*. Recent. Aboral surface. *r*, radial plates; *l*, lateral plates of arms; *d*, dorsal plates of arms. × 1½.

large plate, the *radial* (r). Some forms have a primary circlet of plates similar to that mentioned for the Asteroidea (p. 123).

The mouth-angle plates are fused with the proximal pair of ambulacrals to form stout jaws. A single stout plate, the *torus*, situate at the mouth extremity of each pair of jaws, carries strong spines or teeth which are used for grinding.

The Palæozoic Ophiuroidea differ from recent forms in several respects. All the best known forms are devoid of ventral plates covering the groove. The radial water vessels are protected by outgrowths of the ambulacral ossicles which form a closed canal. The opposite members of each pair of ambulacral ossicles are not fused into single vertebræ. There are no buccal plates, and the madreporite is a separate plate. The vertebræ of some genera possess articulating knobs and prominences similar to those in recent forms. The principal Palæozoic genera are:

Lapworthura. Disc circular, composed of small spicules. The halves of each vertebra (ambulacrals) are opposite. Ludlow Beds. Ex. *L. miltoni*.

Euzonosoma. Disc with concave edges, bordered by a single row of marginal plates. Ambulacral plates alternating. Ordovician to Devonian. Ex. *E. petaloides*, Devonian.

Protaster. Disc composed of overlapping scales. Ambulacral ossicles alternating. Silurian. Ex. *P. sedgwicki*, Ludlow Beds.

Distribution of the Asterozoa

The Asteroidea have a wide distribution in the ocean at the present day; they are most abundant at moderate depths, but also occur in abyssal regions.

The majority of the Ophiuroids live in shallow water, more than half of the known species being found at a depth of less than 30 fathoms, and most of these not extending lower. Other forms occur at greater depths, some species being found below 1000 fathoms.

The earliest representatives of the Asterozoa at present known are found in the Upper Cambrian. Complete specimens are usually rare as fossils since the skeleton readily breaks up after death, but at some horizons and localities numerous examples have been found, viz.: Lower Ordovician of Bohemia; Upper Ordovician of Thraive Glen, Girvan; Wenlock Beds of Gutterford Burn, Pentland Hills; Lower Ludlow of Leintwardine, Herefordshire; Lower Devonian (Budenbach Slates) of the Rhine; Lias of Whitby and Lyme Regis; Corallian (Calcareous Grit) of Yorkshire; Upper Chalk of Bromley, Kent.

The classification of the fossil Asterozoa is not yet settled. The following Palæozoic genera are closely allied to the recent Asteroidea—*Hudsonaster*, *Mesopalæaster*, and *Promopalæaster* (Ordovician and Silurian), *Xenaster* and *Devonaster* (Devonian); these genera show, to some extent, characteristics found in the young of recent forms, for they usually possess a comparatively simple skeleton and have a very distinct primary circlet of plates in the centre of the aboral surface of the disc.

An extinct branch of the Asteroidea is formed by the Palæozoic genus *Urasterella* and its allies; the disc of these forms is small and the arms are long and thin; the adambulacral plates are broad and possess a distinct ridge which bears stout Ophiuroid-like spines.

Some Palæozoic Asterozoa have an Asteroid shape and Asteroid-like ambulacrals, but the madreporite, when known, is on the oral surface, and they show other peculiarities of structure which ally them with the Ophiuroidea rather than with the Asteroidea; these include *Stenaster* (Ordovician), *Helianthaster* (Devonian), *Palasteriscus* (Devonian), *Sturtzaster*, *Rhopalacoma* and *Bdellacoma* (Lower Ludlow).

Well-known Palæozoic Ophiuroids are *Lapworthura* (Ludlow), *Euzonosoma* (Ordovician to Devonian), *Protaster* (Ludlow), and *Onychaster* (Devonian and Carboniferous).

Forms very similar to living Ophiuroids are found in the Jurassic and have been referred to the recent genera *Ophiura*, *Ophiolepis*, and *Ophiocten*. In the Cretaceous *Ophiura* and *Amphiura* occur. A few forms, such as *Ophioglypha*, have been found in the Eocene.

The Asteroidea in the Jurassic formations closely resemble living forms and have been referred to the genera *Astropecten*, *Solaster* and *Plumaster*. The Asteroidea of the Cretaceous are found chiefly in the Chalk where isolated marginal plates are often abundant and can be used for the determination of zonal horizons; the principal genera are *Metopaster*, *Mitraster*, *Crateraster*, *Pycinaster*, *Calliderma* and *Stauranderaster*. In the Cainozoic rocks of England star-fishes are rarely found.

CLASS II. ECHINOIDEA

The echinoids or sea-urchins have usually a globular, heart-shaped, or discoidal body, covered with spines. The shell or test is covered by a layer of ectoderm and consists of numerous calcareous plates, which, in the majority of cases, are immovably united. Nothing corresponding to the ambulacral groove of the starfish is to be seen on the surface, since the water-vascular system is internal to the skeleton, and as a result the tube-feet, in order to reach the exterior, must pierce the plates of the test. The mouth is on the inferior surface, and is either central or in front of the centre. The anus is either at the summit of the test or posterior to it, somewhere along a line drawn from the summit to the centre of the base. In the regular echinoids both anus and mouth are central—being placed at opposite poles of the test; in the irregular echinoids the anus is always, and the mouth often, excentric. In the test we may distinguish three parts: a small patch of plates placed at the summit, known as the *apical disc* or *apical system*; the main part of the test termed the *corona*; and the part between the mouth and the lower margin of the corona, which usually bears plates and is known as the *peristome*.

In a typical echinoid of the regular group (*e.g.* *Echinus*) the anus is placed within the apical disc (fig. 59 B), which then consists of the following parts. Near the centre is the anus (*a*), which is surrounded by a membrane bearing small plates and known as the *periproct* (*p*). The periproct is encircled by a ring formed of ten plates, five are called *genital* (*g*) and five *ocular* (*o*). The genital plates form the inner part of the ring; they are often more or less hexagonal in outline, and are usually provided with a

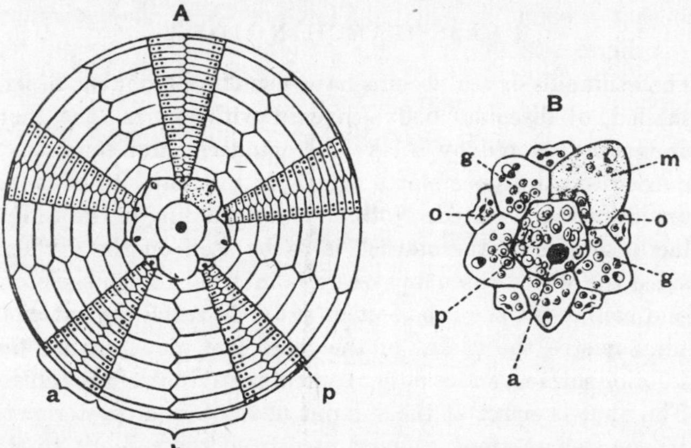

Fig. 59. A. Diagram of the upper surface of a regular echinoid, with the tubercles and spines omitted. *a*, ambulacral areas; *b*, interambulacral areas; *p*, pores in the ambulacral plates.

B. Apical disc of *Echinus esculentus*, Recent. *a*, anus; *p*, periproctal membrane with small plates; *g*, genital plates, each with a pore; *m*, madreporic plate; *o*, ocular plates. $\times 1\frac{1}{3}$.

perforation which serves as the opening for the genital ducts—whence their name; one, the anterior right, is pierced by numerous pores and is the madreporic plate (*m*). Outside the genital plates and alternating with them are the ocular plates; these are smaller than the genital and usually triangular or pentagonal, and each has a perforation through which the terminal tentacle of the radial water-vessel projects; this is pigmented and has sometimes been regarded as a rudimentary visual organ.[1]

In most of the regular echinoids the apical disc is large, but particularly so in *Cidaris*, *Salenia*, *Peltastes*, and their allies. In a few regular forms (fig. 60 D) the genital plates are completely separated from one another by the oculars, so that a single row of ten plates encircles the periproct; in others, some only of the genital plates are separated by oculars. When the oculars separate the genitals and touch the periproct they are said to be *insert* (fig. 60 D); when they do not touch the periproct they are *exsert* (A). Each genital plate has usually one perforation only, but in many Palæozoic forms (fig. 60 D) there are three or more, and in *Cidaris* often two. Similarly the oculars in a few Palæozoic echinoids have two perforations instead of one. In *Salenia* and *Peltastes* there is an extra plate in the apical disc; it is in front of the periproct and is known as the *sur-anal* plate (fig. 60 A, *b*).

In the irregular echinoids the apical disc is small, since it does not enclose the periproct. The madreporic plate may extend to the centre of the disc (fig. 60 E, *m*), and sometimes (*Spatangus*) reaches to the posterior border, separating the posterior oculars (G). The posterior genital is sometimes absent (60 B), and when present may be without a perforation (F). In *Echinocorys* and *Holaster* the apical disc is elongated, and the anterior genitals are separated from the other genitals by two oculars which join in the middle, and the posterior genital is absent (fig. 60 B); in *Collyrites* (C) the apical disc is still more

[1] The genital plates are sometimes termed *basals* and the oculars are also known as *radials*, since, by some authors, they have been considered to represent the plates which bear those names in other groups of the Echinoderma. It is more probable that, although occupying similar positions, they have originated independently in the different groups.

elongated, since the two posterior oculars are separated from the rest of the apical disc by a chain of small plates. In *Clypeaster* (H) the genital plates are fused together.

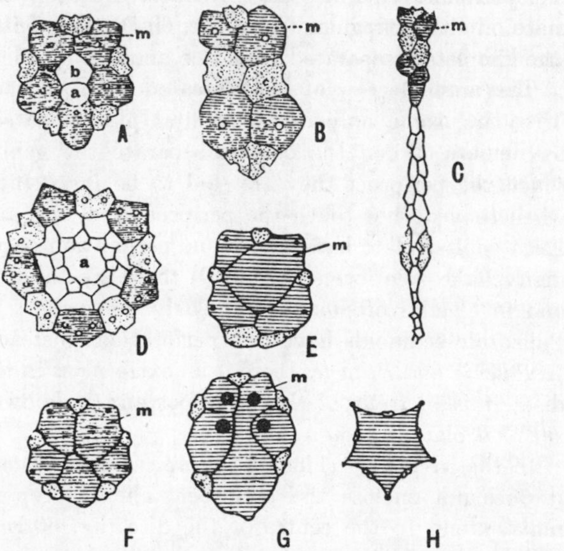

Fig. 60. Some types of apical disc. A. *Peltastes wrighti*, Lower Greensand. B. *Echinocorys vulgaris*, Upper Chalk. C. *Collyrites bicordata*, Corallian D. *Palæechinus*, Carboniferous Limestone. E. *Conulus subrotundus*, Chalk. F. *Holectypus hemisphæricus*, Inferior Oolite. G. *Spatangus purpureus*, Recent. H. *Clypeaster rosaceus*, Recent. In the figures the ocular plates are distinguished by dots, the genital plates by lines. *m*, madreporic plate; *a*, anus; *b*, sur-anal plate. All enlarged.

The corona in a typical echinoid consists of twenty columns of plates, each column extending from the apical disc to the peristome. The plates are of two kinds, *ambulacral* (fig. 59 A, *a*) and *interambulacral* (*b*); there are five double columns of ambulacrals separated by five double columns of interambulacrals; each double column is termed an *area*. The former end against the ocular plates, the latter against the genital, and in each case fresh plates are developed next the apical disc. In each area the plates alternate on either side, and since their inner ends are angular, the line between the two rows is zig-zag.

The ambulacral plates are smaller and more numerous than the interambulacral, and they are perforated by pores (p) for the passage of the tube-feet to the exterior, a radial water-vessel being placed in the middle line under each ambulacral area. The pores are usually round, but sometimes elongated; in most cases they are situated in the outer portion of the plates and are generally in pairs; each pair of pores corresponds to a single tube-foot, since each tube-foot divides at its base into two canals. Frequently each pair of pores is surrounded by an oval raised rim, the *peripodium* (fig. 61); the two pores in each pair are sometimes horizontal, but usually inclined so that the inner pore is lower than the outer pore. In some echinoids, such as *Cidaris*, and all the Palæozoic genera, each ambulacral plate is formed of one piece only (as in fig. 59)—such plates are called *simple* or *primary*. In other cases some of the ambulacral plates are *compound*, consisting of two, three or more small plates which have become fused together; but the original plates are still indicated by the lines of suture between them and also by a pair of pores on each (fig. 61); in some genera (fig. 61 A) the plates which are united are all *primaries*—that is to say, each extends from the margin to the middle line of the ambulacral area; but frequently some of the plates taper away and do not reach the middle line (or inner edge of the compound plate)—such are called *demi-plates* (*e.g.* the middle plates in fig. 61 B, the upper plate in fig. 61 C). Others, termed *occluded* plates, start from the inner margin of the compound plate but taper away before reaching the outer margin. Some known as *included* plates, do not reach either the inner or outer margin (*e.g.* the lower plate in fig. 61 C). This fusion of plates appears to be due to growth-pressure—since each plate of the test is enlarging and new plates are being added next the apical disc; also the perignathic girdle (p. 141) interferes with the passage of the ambulacral plates on to the peristome. The fact that some of the fused primary plates are smaller than others, and also the presence of demi-plates, is attributed

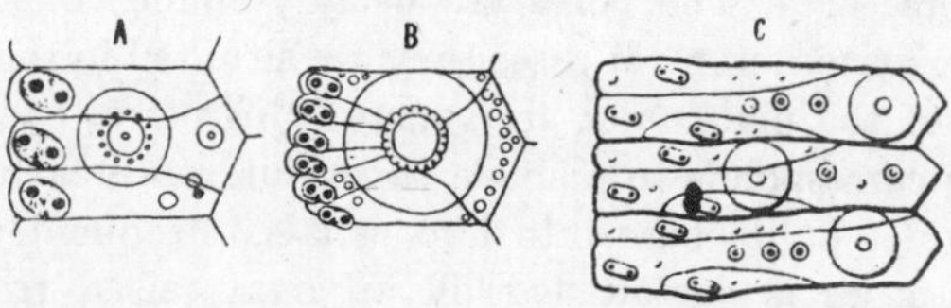

Fig. 61. Compound Ambulacral Plates. A. *Pseudodiadema hemisphæricum*, from the Corallian, formed of three fused plates. B. *Phymosoma koenigi*, from the Chalk, formed of six fused plates. C. *Stomechinus perlatus*, Upper Jurassic, three plates, each formed of three fused plates, with trigeminal pores. The upper plate is a *demi-plate*; the middle one a *primary*; the lower an *included* plate. Enlarged.

to the reduction in size of the original plates by the absorption of material under pressure. The pores in the ambulacra of some echinoids are placed one immediately above the other, so that one vertical row of pore-pairs is seen—such pores are termed *unigeminal* or *uniserial* (figs. 59, 61 A, B); in other cases the pore-pairs are alternately near to, and more distant from, the margin of the ambulacral plate, and consequently two vertical rows are formed, and the pores are said to be *bigeminal* or *biserial*; in a similar way three or more vertical rows of pore-pairs may be produced, when the pores are known as *trigeminal* (fig. 61 C) or *polygeminal*. Sometimes the pores in each pair are united by a groove on the surface of the plate, and are then termed *conjugate*. In some sea-urchins each ambulacral area has a leaf-like or lanceolate form on the upper surface of the test (fig. 71). In such cases the two rows of pores in each area diverge rapidly after leaving the apical disc, and then come together again before reaching the circumference (or *ambitus*), so that the five ambulacra together form a rosette on the upper surface of the corona; the ambulacral areas in such cases are termed *petaloid* (*e.g. Scutella*, fig. 71), but when the rows of pores diverge to only a small extent they are *sub-petaloid* (*e.g. Micraster*, figs. 73, 74). In the petaloid or sub-petaloid part the ambulacral plates are low and numerous, and consequently the tube-feet are likewise numerous. The remainder of each area (mainly on the lower surface of the test) consists of tall plates, few in number, with the pores irregularly developed or sometimes wanting. When, as in *Cidaris*, the distance between the two rows of pores increases uniformly and slowly in passing from the apical disc to the equator, and the pores are as well-developed on the under as on the upper surface of the test, the ambulacra are said to be *simple* (fig. 59).

The advantage gained by the development of compound plates, which appear first in Triassic echinoids, seems to be to give a larger number of tube-feet in each vertical row. The bigeminal or trigeminal arrangement of pores causes the tube-feet to be spread over a larger area, and so increases their mechanical efficiency; the same result was attained by the development of numerous columns of plates in Palæozoic echinoids (see below). Petaloid ambulacra are particularly well developed in flattened or cake-like echinoids, and in such forms the tube-feet have for the most part lost their locomotory function and have become respiratory organs.

With only a few exceptions the corona in the Mesozoic and later echinoids is formed of twenty columns of plates, as described above; but in the Palæozoic echinoids, more than twenty columns of plates are found (fig. 70), except in *Bothriocidaris* (Ordovician) and *Miocidaris*; the former is remarkable in having only one column of plates in each interambulacral area, with the usual two columns in each ambulacral area (fig. 62). In other Palæozoic forms the number of columns is variable and often great, so that the total number of plates in the corona becomes considerable: thus, *Archæocidaris* possesses two columns in each ambulacrum, and four in each interambulacrum (fig. 69); *Oligoporus* has four ambulacral and from four to nine interambulacral columns; *Melonechinus*, six to twelve ambulacral, and from three to eleven interambulacral columns (fig. 70); *Lepidesthes* of from eight to eighteen ambulacral, and three to seven interambulacral columns; whilst *Meekechinus* has twenty ambulacral and three interambulacral columns. In these Palæozoic forms each ambulacral plate possesses one pair of pores.

In most echinoids the plates join by a vertical suture and the test is rigid, but in some genera the plates of the corona overlap to a slight extent, giving some flexibility to the test; such is the case in several Palæozoic genera, and also in a few later forms, especially *Pelanechinus* from the Corallian, *Echinothuria* from the Chalk, and some living species of the deep-sea genera *Asthenosoma* and *Phormosoma*.

The plates of both the ambulacral and interambulacral areas are often provided with rounded elevations known as *tubercles* and *granules*. The tubercles are of various sizes, the largest being the *primary*, and those of smaller size the *secondary*. In a primary tubercle the following parts may be distinguished: at the summit a hemispheroidal piece, sometimes perforated at the top, and known as the *mamelon* (fig. 63 B, *m*). The mamelon rests on the *boss* (*b*), the upper margin of which is sometimes smooth, sometimes crenulated. The base of the boss is frequently surrounded by a smooth excavated space, the *areola* or *scrobicule* (*a*),

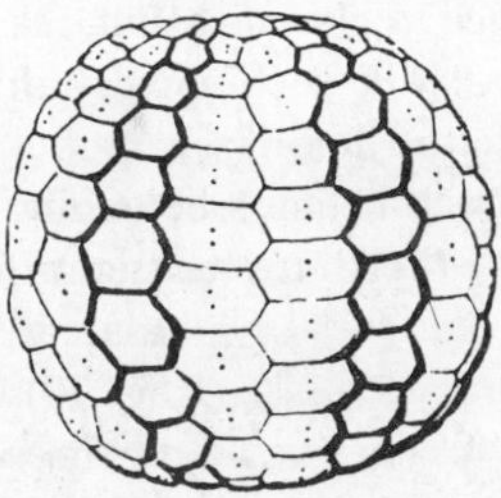

Fig. 62.

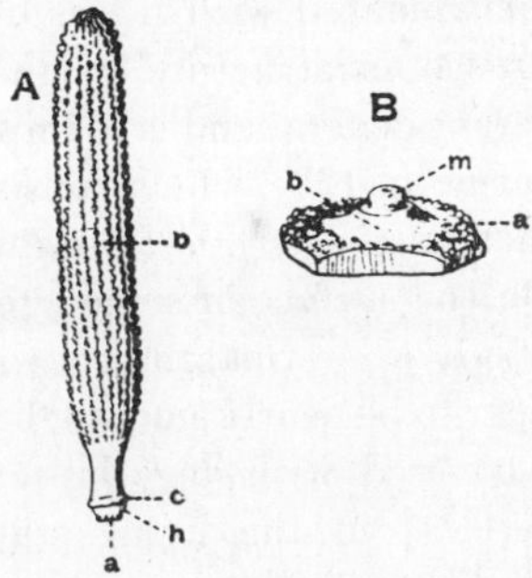

Fig. 63.

Fig. 62. *Bothriocidaris globulus*, Ordovician. Interambulacral plates shown by thick outlines. (After Jackson.) × 1½.

Fig. 63. A. Spine of *Cidaris florigemma*, from the Corallian Rocks. *a*, acetabulum; *h*, head or base; *c*, collar; *b*, shaft or stem. B. Ambulacral plate of *Cidaris* (recent) with a large primary tubercle and secondary tubercles. In the primary tubercle, *m*, mamelon; *b*, boss; *a*, areola. Natural size.

to which muscles from the spine are attached. The granules are smaller than the tubercles and have no distinct mamelon.

Attached to the tubercles are the *spines* or *radioles*; these are of different sizes and shapes in different genera and species and even on the same individual, being needle-like, rod-like, flask-shaped, etc.; the larger spines are attached to the primary tubercles, the smaller to the secondary tubercles. They serve for protection and also assist in locomotion. At the end of the spine, where it articulates with the mamelon, there is a rounded cavity, the *acetabulum* (fig. 63 A, *a*); next comes the *head* (*h*) limited above by a ring or collar (*c*), which may be smooth or crenulated and serves for the attachment of the muscles that move the spine. Beyond the collar and forming the greater part of the spine is the *shaft* or *stem* (*b*), which may be smooth, or ornamented with ridges or rows of spiny processes. The microscopic structure of the spines (fig. 53 A) varies in different genera, and is of importance in classification. Pedicellariæ (p. 123), which consist of a stalk with usually three blades, also occur, but are rarely found fossil.

On the surface of some irregular sea-urchins belonging to the sub-order Spatangina (p. 149) there are bands which appear to be nearly smooth, but are covered with very minute tubercles; in the living state they bear slender spines and their cilia produce a current of water which helps to keep the test clean. These bands are termed *fascioles*, and their position varies in different genera; sometimes they form a ring beneath the anus (*e.g. Micraster*, fig. 64, *c*), when they are said to be *sub-anal*; in other cases they encircle the rosette formed by the petaloid ambulacra (*e.g. Hemiaster*) and are said to be *peripetalous*; or they extend round the margin of the test (*e.g. Cardiaster*).

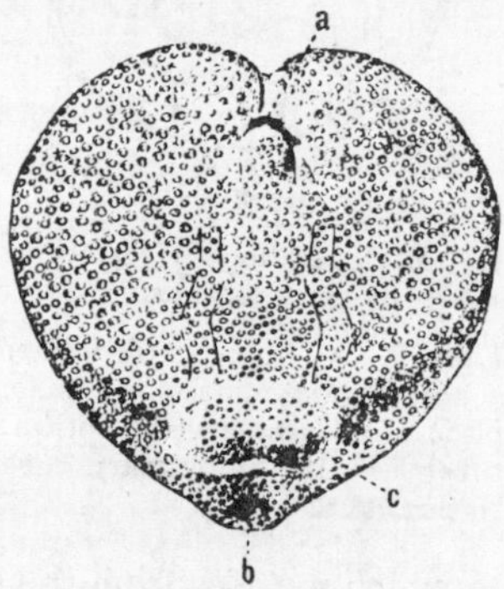

Fig. 64. Under surface of *Micraster cor-anguinum* from the Upper Chalk, showing fasciole. *a*, peristome; *b*, periproct; *c*, fasciole. × ⅔.

On the lower surface of the test is the *peristome* (figs 64, *a*, 65) in the centre of which is the mouth. The peristomal membrane, which extends from the mouth to the edge of the corona, is sometimes (*e.g. Cidaris*, fig. 65) completely covered with rows of thin, overlapping plates, but more usually bears five pairs of plates around the mouth and only small isolated plates on the remainder of the peristome, or is without plates. Some of the plates are perforated and have been derived from the ambulacral areas; others are not perforated. The plates of the peristome

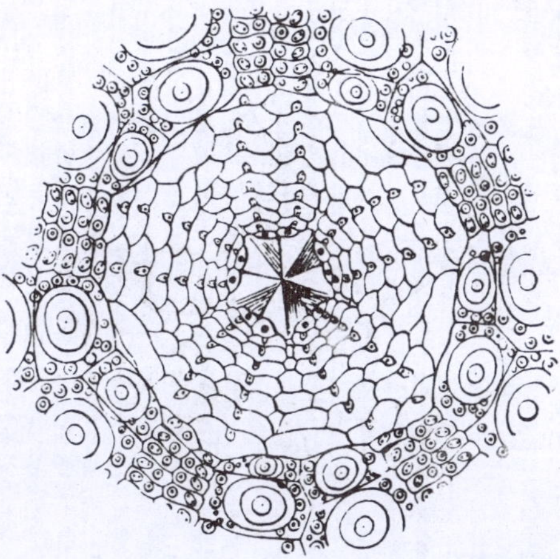

Fig. 65. *Cidaris hystrix*, Recent. Peristome and margin of corona. (After Lovén.)

are usually lost in fossil specimens. The peristome varies in shape, size, and position in different genera; it may be circular, pentagonal, or decagonal when the mouth is central, but becomes transversely oval when the mouth is anterior; its margin is entire in Palæozoic echinoids and in the Cidaridæ (figs. 65, 69), but in other regular echinoids and in the Holectypina there are ten notches or incisions, by which the five pairs of gills or branchiæ pass to the exterior.

The peristome is usually larger in the regular than in the irregular echinoids. In some irregular echinoids belonging to the sub-order Spatangina (p. 149) the parts of the ambulacra near the peristome are depressed and leaf-like, with the pores close together, whilst the intervening interambulacra are convex; this part of the corona has consequently a petaloid appearance, and is known as the *floscelle*. The

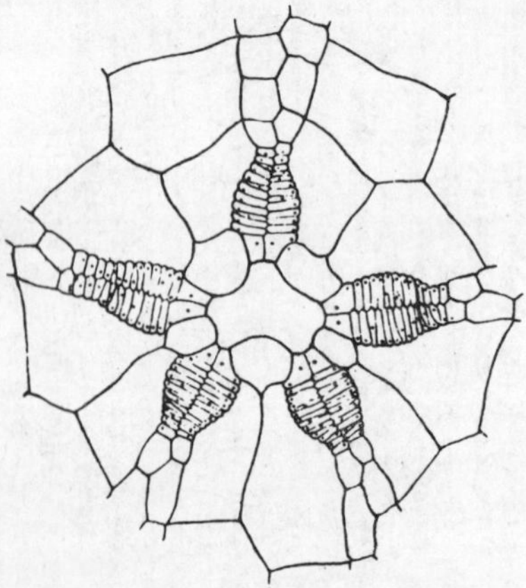

Fig. 66. *Rhyncholampas pacifica*, Recent. Part of the oral surface showing the floscelle. (After Lovén.)

ambulacral plates are low and numerous, each with a pair of pores, so that the tube-feet are numerous and serve to convey food to the mouth (fig. 66).

A pyramidal or conical structure which functions in mastication, and is known as Aristotle's lantern, is found in regular echinoids and in some irregular forms (Holectypina and Clypeastrina). The lantern consists usually of 40 calcareous pieces including five teeth which project through the mouth. Numerous muscles are attached to the calcareous parts, some of which serve to open or close the teeth and are attached to vertical plates at the margin of the peristome—these constitute what is known as the *perignathic girdle* (fig. 67, 68). This may consist of plates arising from the interambulacral areas only, known as *apophyses* (fig. 67), or there may be also processes from the sides of the ambulacral plates, termed *auricles* (fig. 68); and these may remain separate at their summits or unite to form an arch over each ambulacral area at the margin of the peristome.

The first part of the alimentary canal passes through the axis of the lantern. The circular vessel of the water-vascular system forms a ring round the œsophagus at the top of

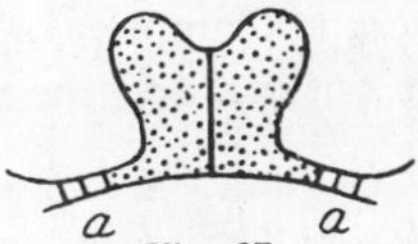

Fig. 67.

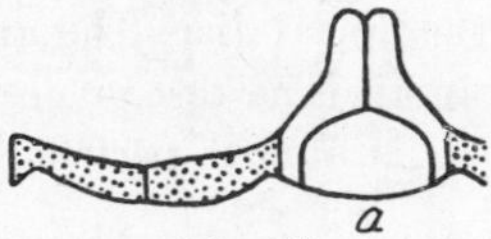

Fig. 68.

Fig. 67. Part of the perignathic girdle of *Dorocidaris* and Fig 68, *Centrechinus*. *a*, margin of ambulacral area. Apophyses dotted; auricles plain.

the lantern, and gives off five radial branches which pass through the auricles and up the middle of the inside of each ambulacral area; lateral branches, which alternate on either side, come off from the radial vessels and open into the tube-feet. The stone canal (p. 124) passes from the circular water vessel to the madreporic plate.

In the irregular echinoids there is a well-marked bilateral symmetry; a plane which passes through the anus (which is in the middle line of the posterior interambulacral area), the apical disc, and the mouth, divides the body into two similar parts. When the mouth is anterior (figs. 64, 73) the ambulacra differ considerably in size, and to some extent in structure; the anterior one is shorter than the others and sometimes, especially in burrowing forms (e.g. *Spatangus*), consists of taller and fewer plates, while the four other ambulacra are paired. The interambulacra are also unlike—the posterior one forming a large part of the base of the test (fig. 72). The bilateral character is inconspicuous in the regular sea-urchins, but the plane of symmetry may be found by means of the madreporic plate, which is always at the summit of the right anterior interambulacral area.

The Echinoidea may be divided into two Orders, (1) Regularia, (2) Irregularia.

ORDER I. REGULARIA

The peristome is at the centre of the base, and the anus within the apical disc. The ambulacra are simple. Lantern present in all. The test is circular in outline, and the radial symmetry is almost perfect.

1. *Endobranchiata*

Peristome entire. No external gills. Ambulacral plates simple. Ordovician to present day.

Palæechinus (fig. 60 D). Test spheroidal or elliptical, rigid. Apical disc with five large genital plates, each with two to five perforations; ocular plates five, small, separating the genitals. Ambulacra narrow, straight, with two columns of plates; one vertical row of pairs of pores on each side of the area. Interambulacra wide, with four to six columns of plates at the ambitus, fewer towards the poles; plates hexagonal, except those next the ambulacral area, which are pentagonal; surface of plates covered with granules. Spines small. Carboniferous. Ex. *P. ellipticus*, Carboniferous Limestone.

Maccoya. Distinguished from *Palæechinus* chiefly by the ambulacra in the middle part of the test consisting of alternate primary and smaller plates—the latter are nearly or quite cut off from contact with the interambulacral margin; the pore-pairs in this part of the test form two vertical rows. Carboniferous. Ex. *M. intermedia*.

Archæocidaris (fig. 69). Test depressed spheroidal, plates overlapping. Ambulacra narrow, sinuous, formed of two rows of plates; pores unigeminal. Interambulacra of four columns of large plates, the middle ones being hexagonal; each plate has a large primary perforated tubercle which bears a long spine, and small tubercles at the margin. Peristome covered with plates. Carboniferous and Permian. Ex. *A. urii*, Carboniferous Limestone.

Melonechinus (= *Melonites*) (fig. 70). Test spheroidal, with melon-like ribs from apex to peristome. Apical disc with five genital plates, each having from two to four pores; oculars without pores, separating the genitals. Ambulacra broad, concave on each side of a median ridge, with six to twelve columns of plates, each plate with a pair of pores; four plates at the peristomal edge of each area. Interambulacra consisting of three to eleven columns of small thick plates, which are pentagonal next the ambulacra, hexagonal elsewhere; tubercles very small. Jaws large. Carboniferous. Ex. *M. multiporus*.

Cidaris (figs. 63, 65). Test spheroidal, the summit and base equally flattened. Apical disc very large, rarely preserved fossil, ocular plates large and exsert. Ambulacra narrow, flexuous or nearly straight; plates numerous, simple, all similar in form, pores unigeminal; between the rows of pores are vertical rows of small tubercles and granules. Interambulacra wide, plates large, each with a primary tubercle which is perforated, and may be crenulated or smooth; areola large, surrounded by secondary tubercles, beyond which may be granules. Peristome large, without incisions, its membrane covered with plates. Spines large, of various forms, generally ornamented with rows of granules. The term *Cidaris* is here used in the extended sense, and includes several divisions usually regarded as genera. Jurassic to present day; allied forms occur in the Trias. Ex. *C. vincenti*, Eocene; *C.* (*Paracidaris*) *florigemma*, Corallian and Kimeridgian. The Cidarids were abundant and widely distributed in Mesozoic times, and some species are found in the Eocene, Oligocene and Miocene. At the present day they live mainly in tropical and sub-tropical seas, especially in the Indo-Pacific region. Two species occur in the Mediterranean Sea. *Cidaris* is the earliest and most primitive of living echinoids.

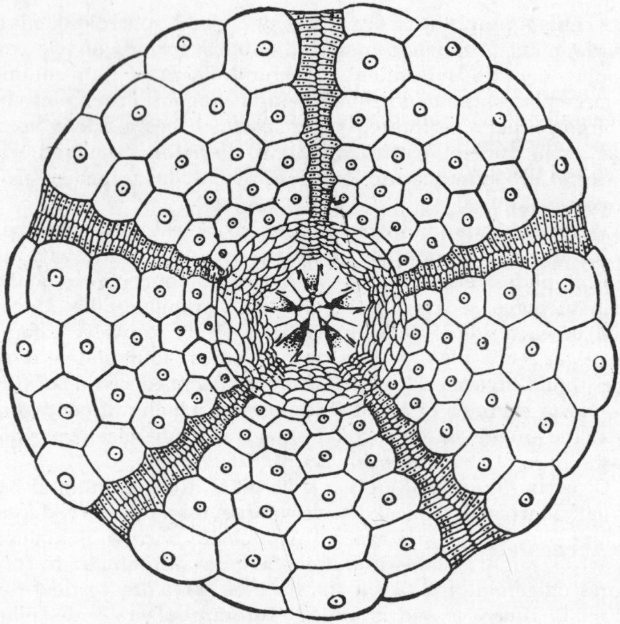

Fig. 69. *Archæocidaris Wortheni*, Lower Carboniferous. Ventral surface. Restoration based on Jackson's figures. ×1⅔.

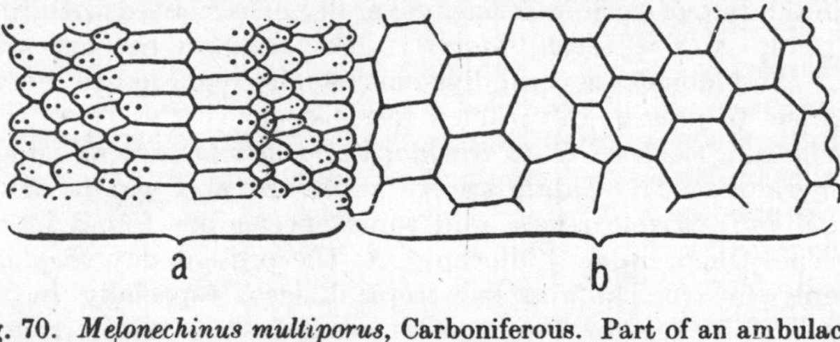

Fig. 70. *Melonechinus multiporus*, Carboniferous. Part of an ambulacral area (*a*) and an interambulacral area (*b*) from the equator of the test. Based on figures given by Jackson. ×2.

2. *Ectobranchiata*

Margin of peristome with notches. External gills present. Generally some or all of the ambulacral plates compound, but sometimes arranged in groups of two or three differing in shape and not fused together. Trias to present day.

Peltastes (fig. 60 A). Test small, circular in outline, depressed. Apical disc very large, prominent, with a sur-anal plate in front of the periproct; the madreporic plate has an oblique fissure. Ambulacra narrow, straight or slightly flexuous, with small tubercles; pores unigeminal except near the peristome; plates, primaries. Interambulacra wide, with large primary tubercles, which are imperforate, but may be crenulate. Peristome slightly notched. Upper Jurassic to Chalk. Ex. *P. wrighti*, Lower Cretaceous.

Salenia. Similar to *Peltastes*, but the periproct is on the right of a median line drawn from the anterior to the posterior margin. Lower Cretaceous to present day. Ex. *S. petalifera*, Upper Greensand.

Acrosalenia. Form similar to *Peltastes*. Apical disc rather large; genital plates large, the posterior smaller than the others and differing in shape. A sur-anal, and sometimes other extra plates, in front of the periproct, which is in the antero-posterior line and situated posteriorly. Ambulacral plates compound at and below the ambitus; pores unigeminal except near the peristome. Interambulacra with large perforate tubercles. Spines smooth or striated. Lias to Lower Cretaceous. Ex. *A. spinosa*, Inferior and Great Oolites.

Hemicidaris. Test spheroidal, inferior surface flattened. Apical disc small. Ambulacra narrow on the upper surface, slightly flexuous, with two rows of tubercles which become smaller on the upper surface; plates at and below the ambitus compound, each formed of two to four fused plates; pores unigeminal, but bigeminal near the peristome. Interambulacra broad; plates large and few, each with a large perforate and crenulate tubercle, and also smaller tubercles and granules. Spines cylindrical, long. Peristome large, with well-developed notches. Inferior Oolite to Cretaceous. Ex. *H. intermedia*, Corallian.

Pseudodiadema (fig. 61 A). Test circular or slightly polygonal, sub-hemispherical, depressed. Apical disc and periproct large. Ambulacra straight, narrower than the interambulacra, with two rows of crenulate and perforate tubercles; plates compound, each consisting of three fused primaries, the middle being largest, usually with three pairs of pores on each plate, unigeminal. Interambulacra with two or more rows of primary crenulate and perforate tubercles. Peristome large, decagonal. Lias to Cretaceous. Ex. *P. pseudodiadema* (= *hemisphericum*), Corallian.

Hemipedina. Test circular or slightly polygonal, depressed. Apical disc rather large. Ambulacra narrow, plates formed of three fused primaries (but simple near the apical disc), pores unigeminal; two rows of tubercles, perforate, not crenulate. Interambulacra with two (sometimes more) vertical rows of primary, perforate, not crenulate tubercles. Spines of moderate length, finely striated. Peristome with slight incisions. Lias to present day. Ex. *H. etheridgei*, Lias.

Diplopodia. Form and tubercles similar to *Pseudodiadema*. Pores bigeminal near the apex and peristome, unigeminal at the ambitus; plates at the ambitus composed of four primaries or sometimes the lowest plate is a demi-plate. Rhætic to Lower Chalk. Ex. *D. versipora*, Corallian.

Stomechinus. Test hemispherical. Genital plates relatively large, projecting outwards; oculars small. Ambulacra wide, plates formed of three primaries—the middle one largest; pores trigeminal. On each ambulacral and interambulacral area are two vertical rows of primary, imperforate, non-crenulate tubercles, of about the same size on each area; also secondary tubercles and granules, usually numerous. Peristome large with ten deep incisions. Inferior Oolite to Lower Cretaceous. Ex. *S. bigranularis*, Inferior Oolite.

Phymosoma (= *Cyphosoma*) (fig. 61 B). Form similar to *Pseudodiadema*. All oculars insert. Ambulacral plates high, compound, each may consist of four, five, or six fused plates (the middle ones being demi-plates) with the same number of pairs of pores; two rows of primary imperforate tubercles; pores unigeminal, but bigeminal near the apical disc. Interambulacra with two or more rows of primary imperforate tubercles. Peristome with small notches. Oxfordian to Eocene; common in the Chalk. Ex. *C. koenigi*, Upper Chalk.

Echinus. Test more or less hemispherical. Apical disc as in fig. 59 B. Ambulacra rather narrow, trigeminal, plates compound consisting of a lower primary, a middle demi-plate, and an upper primary or demi-plate. Two vertical rows of small, primary tubercles on each area, and often numerous secondary tubercles. Peristome rather small, circular, with small incisions. Pliocene to present day. Ex. *E. woodwardi*, Pliocene; *E. esculentus*, Pliocene and living.

ORDER II. IRREGULARIA

The anus is outside the apical disc, in the mid-line of the posterior interambulacral area. The mouth is either central or in front of the centre. The test is bilaterally symmetrical. Ambulacra simple or petaloid. Lantern and perignathic girdle may be present or absent.

SUB-ORDER 1. *HOLECTYPINA*

Peristome central, with notches. Lantern and perignathic girdle present. Ambulacra not petaloid: plates mainly simple, but some may be compound. Lias to Recent. Mainly Mesozoic.

Conulus (= *Echinoconus, Galerites*) (fig. 60 E). Test conical, or almost hemispherical, inferior surface flat, outline pentagonal or oval. Apical disc small, with only four genital plates. Ambulacra narrow, straight, with some demi-plates; pores unigeminal, but trigeminal near the mouth. Interambulacra with broad plates, tubercles very small, perforated and crenulated. Peristome small, central, decagonal. Periproct marginal or submarginal. Upper Greensand to Upper Chalk. Ex. *C. albogalerus* (*conicus*), Upper Chalk.

Holectypus. Test hemispherical, depressed, base excavated. Apical disc small; madreporic plate extending to the centre (fig. 60 F). Ambulacra narrow, straight, with some demi-plates; pores unigeminal, tubercles small. Interambulacra formed of rather large plates, with small tubercles. Peristome central, decagonal, with notches. Periproct large, placed between the peristome and the posterior margin of the test. Upper Lias to Corallian; also foreign Cretaceous. Ex. *H. hemisphericus*, Inferior Oolite.

Discoidea. Form similar to *Holectypus*. On the base of the interior are ten vertical plates extending from the margin of the test towards the mouth, and placed one on each side of the ambulacral areas. Cretaceous. Ex. *D. cylindrica*, Chalk.

Pygaster. Test large, depressed, outline pentagonal or circular, base concave. Apical disc small; madreporic plate large, extending to the front of the periproct; posterior genital absent. Ambulacra straight, simple; pores unigeminal; tubercles in vertical rows. Interambulacra wide, tubercles perforate. Peristome central, large, decagonal. Periproct very large, placed just behind the apical disc. Lias to Lower Cretaceous. *Pygaster* in a more restricted sense is found in the Middle and Upper Oolites; the Liassic and some Middle Jurassic species are separated under the name *Plesiechinus*; the Cretaceous and some Jurassic species are referred to *Macropygus*. Ex. *P. semisulcatus*, Corallian.

SUB-ORDER 2. *CLYPEASTRINA*

Peristome central, without notches. Lantern and perignathic girdle present. Ambulacra petaloid, plates simple. Ocular and genital plates fused together. Upper Cretaceous to present day.

Clypeaster. Outline sub-pentagonal or ovoid, usually truncated posteriorly; base of test flattened but concave around the peristome; upper surface usually convex in the central part and sloping to the margin often forming a thin edge. Apical disc small, pentagonal, the genitals fused together (fig. 60 H). Petaloid parts of ambulacra broad, with the pores widely separated and conjugate. Tubercles small, sunk in depressions; spines very small. Periproct at or near the margin. Peristome central, sunk in a deep depression. Interior with partitions near the edge of the test. Miocene to present day. Ex. *C. rosaceus*, Pliocene and living.

Scutella (fig. 71). Test much flattened, circular or subcircular, broadest posteriorly; base flat, with branching ambulacral furrows radiating from the small central peristome. Apical disc small, central, pentagonal, with central madreporic plate and four genital pores. Ambulacra petaloid, the petaloid parts unequal and nearly closed. Periproct small, inframarginal. Tubercles very small. Interior of test with supports near the margin. Eocene, mainly Oligocene and Miocene. Ex. *S. subrotunda*, Oligocene; *S. leognanensis*, Miocene.

Fig. 71. *Scutella leognanensis*, Miocene. (From Nicholson.) × $\frac{1}{3}$.

SUB-ORDER 3. *SPATANGINA*

Peristome excentric, without notches. Lantern and perignathic girdle absent. Ambulacra commonly petaloid or sub-petaloid, the plates simple; anterior ambulacrum often different from the others. The bilateral symmetry of the test is particularly well-marked. Lias to present day.

Hyboclypeus. Test oval, depressed, anterior part usually more elevated. Apical disc elongated—the two anterior genitals separated from the two posterior by two oculars. Ambulacra simple, pores unigeminal. Interambulacra wide. Tubercles very small. Periproct next the apical disc, in a long groove on the upper surface. Peristome a little in front of the centre. Inferior Oolite to Corallian. Ex. *H. gibberulus*, Inferior Oolite.

Nucleolites (=*Echinobrissus*). Test depressed; outline oval or quadrilateral, rounded anteriorly, truncated and broadest posteriorly; inferior surface concave. Apical disc compact, four perforate genital plates, and one imperforate. Ambulacra sub-petaloid, pores unigeminal, the outer pore elongated in the sub-petaloid part. Interambulacral plates wide, tubercles small.

Peristome oval or pentagonal, excentric, a little anterior. Periproct placed in a sulcus on the upper surface. Inferior Oolite to Lower Chalk. Ex. *N. scutatus*, Corallian.

Clypeus. Test large, flattened, more or less discoidal, with circular or pentagonal outline, and flat or concave base. Apical disc small. Ambulacra large, petaloid, pores unigeminal (except near the peristome), outer pore elongated and in a long groove. Peristome nearly central, with a floscelle. Periproct on the upper surface, often in a sulcus. Tubercles very small. Inferior Oolite to Corallian. Ex. *C. ploti*, Inferior Oolite.

Echinolampas. Test variable in form, more or less ovoid, often inflated above, sometimes hemispherical or conical. Apical disc small, a little in front of the centre; genitals fused. Ambulacra petaloid, pores conjugate; poriferous zones often of unequal length. Tubercles small, perforate, not crenulate. Periproct oval, transverse, inframarginal. Peristome nearly central, transverse; floscelle present, not much developed. Lower Eocene to present day. Ex. *E. ellipsoidalis*, Eocene.

Catopygus. Test small, oval, elevated, truncated behind with flat base. Apical disc small. Ambulacra sub-petaloid unigeminal, outer pore elongated in the sub-petaloid parts. Tubercles very small. Periproct high up on the posterior end. Peristome a little excentric, small, with a floscelle. Cretaceous to present day. Ex. *C. columbarius*, Upper Greensand.

Collyrites (fig. 60 C). Test ovoid, inflated. Apical disc greatly elongated; at the anterior end are four perforated genital plates separated by two oculars, at the posterior end are two oculars; these two groups of plates are connected by numerous small plates. Ambulacra simple, pores unigeminal. The three anterior ambulacra meet at the anterior end of the apical disc, the other two meet at the posterior end. Interambulacra broad, tubercles small. Peristome excentric. Periproct above the posterior margin. Lias to Cretaceous. Ex. *C. bicordata*, Corallian.

Echinocorys (=*Ananchytes*) (fig. 60 B). Test very convex above, inferior surface flattened, outline oval. Apical disc elongated; only four genital plates, the two anterior separated from the two posterior by two large ocular plates. Ambulacra simple, pores unigeminal. Interambulacral plates large, tubercles small. Peristome anterior. Periproct oval, infra-marginal. Upper Chalk. Ex. *E. vulgaris*.

Holaster (fig. 72). Test heart-shaped, inferior surface more or less flattened, superior surface with a broad shallow groove in front. Apical disc elongate, the two pairs of genital plates separated by two oculars. Ambulacra large, simple; pores unigeminal, round or elongate; the anterior ambulacrum in the groove. Interambulacra with small tubercles and granules. Peristome near the anterior margin, elliptical. Periproct supra-marginal. Upper Greensand and Chalk; also Tertiary in Australia. Ex. *H. subglobosus*, Lower Chalk.

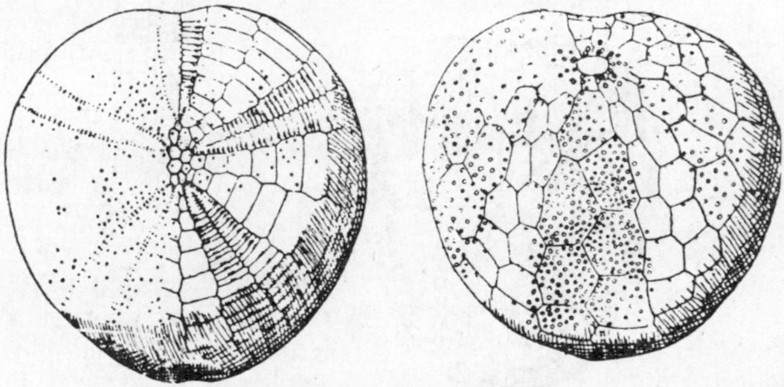

Fig. 72. *Holaster subglobosus*, Lower Chalk. Upper and lower surfaces. $\times \frac{2}{3}$.

Cardiaster. Form similar to *Holaster*, but anterior groove usually with sharp borders. Apical disc similar to *Holaster*. Pores elongate, unigeminal. Small perforate and crenulate tubercles. Peristome near the anterior margin, with a projecting lip. Periproct on the posterior truncated end. Fasciole passes beneath the periproct and round the margin of the test. Cretaceous. Ex. *C. ananchytis*, Chalk.

Micraster (figs. 64, 73, 74). Test heart-shaped or oval, truncated behind. Apical disc small, excentric; madreporic plate extending to the centre; posterior genital absent. Ambulacra sub-petaloid, placed in sunken areas, the sub-petaloid parts of the two anterior lateral longer than those of the two posterior lateral; pores unigeminal. The anterior unpaired ambulacrum in a deep groove, with its pores circular. Interambulacra with large plates; tubercles small, perforate and crenulate. Fasciole below the anus. Peristome near the anterior border, with a projecting lip (labrum). Periproct on the upper part of the posterior end. On the under surface the posterior interambulacrum bulges out forming a *plastron*. Middle and Upper Chalk. Sub-genera in the Tertiary. Ex. *M. cor-anguinum*. Upper Chalk.

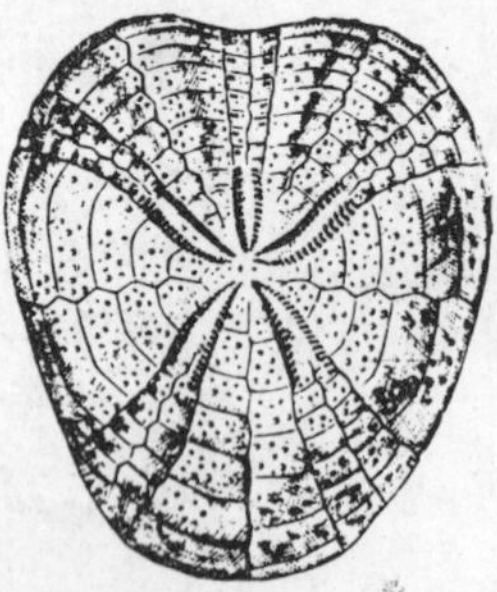

Fig. 73.

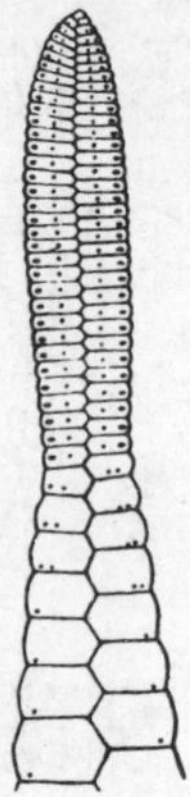

Fig. 74.

Fig. 73. *Micraster cor-bovis*. Upper Chalk. × ½.

Fig. 74. *Micraster cor-anguinum*, Upper Chalk. Part of the right anterior ambulacral area. × 1½.

Epiaster. Form similar to *Micraster*, but usually more elevated. No fasciole. Upper Cretaceous. Ex. *E. gibbus*, Upper Chalk.

Hemiaster. Form similar to *Micraster*. A peripetalous fasciole only. Pores slit-like in the petaloid parts of the ambulacra, except in the anterior ambulacrum. Cretaceous to present day. Ex. *H. bailyi*, Gault.

Schizaster. Test heart-shaped, highest behind, with apex posterior to the centre. Anterior ambulacrum long, placed in a groove; other ambulacra petaloid and in deep grooves—the posterior pair much shorter than the antero-lateral pair. Peristome near the anterior margin, with projecting lip. Periproct on the posterior truncated end of the test. A peripetalous fasciole, and usually also a lateral fasciole diverging from the former and passing beneath the periproct. Eocene to present day. Ex. *S. d'urbani*, Bracklesham Beds.

Toxaster. Test sub-cordate, broadest anteriorly, highest posteriorly. Apical disc behind the centre. Anterior ambulacrum in a broad, shallow groove; other ambulacra level with the interambulacra or only slightly sunk, sub-petaloid, flexuous; posterior pair shorter than the anterior pair; pores elongate. Peristome anterior, transverse. Periproct on the upper part of the posterior truncated end. Tubercles small, perforate. No fasciole. Lower Cretaceous. Ex. *T. complanatus*.

Distribution of the Echinoidea

Some echinoids live at great depths in the ocean, no less than a dozen species having been found below the 2000 fathom line, and one even at 2900 fathoms; but by far the larger number occur near the coasts in shallow water; thus, of the 297 existing species recorded by Agassiz, 201 are found in water of less than 150 fathoms in depth. Echinoids are most abundant where the sea-bottom is rocky, sandy, or calcareous, and less common where it is muddy; consequently fossil forms are rare in clayey strata. Those found in deep water have a much wider range in space than those found in shallow water. Many genera, especially those with a considerable range in depth, have also a long range in time, some extending back to the Cretaceous or even to the Jurassic period, *e.g. Hemipedina, Catopygus, Salenia, Hemiaster.*

All the Palæozoic Echinoids belong to the Endobranchiate division of the Regular group. With two exceptions the corona is characterised by consisting of more than 20 columns of plates. In each ambulacral area the number ranges from 2 to 20; each plate has one pair of pores, so that in some cases the number of tube-feet in each area was very large. In each interambulacral area the number of columns varies from 3 to 14. Some forms have a rigid test, but in many genera the plates of both areas overlap, so that the test was flexible. The ocular plates are generally insert, and the genital plates commonly have 3 or more perforations.

Echinoids are rare in Palæozoic formations, especially in those of pre-Carboniferous age. The earliest representatives are found in the Ordovician. *Bothriocidaris* (fig. 62), from the Upper Ordovician of Russia, has been regarded as a Cystid, but is more probably an echinoid of a peculiar type, differing from all others in having a single column of plates in each interambulacral area, and in the ambulacral plates being as large as the interambulacrals with which they alternate. Also the pairs of pores in the ambulacrals are placed vertically, and the test was rigid. In *Aulechinus* from the Upper Ordovician of Girvan, and in *Myriastiches* from the Middle Ordovician, there are numerous columns of ambulacral plates and the test was flexible.

Three genera are known in the Silurian. All have flexible tests, with several columns of interambulacral plates. *Palæodiscus* and *Koninckocidaris* have two columns of ambulacral plates, *Echinocystis* has four. In the Devonian Echinoids are still very rare, the genera represented being *Eocidaris*, *Lepidechinoides*, *Lepidocentrus*, and *Nortonechinus*.

In the Carboniferous the echinoids with numerous columns of plates in the corona reach their maximum development. The genera with only two columns of ambulacrals are *Archæocidaris* (fig. 69), *Maccoya*, *Palæechinus* and *Perischodomus*; those with more than two columns of ambulacrals are *Lepidesthes*, *Lovenechinus*, *Melonechinus* and *Oligoporus*. *Miocidaris*, the earliest representative of the Cidarids, appears in the Carboniferous and Permian but is more abundant in the Trias and Lower Jurassic; it resembles *Archæocidaris* but has two instead of four columns in each interambulacral area, and is the earliest echinoid known in which the corona consists of 20 columns of plates. In the Permian there are few echinoids; *Archæocidaris* and *Miocidaris* are represented; *Meekechinus* with 20 columns of ambulacrals and 3 columns of interambulacrals is the last representative of the type of echinoid characteristic of the Palæozoic.

In the Trias echinoids are found in St Cassian, Bakony and Timor. All the characteristic Palæozoic types have disappeared, in most cases without leaving any descendants, and their place is taken by genera with only 20 columns of plates in the corona and a rigid test. The Cidarids are the chief forms, and they differ from *Miocidaris* in having a rigid instead of a flexible test. Associated with the Cidarids are some other regular forms, the first representatives of the Ectobranchiates—forms with notches in the peristome, indicating the appearance of external gills, and with compound plates in the ambulacra. Some of these early Ectobranchiates differ but little from Cidarids, from which they appear to have been derived.

In the Jurassic rocks the echinoids are much more numerous, relatively to the other groups of animals, than in the earlier formations; they are comparatively rare in the Lias and the other clayey divisions, but very abundant in the calcareous beds, especially in the Inferior Oolite and the Corallian. *Cidaris* is abundant throughout, and the Ectobranchiates develop rapidly and are abundant in the Middle and Upper Jurassic, *e.g.* *Acrosalenia*, *Diademopsis*, *Diplopodia*, *Hemipedina*, *Pedina*, *Pseudodiadema*, *Stomechinus*. In the Lias Irregular Echinoids, belonging to the Holectypina, make their appearance. One of the earliest of these is *Pygaster* (*Plesiechinus*) which differs in structure but little from the Ectobranchiates, from which it has probably been derived. It is only just irregular, since the periproct touches the posterior end of the apical disc. In other genera (*e.g. Holectypus*) the irregularity becomes more marked owing to the shifting of the periproct further from the apical disc. *Galeropygus* is another genus of the Holectypina which appears first in the Lias and is only just irregular and resembles in structure the early Ectobranchiates The three genera mentioned all become abundant in the Middle and Upper Jurassic. The more irregular group, the Spatangina, also appears in the Lias. In most of the early forms the mouth is only a little in front of the centre, but later, in relation to a burrowing mode of life, it tends to move towards the anterior margin and then the test becomes more distinctly elongate and the bilateral symmetry more marked. In these, owing to a change in feeding habits, the lantern and perignathic girdle disappear. Further, the ambulacral areas are now formed of simple plates. The principal genera of the Spatangina which are common in the Middle and Upper Jurassic are *Clypeus*, *Collyrites*, *Hyboclypeus*, *Nucleolites* and *Pygurus*.

In the Cretaceous the echinoids are even more abundant than in the Jurassic, and attain a great development in the upper division of the system; many of the genera found in the Lower Cretaceous occur also in the Upper Jurassic, but the irregular forms are more numerous than hitherto, and show a still greater development in the Upper Cretaceous, where *Micraster*, *Echinocorys* and other allied genera are characteristic. The most important genera are: (1) regular, *Cidaris*, *Pseudodiadema*, *Phymosoma*, *Peltastes*, *Salenia*; (2) irregular, *Discoidea*, *Conulus*, *Catopygus*, *Hemiaster*, *Micraster*, *Epiaster*, *Cardiaster*, *Holaster*, *Echinocorys*.

The Clypeastrina make their first appearance in the Upper Chalk.

Between the Cretaceous and Eocene there is, in Britain, a great break in the succession of the echinoids; not a single species is common to the two systems, and most of the genera also are different. This change is due in part to the great difference in the conditions under which the deposits were formed, the Chalk being a comparatively deep-water formation, and the Eocene beds, shallow water; but the Eocene forms differ more from those of the Upper Chalk than from those of the Chalk Marl, the latter deposit having been formed in water of less depth. Throughout the English Tertiaries the echinoids are much rarer than in the Cretaceous; in the Eocene this can be accounted for largely by the fact that the sea-bottom was for the most part muddy; in the Oligocene by the prevalence of fresh-water and estuarine conditions; and in the Pliocene, by the lower temperature of the ocean. The London Clay echinoids belong to tropical or sub-tropical genera. The commonest forms in the Eocene of England are *Hemiaster* and *Schizaster*. In the Eocene of the South of Europe, North Africa, India, etc., echinoids are numerous; the regular forms are relatively less important than in earlier formations, but the Clypeastrina and Spatangina, in this and subsequent deposits, become increasingly abundant; the first group is represented by *Clypeaster*, *Scutella*, etc., the second by *Echinolampas*, *Schizaster*, *Hemiaster*, etc. The Pliocene echinoids found in East Anglia include some forms similar to those found in the North Atlantic, and others which show considerable affinity to species now living in the West Indian seas; the principal genera represented are *Echinus*, *Strongylocentrotus*. *Echinocyamus*, *Spatangus*, and *Temnechinus*.

CLASS III. HOLOTHUROIDEA

This Class includes the sea-cucumbers. They possess an elongated and usually cylindrical body with the mouth at one end and the anus at the other; around the mouth is a circle of tentacles, which are really modified tube-feet. From the water-vascular ring five radial vessels are given off and end near the anus; branches also go to the tentacles. In *Synapta* and its allies tube-feet (with the exception of the tentacles), as well as radial vessels, are absent. The stone-canal in almost all cases opens into the body-cavity. The integument is leathery, and the skeleton is very poorly developed, consisting of minute isolated pieces of various shapes, such as spicules, anchors, and wheels (fig. 75).

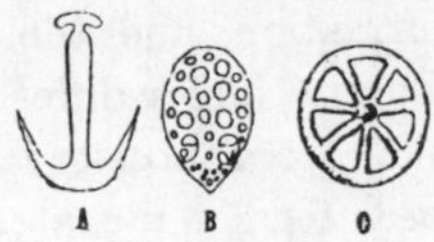

Fig. 75. A, B, anchor and plate of *Synapta tenera*, Recent. C, wheel of *Chirodota convexa* from the Inferior Oolite. Enlarged.

At the present day the Holothurians are widely distributed, but owing to the nature of their hard parts they are rarely found fossil. Specimens found in the Middle Cambrian of British Columbia were regarded by Walcott as Holothurians and may possibly belong to this Class. The earliest known European forms, represented only by skeletal structures, occur in the Carboniferous rocks of Scotland, and in the Permian of Germany. Some specimens have been recorded from Jurassic, Cretaceous and later formations. An impression of the body of a Holothurian has been found in the Upper Jurassic of Solenhofen. *Synapta* has been recorded from the Oligocene.

II. PELMATOZOA

The Pelmatozoa, unlike the Eleutherozoa, are generally sedentary, being attached to the sea-floor or some foreign object by the aboral surface, usually by means of a jointed stem; in most cases the attachment is permanent, but it may be temporary only. The group is essentially distinguished by the ciliated grooves which radiate from the mouth; the cilia produce a current of water which carries small organisms to the upwardly directed mouth. The Classes of the Pelmatozoa are: (1) Crinoidea, (2) Cystidea, (3) Blastoidea, (4) Edrioasteroidea.

CLASS I. CRINOIDEA

The Crinoidea include the sea-lilies and feather-stars. The body consists of a *stem*, a *calyx*, and movable *arms* given off from the margin of the calyx (fig. 76).

The calyx is more or less globular, or cup- or basin-shaped, and contains the digestive and other important organs. The mouth is either at or near the centre of the ventral or oral surface, and the anus, which is in the posterior inter-radial area, is also on the oral surface, and is usually situated at the end of a tubular process; the alimentary canal is tubular and makes a complete coil around the cavity of the calyx. There is a groove on the ventral surface of each arm, and these grooves—the *food-grooves*—are continued over the oral surface to the mouth; they are lined with cilia, by the movements of which food is conveyed to the mouth. There are five arms, but each may branch repeatedly. Immediately under the groove of each arm there is a radial nerve-cord; these cords unite to form larger trunks and ultimately join as a ring round the mouth. Beneath the nerve of each arm is a radial vesssel of the water-vascular system, which is continued over the oral surface and joins a ring round the mouth; from this ring tubes (stone canals) hang down and open into the body-cavity, which communicates with the water of the exterior by means of pores. From the radial vessels tubes are given off to the tube-feet, which form a row on each side of the food-grooves, and function in respiration. In addition to the nervous system already mentioned, there is another supplying the aboral elements of the skeleton; from a centre at the aboral pole of the calyx nerve cords are given off, which pass through canals in the plates of the calyx to the arms and pinnules, and also into the stem when present. All the organs of the body are thus radially symmetrical with the exception of the alimentary canal.

The stem (fig. 76) in the crinoids is more or less flexible, and is sometimes several feet in length. It consists of a number of segments, known as *columnals*, which may be disc-like or pentagonal (occasionally square or elliptical); or they may be higher than broad, forming cylinders; these columnals articulate by their flat surfaces, which are often provided with radiating striæ or with ridges in the form of a rosette. Each columnal is pierced at the centre by a canal which is circular or pentagonal and contains a prolongation of the aboral nervous system and vascular organ. The columnals are generally of different heights—larger plates being separated by smaller; the former are first developed, and the latter are those which are subsequently introduced between them. The lower (or distal) end of the stem may taper, but usually branches or expands and serves to fix the animal; when a crinoid lives on a soft sea-floor it is fixed by a branching root-like structure (fig. 76), but when it is attached to a rock or other firm substance solid calcareous material is secreted forming an encrusting plate or mass (*e.g. Apiocrinus*). In the course of evolution, and sometimes in the development of the individual (fig. 76), the stem at first consists of cycles of five plates (pentameres) alternating in position; subsequently the five plates of a cycle become horizontal, and afterwards fuse to form a single columnal. From the stem small branches known as *cirri* are sometimes given off; these have a structure similar to that of the stem, and are also pierced by a central canal. In the course of evolution cirri originate as root-like branches at the distal end of the stem, but subsequently they appear at higher levels and are then borne on some of the larger columnals.

The part of the calyx below the origin of the arms is called the *dorsal cup* (fig. 76); the part above them is the *tegmen*. The dorsal cup consists at its base of a cycle of five plates, known as *basals* (figs. 81, *b*; 82, *c*); but, owing to fusion, the number of basals is sometimes reduced to four, three, or rarely two. In some forms there is below the basals and alternating with them another row of plates (five or three), termed *infra-basals* (fig. 82, *b*), and the base is then said to be *dicyclic*; when basals only are present it is *monocyclic*. Above the basals, and alternating with them, is a cycle of five *radial* plates (figs. 81, *r*; 82, *d*), which usually form the sides of the dorsal cup; each radial is in a direct line with one of the arms. In some genera there are, between the two posterior radials, other plates, the *anal inter-radials* (figs. 81, *a*; 82, *e*). Sometimes there are inter-radial plates between the other radials as well.

The arms are characteristic of the Crinoidea; they come off directly from the radials, and are formed either of a single or of a double row of plates, the *brachials*; when there is a single row the arm is termed *uniserial* (fig. 82);

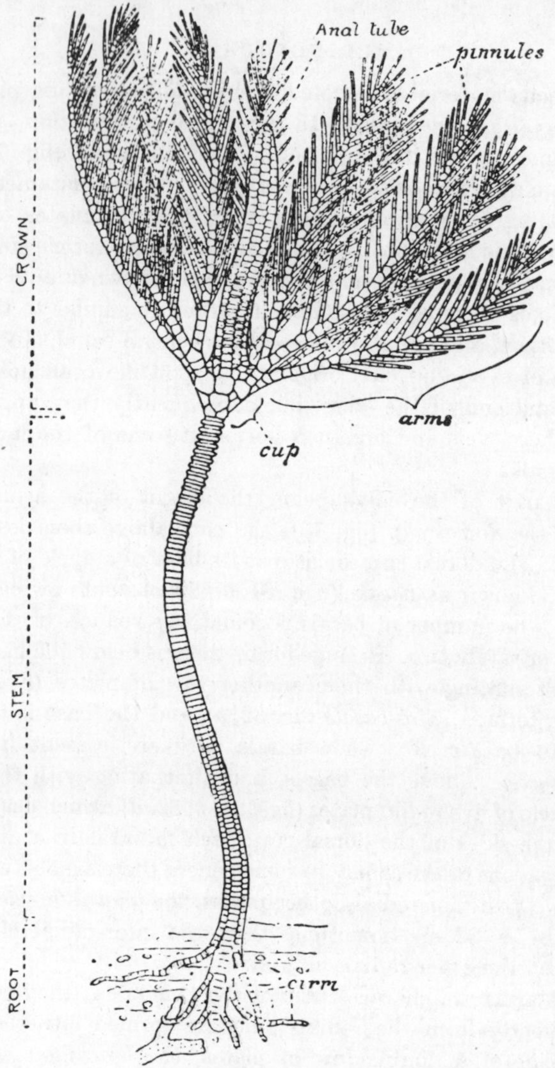

Fig. 76. *Botryocrinus decadactylus*, from the Wenlock Limestone—a simple form of Crinoid, seen from the posterior inter-radius. (From the *Guide to the Geol. Dept., Brit. Mus.*) Natural size.

when there are two rows it is *biserial*. In biserial arms the plates alternate with one another. The brachial plates are connected by muscles by means of which the movements of the arms are effected. The dorsal or outer surface of the brachial plates is rounded; on the ventral or inner surface there is a groove in which the soft parts, above described, are placed; and there is usually also a perforation below the groove, in which the dorsal nerve-cord is situated. The groove in the arms is covered over by a series of plates—the *covering plates*, which can be opened and closed, and serve for the protection of the soft parts. Where an arm branches, the brachial which supports two branches (figs. 76, 82, 83, *a*) has sloping sides, and is known as an *axillare*. Small unbranched appendages called *pinnules* occur on the arms of many crinoids (fig. 76); they are similar in structure to the arms, and are given off alternately on opposite sides. In living crinoids the genital products mature in the pinnules.

The arms in simple types of crinoids are short, uniserial and unbranched, but in more advanced types they branch several or many times and may attain a great length, in this way the food supply is increased. The two branches formed at each division may be of equal thickness; or all the branches on one side may be thin and those on the other thick; or the branches may be alternately thick on one side and thin on the other. The character of the arms also shows a relation to habitat. Crinoids which live in deep or quiet water have long, thin arms, while those found in rough water have short and thick arms.

The biserial arm arose from the uniserial type by the development of pinnules (fig. 77). In uniserial arms a pinnule is borne at the distal end of each brachial, alternately on the two sides of the arm. Each plate is thicker on the side bearing the pinnule than on the opposite side. This difference increases until the plates become wedge-shaped. Then the thin edge of the wedge ceases to reach the margin of the arm, and ultimately two rows of short brachials, meeting in the middle line of the arm, are developed, thus greatly increasing the number of pinnules in a given length of arm.

In some forms, the earlier rows of brachial plates become firmly united to one another and to the radials (figs. 81, 85, 2, 3, *br*); these *fixed brachials* have often been regarded as radials, but morphologically they are only brachials which have become incorporated into the calyx. The fixed brachials may be in contact at the sides, or, as in most Palæozoic

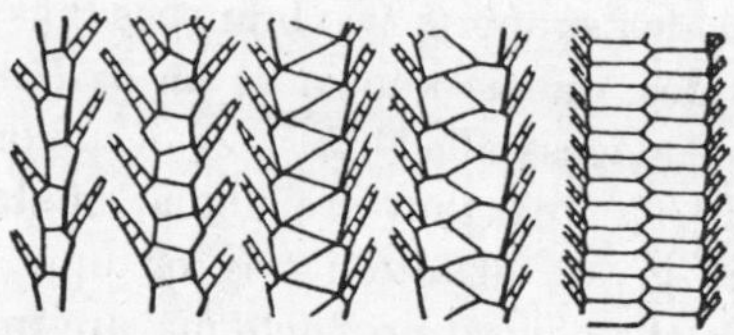

Fig. 77. Evolution from uniserial to biserial arms. (After Bather.)

crinoids, they may be separated by other plates which are termed *inter-brachials* (fig. 81, *ir*). In the posterior interradial area (that which leads up to the anus) the interbrachial plates are often more numerous than in the other areas, so that the radial symmetry of the dorsal cup is no longer perfect.

In several groups of crinoids a tendency for the root to disappear is seen, so that the animal was no longer permanently fixed but could moor itself by means of the cirri or by the distal end of the stem coiling around some object. In some cases this led to a complete or almost complete loss of the stem and the adoption of a free-swimming mode of life. In *Millericrinus pratti* (Jurassic) all stages in the process of reduction can be seen in different individuals, from those with a long stem consisting of 70 columnals to those with a single ossicle fused with the base of the calyx. In *Antedon* a portion of the stem remains but is compressed and fused with the infra-basals into a single ossicle which bears cirri. In *Saccocoma* (Upper Jurassic to Chalk), *Marsupites* and *Uintacrinus* (Upper Chalk) there is no stem, but a central pentagonal plate at the base of the calyx (fig. 84, *c*). In these genera the calyx is large, with thin walls, making a light body. In *Saccocoma* it is formed

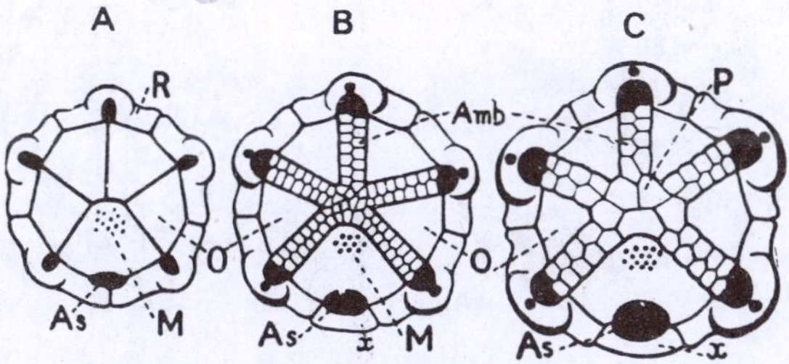

Fig. 78. Three stages in the evolution of the Tegmen. A, orals only; B, orals and ambulacrals; C, ambulacrals enlarged near the mouth (*p*). *As*, anus; *P*, peristomal plates; *M*, madreporic pores; *O*, orals; *R*, radial plate; *x*, anal plate; *Amb*, ambulacral plates. (After Bather.)

mainly of the radial plates; in *Marsupites* of large infra-basals, basals and radials; while in *Uintacrinus* the calyx consists of a large number of small plates, owing to the incorporation of numerous brachials, inter-brachials and pinnulars. In *Saccocoma* and *Uintacrinus* the arms were long, suggesting adaptation for a pelagic mode of life; this accords with the wide geographical distribution of *Uintacrinus*, which has been found in Europe, North America and Australia.

The tegmen or oral surface of the calyx (figs. 78, 79) is usually more or less completely covered by plates. Sometimes (fig. 78 A) five large triangular plates (*orals*) only occur, between which are the food-grooves leading to the mouth; but usually the oral plates become reduced in size, and other smaller plates appear—the food-grooves being usually covered by plates, sometimes called 'ambulacrals' (fig. 78 B, C), and between them there may occur numerous

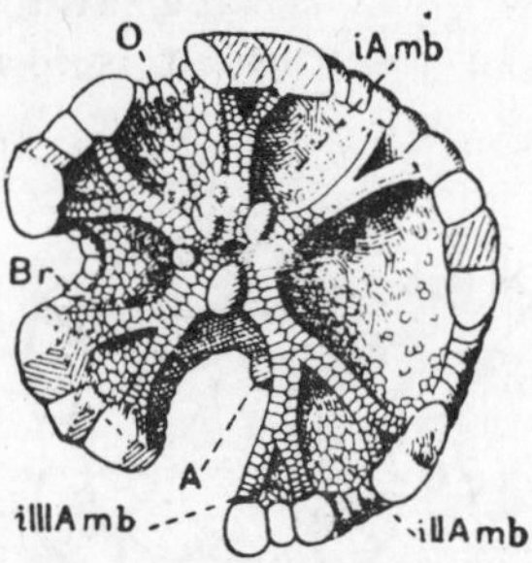

Fig. 79.

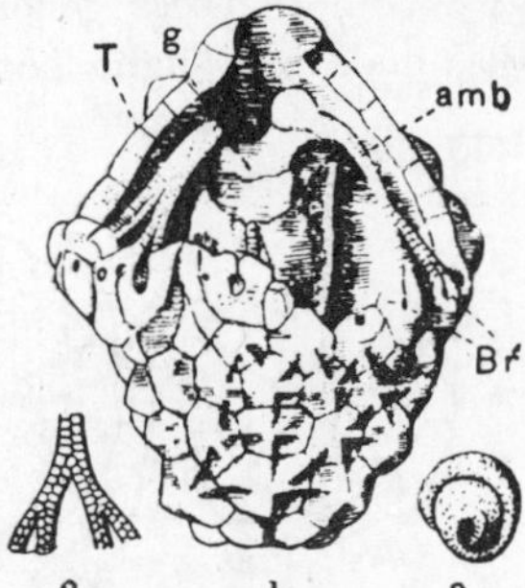

Fig. 80.

Fig. 79. Tegmen of *Taxocrinus intermedius*, Silurian. *A*, anal ridge; *Br*, edges of brachial plates; *iAmb.*, *iIIAmb.*, *iIIIAmb.*, interambulacrals; *O*, oral plate. (From Bather, after Wachsmuth and Springer.)

Fig. 80. *Cactocrinus proboscidialis*, Carboniferous. 1, specimen with one side of tegmen broken away; 2, food-canal from above; 3, convoluted organ from below. *T*, tegmen; *amb.*, tube formed of ambulacral and side plates enclosing food groove and water vessel; *Br'*, arm openings; *g*, convoluted skeleton of gut. (From Bather, after Meek.) × 1⅓.

'interambulacral' plates (fig. 79). In the Camerate type of crinoid, which flourished in the Palæozoic (*e.g. Actinocrinus*), the tegmen consists of a complete vault or dome of stout plates concealing the mouth as well as the food-grooves and their covering plates (fig. 80); commonly this plated tegmen extends upwards around the anal process forming a tube-like covering (fig. 76).

The various plates of the crinoid skeleton are joined together by fibres of connective tissue continuous with those which form the organic basis of the plates. In some cases adjacent plates become fused owing to the deposition of calcareous material between them.

In the genera described below, the basals, radials, and arms are five in number unless otherwise stated.

A. *Monocyclic Crinoids*

Platycrinus. Basals three, unequal. Radials large. Some fixed brachials. One inter-brachial in each area—more in the posterior (anal) area. No inter-radial. Arms bifurcating once to

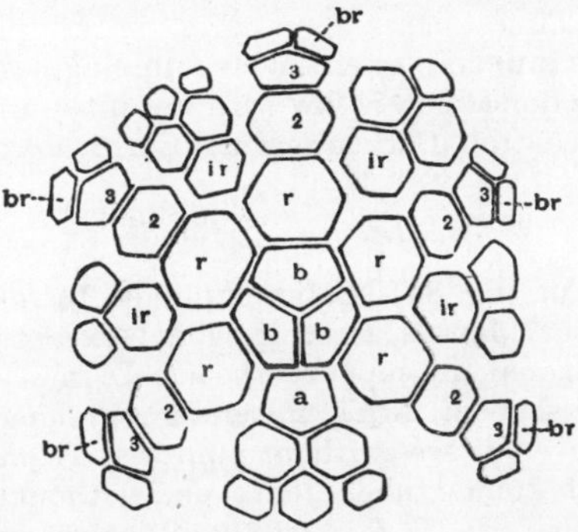

Fig. 81. Diagram of the plates of *Actinocrinus triacontadactylus*, Carboniferous Limestone. *b*, basal plates; *r*, radials; 2, 3, fixed brachials; *br*, brachial plates; *ir*, inter-brachials; *a*, anal inter-radial.

thrice, uniserial at the lower end, biserial above; pinnules long. Tegmen with small plates; anus sub-central, sometimes at the end of a long process. Stem long, section often elliptical. Devonian, but mainly Carboniferous. Ex. *P. lævis*, Carboniferous Limestone.

Eucalyptocrinus (=*Hypanthocrinus*). Calyx deeply concave at the base; at the bottom of the cavity four basals, at the sides five radials; several cycles of fixed brachials, and some inter-brachials. Tegmen elevated, and forming a central anal tube composed of five rows of large plates. Ten vertical partitions spring from the outside of the tegmen, forming compartments in which the ten arms rest. Arms biserial except at the base. Mainly Silurian; one Devonian species. Ex. *E. decorus*, Wenlock Limestone.

Actinocrinus (fig. 81). Calyx pear-shaped, ovoid, or more or less spherical. Basals three, equal, forming a hexagon. Radials generally higher than wide. The first two rows of brachials firmly united. Inter-brachials numerous; and also one (posterior) inter-radial, above which the inter-brachials are more numerous than in the other areas. Tegmen formed of thick, tubercled, hexagonal plates, produced into a tube with the anus at the end. Arm-branches ten to thirty, biserial. Stem circular, canal pentagonal. Carboniferous. Ex. *A. triacontadactylus*, Carboniferous Limestone.

Amphoracrinus. In essential structure agrees with *Actinocrinus*, but the dorsal cup is low with few inter-brachials. Anal tube short, excentric. Carboniferous. Ex. *A. amphora.*

B. *Dicyclic Crinoids*

Cyathocrinus (fig. 82). Calyx cup-like. Infra-basals small, equal, pentagonal. Basals large, hexagonal (except the posterior, which is heptagonal and supports the square inter-radial plate). Radials shield-shaped. Arms uniserial, very long, bifurcating from five to seven times, without pinnules. Tegmen produced into a long or short anal tube. Stem round, without cirri. Silurian to Carboniferous. Ex. *C. longimanus*, *C. acinotubus*, Silurian.

Crotalocrinus. Dorsal cup similar to that of *Cyathocrinus.* Some fixed brachials present. Arms uniserial, dichotomous, the branches uniting so as to form lamellar expansions or networks; pinnules absent. Tegmen nearly flat, formed of small plates with five large plates at the centre. Anus near the posterior margin. Stem thick, circular; canal pentagonal; root thick, branching. Wenlock Limestone. Ex. *C. rugosus.*

Botryocrinus (fig. 76). Calyx small, cup-shaped. Infra-basals pentagonal; basals hexagonal (except the two posterior, which are pentagonal); radials with the articular surface occupying $\frac{1}{2}$ to $\frac{2}{3}$ of the width; two anal inter-radials, one as in *Cyathocrinus*, another below it on the right. Arms divide, giving ten main branches which often bear smaller branches or pinnules. Anal tube large, sometimes coiled, anus near its base. Stem formed of low plates, often in five pieces. Silurian and Devonian. Ex. *B. decadactylus*, Wenlock Limestone.

Poteriocrinus. Calyx with thin plates. Infra-basals equal. Basals high. Three anal inter-radials present. Radials with well-

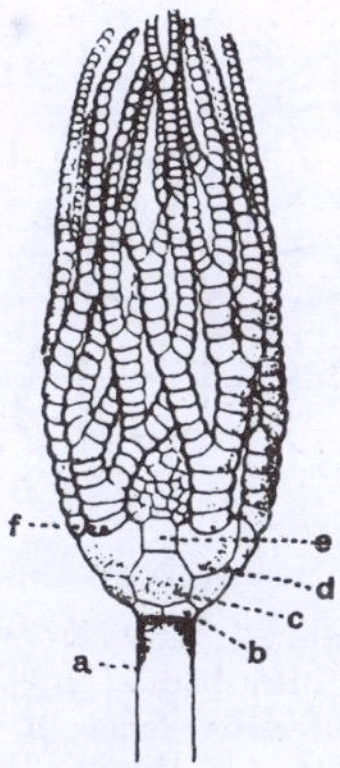

Fig. 82.

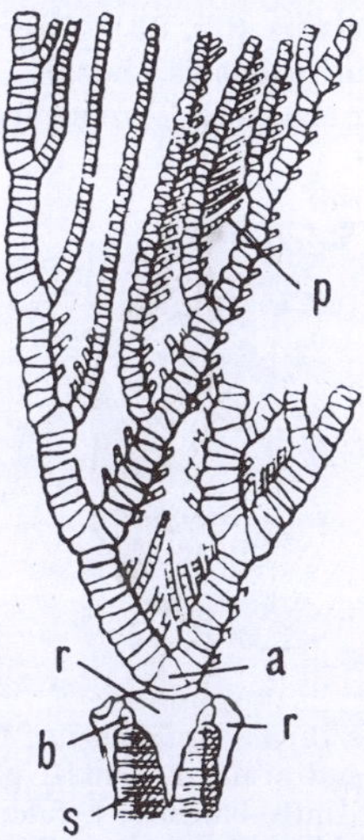

Fig. 83.

Fig. 82. *Cyathocrinus longimanus*, from the Silurian. *a*, portion of stem; *b*, infra-basal plates; *c*, basals; *d*, radials; *e*, anal inter-radial; *f*, first brachial. Reduced.

Fig. 83. *Pentacrinus fossilis*, Lias. Calyx and part of stem and arm. *s*, stem; *b*, basal plate; *r*, radials; *a*, axillare; *p*, pinnules. (After Bather, 1898.)

marked concave articular surfaces which do not occupy the entire width of the plates. Anal tube long. Arms long, branching, with pinnules. (Devonian?), Carboniferous. Ex. *P. crassus*, Carboniferous.

Woodocrinus. Like *Poteriocrinus* but calyx and arms usually shorter; anal tube inconspicuous. The arm-facet occupies the full width of the radial. Carboniferous. Ex. *W. macrodactylus*.

Encrinus. Calyx saucer-shaped. Infra-basals very small, generally concealed by the stem. Basals rather large, hexagonal. Radials large, pentagonal. Two fixed brachials in each ray, the upper being axillary. No inter-brachials, no anal inter-radials. Arms bifurcating, the branches uniserial at first, then alternating, finally biserial; with pinnules. Tegmen covered with plates. Stem long, with small canal. Trias. Ex. *E. liliiformis*, Muschelkalk.

Pentacrinus (fig. 83). Calyx small, bowl-shaped, consisting of small infra-basals, basals, and radials which project like spines over the stem. Arms very long, much branched, uniserial;

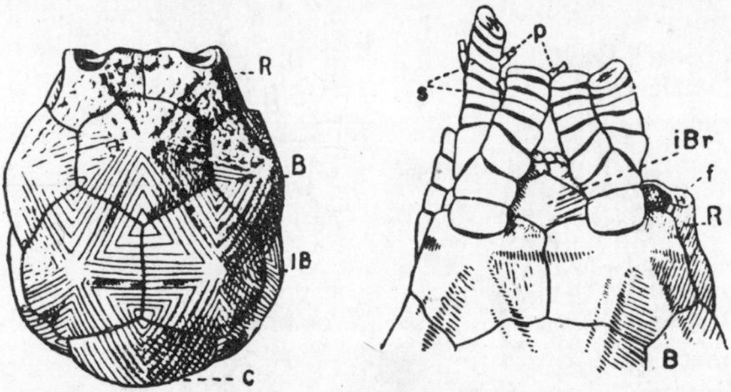

Fig. 84. *Marsupites testudinarius*, Upper Chalk. 1. Calyx from the side. 2. Radials and arms. *c*, central plate; *IB*, infra-basals; *B*, basals; *R*, radials; *iBr*, inter-brachial; *f*, fulcral ridge of radial facet; *p*, pinnules; *s*, junction of brachial plates. (From Bather.)

the small branches all come off on the same side of each main branch. The arms bear pinnules. Stem long, pentagonal, with cirri coming off in whorls; the articular surfaces of the columnals with five raised, crenulate, petaloid parts which are narrow and quite distinct from one another. Jurassic. Ex. *P. fossilis*, Lias.

Marsupites (fig. 84). Calyx large, globular; plates large and thin. Stem absent. Base formed of a large central pentagonal plate (*c*). Infra-basals pentagonal. Basals hexagonal. Radials pentagonal, with crescentic depressions for the articulation of the arms. Arms relatively short, bifurcating, uniserial; first brachial much narrower than the radial. Upper Chalk. Ex. *M. testudinarius*.

Ichthyocrinus. Three very small infra-basals; five small basals; five radials; two or three cycles of fixed brachials. Anal inter-radial small, below the right posterior radial. Arms in contact all round, interlocking, uniserial; no pinnules. Silurian. Ex. *I. piriformis.*

Sagenocrinus. Infra-basals small. Anal inter-radial sunk between basals; radials large. Numerous cycles of fixed brachials, separated by very numerous inter-brachials. Arms dividing, uniserial; no pinnules. Silurian. Ex. *S. expansus.*

Apiocrinus (fig. 85). Calyx-large. Infra-basals enclosed by, and often fused with, the thick basals. Radials low, excavated on their upper surfaces. Four cycles of fixed brachials. Arms ten, bifurcating once or twice, uniserial. Stem long, cylindrical, base expanded; the articular surfaces of the columnals radiately striated. The upper columnals are in contact at the periphery only. The upper part of the stem expands and passes gradually into the calyx; the upper surface of the last columnal is provided with five radiating ridges between which the basals lie. Jurassic (Lower Cretaceous?). Ex. *A. parkinsoni*, Bradford Clay.

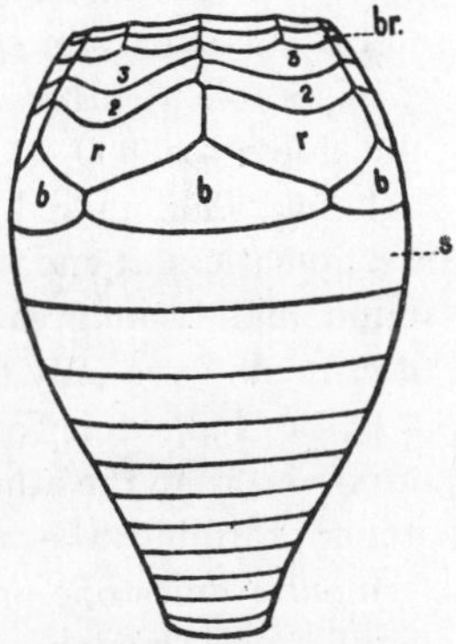

Fig. 85. *Apiocrinus parkinsoni*, from the Bradford Clay. *s*, top columnal of the stem; *b*, basal plates; *r*, radial plates; 2, 3, and *br*, fixed brachial plates. × ¾.

Millericrinus. Allied to *Apiocrinus.* Usually the top columnal only is widened. Articular facets of radials and brachials well developed. Lias (? also Trias) to Lower Cretaceous. Ex. *M. pratti*, Inferior and Great Oolite.

Bourgueticrinus. Calyx small, with vertical or inwardly-sloping sides; basals about half the height of radials; two rows of fixed brachials; no inter-brachials. Free arms unknown. Stem long, the top columnal very large, as wide as the calyx; upper columnals with circular, others with elliptical, articular faces and a transverse ridge across the longer diameter. Cretaceous. Ex. *B. ellipticus*, Chalk.

Distribution of the Crinoidea

Although not so numerous and varied as in the Palæozoic period, the Crinoidea are represented at the present day by a large number of species belonging to about 100 genera. The unstalked forms are the most important (the Antedonidæ, Actinometridæ, etc.); these are widely distributed, and occur chiefly in shallow water, but some are found at considerable depths—*Antedon* extending from the shore-line down to 2900 fathoms, and *Actinometra* down to 800 fathoms. The stalked crinoids (*e.g. Isocrinus*, *Rhizocrinus*) are much less abundant than the unstalked forms, and are found mainly at great depths. In some cases the species of crinoids have only a limited distribution in space.

In the Palæozoic formations the crinoids are much more numerous than the other Echinoderms, their remains (chiefly stems) forming the main part of some limestone beds (crinoidal limestone or marble), as for instance in the Carboniferous. The other Echinoderms are seldom sufficiently numerous to be of importance as rock-builders. The majority of fossil crinoids are stalked forms, and appear to have lived in fairly shallow water, since they are found in association with reef-building corals and other shallow-water organisms.

Crinoids occur first in the Tremadoc Beds. In the Ordovician, *Glyptocrinus*, *Dendrocrinus*, and a few others have been found. In the Silurian, crinoids become very much more abundant, and attain their maximum development; the Camerate type, of which both monocyclic and dicyclic forms occur, are important from now until the close of the Palæozoic; the principal genera are *Botryocrinus*, *Calceocrinus*, *Crotalocrinus*, *Eucalyptocrinus*, *Gissocrinus*, *Ichthyocrinus*, *Marsipocrinus*, *Periechocrinus*, *Pisocrinus*, *Sagenocrinus*, *Taxocrinus*. In the Devonian, *Cyathocrinus*, *Cupressocrinus*,

Haplocrinus, *Hexacrinus* and others are common; in the Carboniferous, *Actinocrinus*, *Amphoracrinus*, *Poteriocrinus*, *Platycrinus*, *Rhodocrinus*, and *Woodocrinus*. Crinoids are found in the Permian of Sicily and Timor; *Eutelocrinus*, *Timorocrinus*, etc. The Palæozoic genera do not survive into the Mesozoic, and throughout the Mesozoic formations crinoids are much less abundant than in the Palæozoic. In the Trias the characteristic form is *Encrinus*; *Isocrinus* is also present. In the Jurassic, *Antedon*, *Isocrinus*, *Pentacrinus*, *Saccocoma*, *Apiocrinus*, and *Millericrinus* are found, the first two living on to the present day. The Cretaceous is characterised by *Bourgueticrinus* and the free-swimming *Marsupites* and *Uintacrinus*—the last two being confined to the Upper Chalk. In the Cainozoic, crinoids are very rare, but *Antedon* has been found in the Lower Eocene.

CLASS II. CYSTIDEA

The Cystidea vary considerably in structure. In some genera there is scarcely any sign of radial symmetry, but in others it is indicated by the food-grooves which radiate from the mouth over the surface of the calyx, and bear simple arm-like structures called brachioles (fig. 89). In forms with fewer plates in the calyx the radial symmetry becomes more distinct. The plates of the Calyx are perforated by canals. The stem is short, and in some genera absent.

The most primitive type is seen in *Aristocystis* from the Ordovician (fig. 86). In this genus the body is ovoid or pear-shaped, and is formed of numerous polygonal plates without any regular arrangement except at the point of attachment; it is without food-grooves, brachioles or stem. The plates are perforated by canals perpendicular to the surface and distributed irregularly. The mouth (*O*) is at the summit and the anus (*As*), covered by a pyramid of plates, is on one side. *Dendrocystis*, from the Ordovician

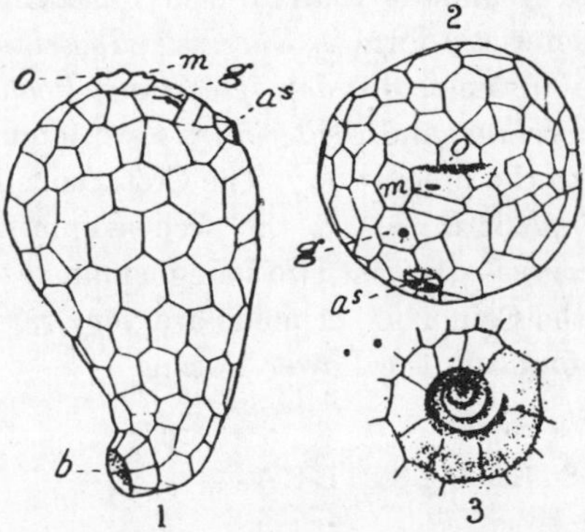

Fig. 86.

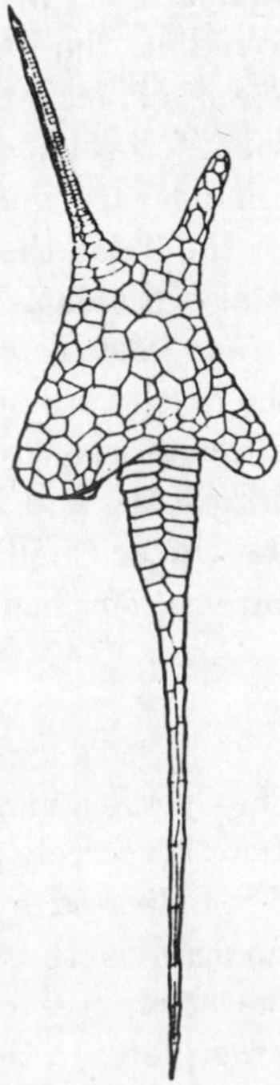

Fig. 87.

Fig. 86. *Aristocystis bohemicus*, Ordovician. 1, side view. 2, oral view, $\times \frac{1}{2}$. 3, base, showing impression of gasteropod, $\times \frac{2}{3}$. *as*, anus; *b*, surface of attachment; *g*, genital opening; *m*, hydropore; *o*, mouth. (After Bather.)

Fig. 87. *Dendrocystis scotica*, Upper Ordovician. $\times \frac{2}{3}$. (After Bather.)

(fig. 87), is similar to *Aristocystis* but a tapering stem, formed of two rows of plates, is developed; and the calyx is laterally compressed, probably indicating that the animal lived in a horizontal position attached by the stem. Other genera, such as *Mitrocystis* and *Placocystis*, show a further modification since the two sides of the calyx differ in structure.

In *Glyptosphæra* (fig. 88 A), from the Ordovician, the calyx is spherical, and composed of a very large number of polygonal plates, which are without any radial arrangement. The mouth is at the summit of the calyx, and is

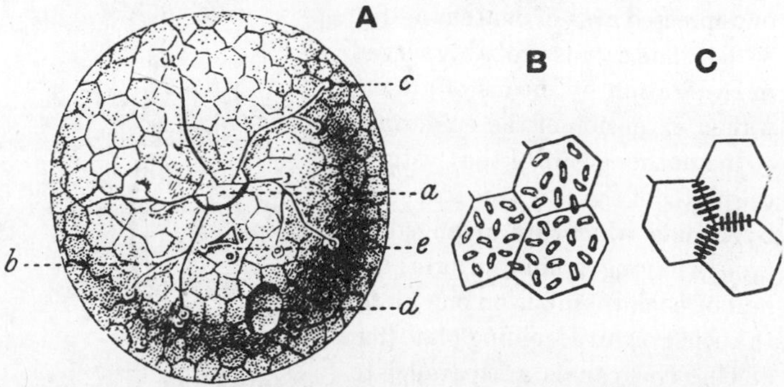

Fig. 88. A. *Glyptosphæra leuchtenbergi*, from the Ordovician of Russia. *a*, mouth covered by oral plates; *b*, food-grooves; *c*, facet for the brachiole; *d*, anus; *e*, just above this is the triangular madreporite, just below is the circular genital aperture (after Volborth). B. A few plates of the same enlarged, showing the pairs of pores. C. Plates of *Echinosphæra*, with pore-rhombs, enlarged.

covered by five oral plates (*a*), between which the five food-grooves start and extend in a radial manner over the upper part of the calyx sometimes giving off branches (*b*); at the ends of these grooves are facets (*c*) to which the brachioles were articulated. The grooves were protected by small covering-plates. On one side of the calyx is the anus (*d*), which in perfect specimens is covered by a pyramid of small triangular plates. Between the mouth and the anus is the madreporite (*e*), which is the external opening of the water-vascular system; just below it is the small, circular genital aperture. All the plates of the calyx are pierced by pairs of canals (diplopores) running perpendicularly to the surface; the canals of each pair are joined by a U-shaped tube, and their external openings are enclosed in a raised or depressed area of oval shape (fig. 88 B). The canals probably served in respiration by means of a thin-walled extension of the epidermis.

In another group of the Cystidea the plates of the calyx are traversed by canals which are arranged in groups having a rhombic form; one half of each rhomb is on one plate, the other on an adjoining plate (fig. 88 C). The canals are parallel to the surface of the plates, and perpendicular to the sutures between the plates. These groups of canals are known as *pore-rhombs*. *Echinosphæra*, from the Ordovician, is a form which possesses many pore-rhombs; it has a spherical calyx, consisting of numerous plates, some of which project at the base and probably served to fix the calyx, there being no stem; around the mouth are from three to five small arms. In most genera belonging to this group, the plates of the calyx are much fewer than in *Echinosphæra*, and have a distinctly radial symmetry—being arranged in cycles,

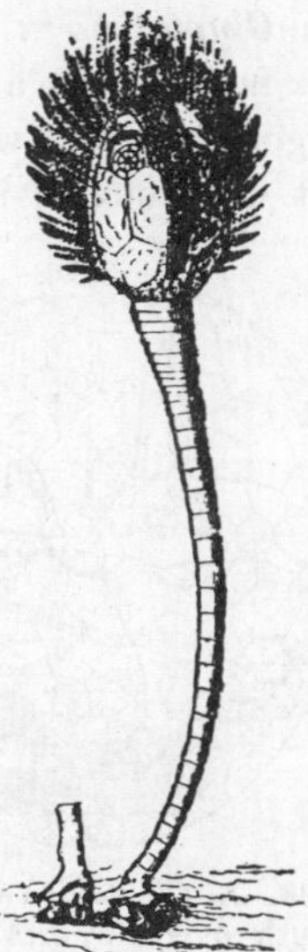

Fig. 89. *Lepadocrinus quadrifasciatus*, from the Wenlock Limestone. Restored figure. The brachioles of the outer rows are erect; those of the middle row depressed. Near the top of the left-hand quarter is the anus; near the top of the right-hand quarter is a pectinirhomb. (From the *Guide to the Geol. Dept., Brit. Mus.*) Natural size.

the plates of each cycle alternating with those immediately below; for example, the calyx of *Lepadocrinus*, from the Silurian (fig. 89), is formed of five cycles of plates; at the base is a cycle of four plates, followed by four cycles of five plates each; from the summit of the ovoid calyx four food-grooves stretch toward the base; they do not rest directly on the calyx, as is the case in *Glyptosphæra*, but on specially-developed plates. Numerous brachioles come off from each side of the food-grooves. In this genus there are only three rhombs, and they are of the more highly-developed type called *pectini-rhombs*, which differ from pore-rhombs in being surrounded by a raised rim and in having the folds of the plate more pronounced. In some other Cystideans of this group the brachioles are found near the mouth only.

Distribution of the Cystidea

The Cystideans are comparatively rare fossils. They range from the Middle Cambrian to the Devonian, and attain their maximum development in the Upper Ordovician. In the Menevian, *Protocystis* is found; this also occurs in the Tremadoc Beds, and with it *Macrocystella*. In the Ordovician, *Aristocystis*, *Dendrocystis*, *Echinosphæra*, *Pleurocystis*, *Glyptosphæra* and others are present; in the Silurian, *Lepadocrinus*, *Pseudocrinus*, and *Placocystis*. In the Devonian there are fewer forms (*Pseudocrinus*, *Jaekelocystis*).

CLASS III. BLASTOIDEA

In the Blastoids (fig. 90) the body consists of a calyx, usually with a stem; but the latter is rarely found attached to the calyx. The calyx may be spherical, oval, pear-shaped, or bud-like; in most cases it is formed almost entirely of thirteen plates, arranged in a regular manner. True arms are not present.

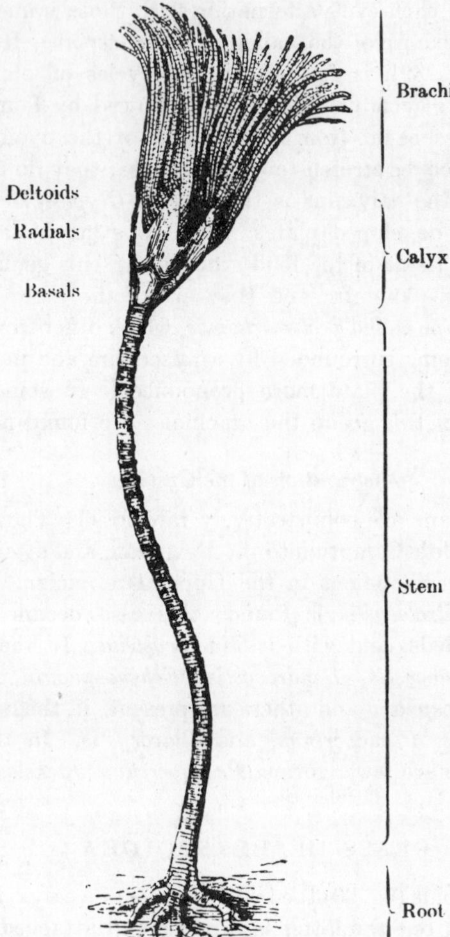

Fig. 90. *Orophocrinus fusiformis*, from the Carboniferous of Iowa. Restored figure. (From the *Guide to the Geol. Dept., Brit. Mus.*) Natural size

Pentremites is the commonest Blastoid, and may therefore conveniently be taken as an example of the group. Its calyx (fig. 91) has the following structure. The aboral part is formed of a cycle of three plates—the *basals* (*b*), two of which are alike, and the third smaller. Above the basals is a cycle of five *radial plates* (*r*); these are larger than the basals, and form the main part of the calyx. At the upper end of each there is a deep incision, which serves for the reception of the food-carrying area (*a*); this is usually

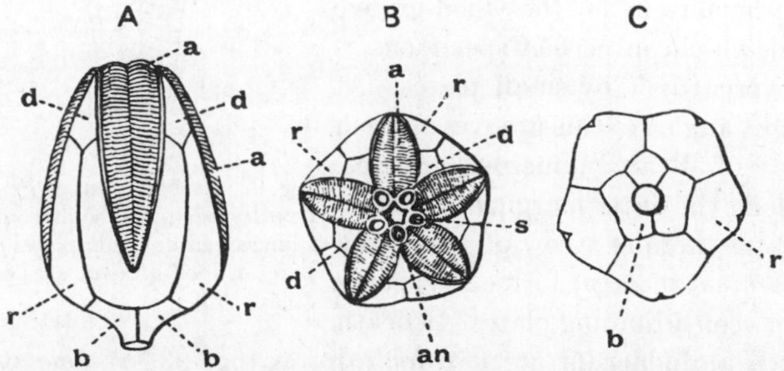

Fig. 91. *Pentremites godoni*, Carboniferous. A, side; B, upper surface; C, under surface. *a*, ambulacra; *b*, basal plates; *r*, radials; *d*, deltoids; *s*, spiracles around the mouth; *an*, anus. ×2.

spoken of as an 'ambulacrum', but there is no evidence of the existence of a radial water vessel, and it is doubtful whether this area is really homologous with the ambulacrum of an Echinoid. Above the radials and alternating with them occur five smaller plates—the *deltoids* (*d*) or inter-radials. The mouth is placed at the summit of the calyx, in the centre, and around it are five other openings termed *spiracles* (*s*), one of which is larger than the others and includes the anus (*an*). From the mouth the five ambulacra (*a*) radiate towards the aboral surface, and are bordered partly by the deltoids but mainly by the radials. Each ambulacrum (fig. 92) consists of the following plates: in the middle is a long pointed plate (l), the *lancet-plate*, which is traversed by a longitudinal canal, in which a nerve may have been present. On each side of the lancet-plate is a row of small plates, the *side-plates* (s). Extending down the middle of each ambulacrum is the food-groove (a), which, in perfect specimens, is covered over by small plates. At right angles to this groove, on each side of it, are numerous grooves. Along the outer margin of the side-plates there is a row of pores, the *marginal pores* (p), formed by spaces between adjoining plates. Beneath each ambulacrum are two *hydrospires* (fig. 93, *h*), one on each side. The hydrospire (fig. 94) is a flattened and folded organ, communicating with the exterior by means of the marginal pores, and also by the spiracles on the oral surface of the calyx. A current of water probably passed in through the former openings and out by the latter. In well-preserved specimens the mouth, as in many crinoids, is not visible externally, but is covered over by a roof of small plates. From the margins of the ambulacra pinnule-like appendages known as *brachioles* (fig. 90) are given off; these are seldom preserved, but pits or facets to which they were attached are seen on the side-plates.

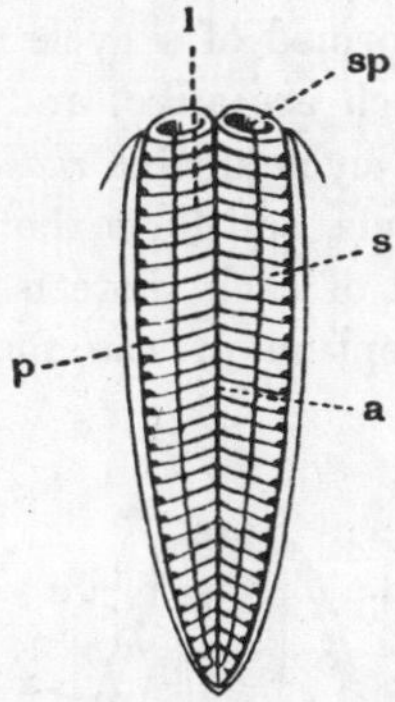

Fig. 92. Ambulacrum of *Pentremites godoni*, Carboniferous. *l*, lancet-plate; *s*, side-plate; *p*, pore; *a*, food-groove; *sp*, spiracle. ×3.

The hydrospires are really folded parts of the radial and deltoid plates—the folds being parallel to the margin of the ambulacra. This is clearly seen in *Codaster*, which is a more primitive form than *Pentremites*; in that genus (fig. 95) the folds open directly to the exterior by slits, owing to the fact that they are not covered by the lancet-plate and side-plates; and on account of this circumstance spiracles are not developed. In some genera, in which the folds are concealed (fig. 96), the space below the lancet-

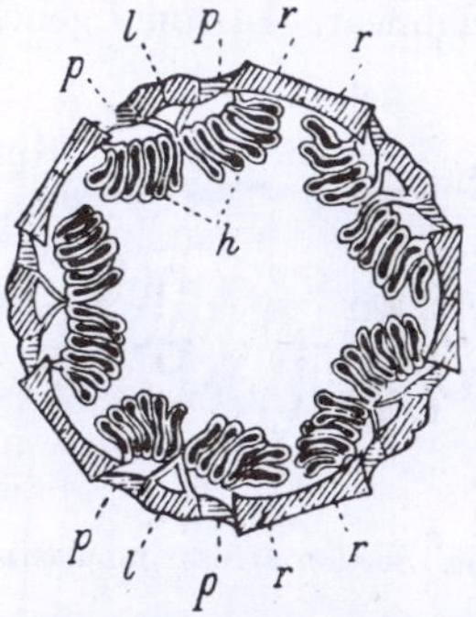

Fig. 93.

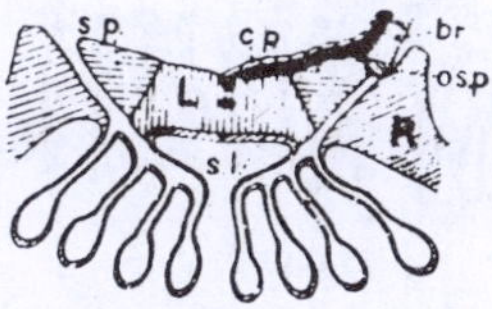

Fig. 94.

Fig. 93. *Pentremites sulcatus*, Carboniferous. Horizontal section of the calyx. *l*, lancet-plate; *p*, side-plates; *r*, radial plates; *h*, hydrospires (not quite correctly drawn, see fig. 94). (After Zittel.) Enlarged.

Fig. 94. *Pentremites*, Carboniferous. Section across ambulacrum. *br*, brachiole; *c.p.*, covering plates; *L*, lancet-plate; *o.s.p.*, outer side-plate; *R*, radial; *s.l.*, sub-lancet-plate; *s.p.*, side-plate. (After Bather.) ×5.

plate and side-plates, into which the folds open, communicates with the exterior at the oral end by slits or incipient spiracles. A further modification is seen in *Pentremites* (fig. 94) in which, owing to the hydrospire being pushed further into the cavity of the calyx, the folds open into a common canal instead of into the space between the summits of the folds and the overlying lancet-plate and side-plates; this canal opens orally by true spiracles (fig. 91 B, *s*).

The number of folds in each hydrospire varies from one to nine. In a few primitive types hydrospires are absent. In many Blastoids there are five pairs of spiracles and an independent anus, but in some genera (*e.g. Pentremites*) the pairs are confluent so that only five spiracles are present, of which the posterior encloses the anus.

The ambulacra vary in width and length; they may be broad and petaloid or narrow and linear. In some genera

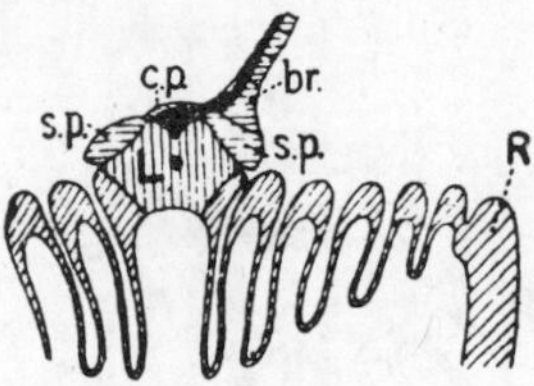

Fig. 95.

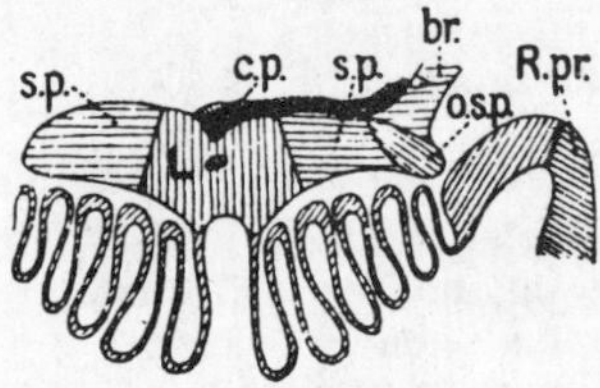

Fig. 96.

Fig. 95. *Codaster trilobatus*, Carboniferous. Section across ambulacrum. (After Bather.) ×5.

Fig. 96. *Phænoschisma verneuili*, Carboniferous. Section across ambulacrum. (After Bather.) Enlarged.

br., brachiole; *c.p.*, covering-plate; *L*, lancet-plate; *o.s.p.*, outer side-plate; *R*, radial; *R.pr.*, part of radial; *s.p.*, side-plate.

the alternate side-plates become squeezed towards the outside of the ambulacrum; here they form an outer row, known as the *outer side-plates*, and are smaller than the plates of the inner row. The side-plates may be entirely at the sides of the lancet-plate (fig. 96), or they may rest on it and partly, or even completely, conceal it (fig. 95). The basals, radials, and deltoids vary considerably in relative size—thus the deltoids may be very small (as in *Troostocrinus*), or they may form a considerable part of the calyx (as in some species of *Orbitremites*).

The most important characters of the Blastoidea as a Class are found in the ambulacra and hydrospires, the absence of true arms, the monocyclic base consisting of three basals only, and the five incised radials. In a few rare cases, hydrospires have been found to be present in the Crinoidea (*Carabocrinus, Hybocrinus*).

Codaster. Calyx in the form of an inverted cone or pyramid. Basals forming a conical and usually deep cup; radials large, with the forked parts sharply bent, forming part of the flattened upper surface of the calyx; deltoids and ambulacra confined to upper surface. A long lancet-plate, with side-plates, occurs between the deltoids and radials. Hydrospires consist of sharp folds of the calyx where the radials and deltoids meet, and open at the surface by slits. Mouth pentagonal, originally plated over; no spiracles; anus between the posterior deltoid and radials. Silurian to Carboniferous. Ex. *C. trilobatus*, Carboniferous.

Orbitremites (= *Granatocrinus*). Calyx elliptical, ovate, or more or less spherical, in section pentagonal or round; with concave base. Basals small, not seen in a side view. Radials of variable size and forming part of the base. Deltoids generally rhombic, large in some species, small in others. Ambulacra narrow, straight, with nearly parallel sides. Lancet-plate narrow. Hydrospires simple, usually with two or three folds only, dilated at the free ends; the inner fold forms a plate next to the lancet-plate. Spiracles five, round or oval, piercing the apices of the deltoids, the posterior one including the anus. Carboniferous Limestone. Ex. *O. derbiensis*.

Distribution of the Blastoidea

In England the Blastoids range from the Devonian to the Carboniferous, being most abundant in the latter. A few primitive types (*Asteroblastus, Blastoidocrinus*) occur in the Ordovician of Russia and Canada; and some others (*Troostocrinus, Codaster*) are found in the Silurian of North America

The English Devonian forms are rare and but little known. In the Carboniferous Limestone the blastoids attain their maximum development; ten genera are represented, the most important being *Codaster*, *Orophocrinus*, *Schizoblastus*, *Orbitremites* and *Mesoblastus*. *Pentremites* is common in the Carboniferous of America, but is not found in Britain. A number of genera have been found in the Permian of Timor.

CLASS IV. EDRIOASTEROIDEA

The calyx in the Edrioasteroids (fig. 97 A) is usually composed of a large number of irregular plates, and in most

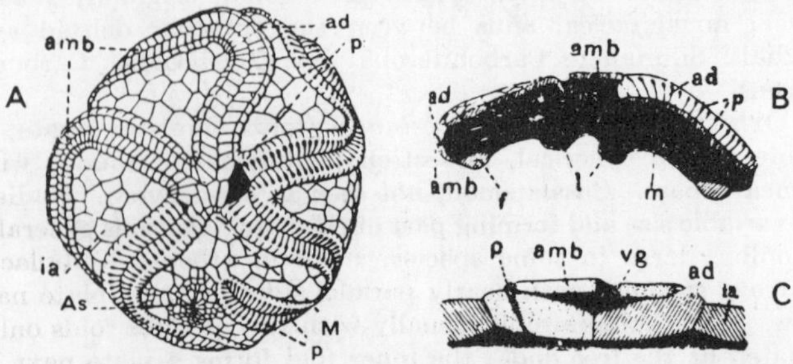

Fig. 97. *Edrioaster bigsbyi*, Ordovician of Canada. (From Bather.)

A. Oral surface. *amb*, covering-plates over the anterior and left-anterior ambulacral grooves, but removed from the other grooves; *ad*, floor-plates of ambulacral grooves; *p*, pores between floor-plates; *ps*, peristome, the greater part of which is roofed by enlarged covering-plates; *ia*, interambulacrum; *M*, madreporite; *As*, anus. Natural size.

B. Section across the same specimen through the right anterior ambulacrum and the left posterior interambulacrum. Lettering as in A. *f*, frame of stouter plates; *m*, membrane with overlapping plates thrown into five lobes (*l*). Natural size.

C. Section across an ambulacrum with covering-plates (*amb*) over the groove (*vg*). Enlarged.

cases is flattened and more or less circular in outline; it is attached to some foreign body by the under part—a stem being rarely if ever present. The mouth is at the centre of the upper surface (*ps*), and is covered by plates; from it five ambulacra extend outwards over the upper surface of the calyx, and sometimes over part of the lower surface also. The ambulacra do not branch as a rule, but are frequently curved. The ambulacral grooves are covered by two rows of alternating plates (*amb*), similar to the covering-plates of crinoids. In *Edrioaster* and its near allies the floor of each groove is formed of special plates (*ad*), between or at the outer margins of which are pores (*p*) which may indicate the existence of tube-feet. Neither brachioles nor arms are developed in connexion with the ambulacra. The anus, which is covered by a pyramid of plates, is on the upper surface—in the area between the two posterior ambulacra (*As*). The calyx was more or less flexible in some cases; and frequently around its border on the upper surface (but sometimes on the lower, fig. 97 B, *f*) there is a series of larger marginal plates, forming a framework, which, in combination with the five conspicuous ambulacra, gives the upper surface something of the appearance of a starfish in which the rays are not prolonged.

The Edrioasteroids include only a few genera, and have usually been regarded as Cystidea, but differ in the absence of brachioles, and in the occurrence of pores between the flooring-plates, suggestive of the presence of a radial water vessel with tube-feet.

Distribution of the Edrioasteroidea

The Class ranges from the Cambrian to the Carboniferous, and is best represented in the Ordovician. The principal genera are: *Stromatocystis* in the Cambrian; *Cyathocystis*,

Edrioaster and *Steganoblastus* (?) in the Ordovician; *Pyrgocystis* in the Silurian; and *Agelacrinus* and *Lepidodiscus*, ranging from the Ordovician to the Carboniferous.[1]

[1] The genera *Turrilepas* (Silurian), *Lepidocoleus* (Ordovician to Devonian) and *Plumulites* (Ordovician to Devonian) have commonly been regarded as early forms of Cirripedes, but later work shows that this view cannot be maintained. It has been suggested that these and other allied genera may represent an early offshoot of the Echinoderm stock, for which the name *Machæridia* has been proposed.

PHYLUM ANNELIDA

CLASS CHÆTOPODA

The Chætopoda include various forms of worms. The body is segmented and generally the segments are numerous and similar. There is a ventral nerve-cord with ganglia, and a nerve-ring round the œsophagus connected with a pair of ganglia above it. A vascular system and a body-cavity (cœlom) are present. The cuticle is thin and flexible. The majority of the Chætopoda possess bristle-like processes termed setæ or chetæ which assist in locomotion. There are two orders: (1) the Polychæta, (2) the Oligochæta, *e.g.* the common earthworm *Lumbricus*; *Protoscolex* (Ordovician and Silurian) probably belongs to this Order.

ORDER I. POLYCHÆTA

The members of this Order are nearly all marine, and are characterised by the possession of numerous setæ arranged in bundles on each segment; the setæ are usually placed on lobes or flaps on the sides of the segments termed *parapodia*. Tentacles are usually present on the head. Many forms live in tubes, which may consist of carbonate of lime, of chitinous material, or of grains of sand cemented together by a secretion; the tubes are sometimes free, but often attached to some foreign object. On account of the possession of this tube the polychætous worms are often found fossil. Other forms, which do not live in tubes, are provided with minute chitinous jaws, and in some formations, especially the Ordovician and Silurian, these are abundantly preserved.

Serpula. Tube calcareous, long, round, angular or flattened; straight, curved irregularly or sometimes spirally, closed at one end; generally attached to some foreign object by a portion of its surface. Silurian to present day. Ex. *S. gordialis*, Chalk.

Spirorbis. Tube calcareous, small, spiral, attached by one side. The spiral either left-handed or right-handed, the last whorl often produced into a free tube. Ordovician to present day. Ex. *S. pusillus* (= *carbonarius*), Carboniferous.

Distribution of the Chætopoda

Nearly all the worms which are found fossil belong to the Order Polychæta; the earliest examples occur in the Cambrian Beds. In addition to worm-tubes and jaws, there are, in various rocks, numerous trails and burrows, which are considered by some authors to have been formed by worms, but in many cases it is probable that they were made by other animals such as crustaceans and gasteropods.

PHYLUM BRACHIOPODA

Classes	*Orders*	*Sub-orders*
1. Inarticulata	1. Atremata	
	2. Neotremata	
2. Articulata	1. Protremata	1. Strophomenacea
		2. Orthacea
		3. Pentameracea
	2. Telotremata	1. Spiriferacea
		2. Rhychonellacea
		3. Terebratulacea

In the Brachiopods the soft parts of the animal are enclosed in a shell which is formed of two parts termed *valves*, one placed on the dorsal surface, the other on the ventral. Generally the main part of the body occupies only the posterior portion of the shell. The interior of the shell is lined by the body-wall, and by the *mantle*, which is a prolongation of the body-wall and is divided into two lobes, one occurring in each valve; the space between the two is known as the *mantle-cavity*. The shell is secreted by the mantle. In most genera the margin of the mantle is thickened, and carries numerous chitinous setæ. The mouth (fig. 98, *v*) is near the centre of the anterior surface of the body, and leads into an œsophagus, followed by a stomach, and an intestine. In the articulate brachiopods the intestine is short and ends blindly, in the inarticulate forms it is long and ends in an anus which opens into the mantle-cavity. The nervous system consists of a ring round the œsophagus, with ganglionic enlargements from which nerves are given off to the arms, mantle, etc. The part of the body-cavity which surrounds the alimentary canal communicates with the mantle-cavity by means of two, or rarely four, funnel-shaped canals, which serve as excretory organs. The body-cavity extends into the mantle as a series of spaces or sinuses; these produce slight depressions on the interior of the valves, and can often be traced as ridges on the internal casts of fossil specimens (fig. 116). The body-cavity is filled with a fluid which is kept in motion by means of cilia. The heart is on the dorsal surface of the stomach.

Reproduction takes place sexually, and the sexes are usually separate. The genital organs are placed in the body-cavity, and in the sinuses of the mantle.

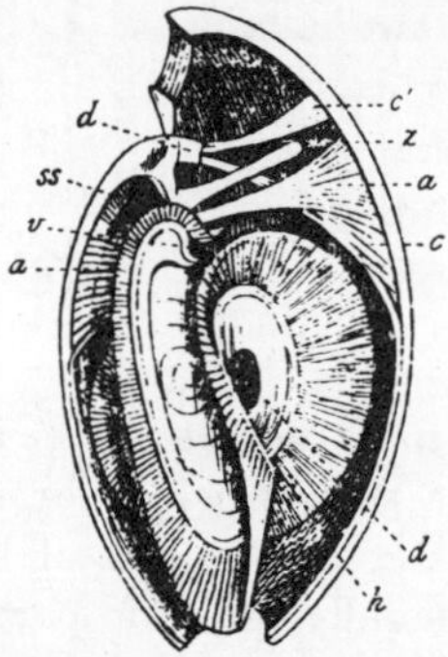

Fig. 98. *Magellania* [= *Waldheimia*] *flavescens*, Recent. Longitudinal section. *d* (upper), cardinal process; *d* (lower), arms; *h*, cirri (tentacles); *a*, adductor muscles; *c*, *c′*, divaricator muscles; *ss*, septum; *v*, mouth; *z*, terminal part of alimentary canal. (After Davidson.) × 1½.

Generally the greater part of the mantle-cavity is occupied by the *lophophore*, consisting of two long processes, given off from the sides of the mouth, known as *arms* or *brachia* (fig. 98, *d*). The arms are covered with cirri (or tentacles) (*h*), the cilia on which produce a current of water conveying food to the mouth. Respiration is carried on mainly by the mantle, but possibly also to some extent by the arms. In some brachiopods spicules of calcite are found in the mantle, and sometimes also in the arms and cirri.

Of the two valves of the brachiopod, the ventral is nearly always larger than the dorsal; each is produced into a beak or *umbo* (fig. 99). The ventral umbo is more prominent than the dorsal, and has generally, either at its apex or just beneath it, an opening. With a very few exceptions the shell of the brachiopod is equilateral, that is to say, a line drawn from the umbo to the opposite margin divides it into two equal and similar parts. This character, combined with the inequality in the size of the valves and the perforation at the umbo, renders it easy to distinguish the shell of a brachiopod from that of a lamellibranch. In many forms the two valves are joined together by means of a hinge, these constitute the group *Articulata*; in others they are held together by the muscles and the mantle only, these form the *Inarticulata*. The hinge consists of two short curved processes or teeth given off from the ventral valve near the

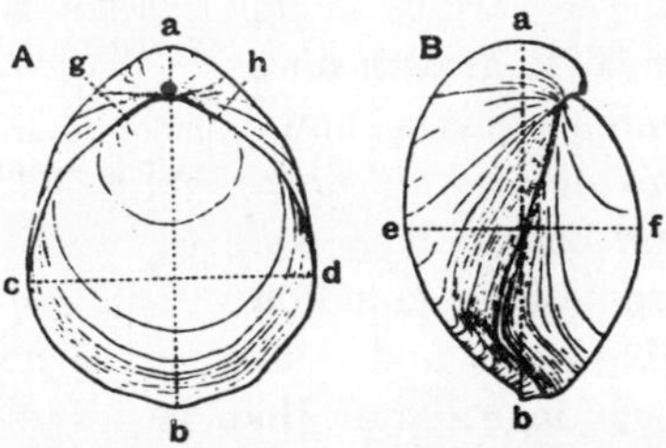

Fig. 99. *Terebratula semiglobosa*, Upper Chalk. A. Dorsal view. B. Lateral view. *a*, posterior; *b*, anterior; *a—b*, length; *c—d*, breadth; *e—f*, thickness; *g—h*, hinge-line. × $\frac{2}{3}$.

umbo, which fit into corresponding sockets in the dorsal valve. In some genera (*e.g. Orthis*) the teeth are supported by plates (the *dental plates*) which are fixed to the inside of the ventral valve. The part of the margin of the valves where the teeth occur and on which the two valves move in the opening and closing of the shell is termed the *hinge-line* (fig. 99, *g—h*). In some genera (*Terebratula*) this is short and curved, in others (*Spirifer*, fig. 113) it is long and straight. The posterior part of the shell is that near the hinge (fig. 99, *a*), the anterior is the opposite margin (*b*). The length of the shell is measured from the anterior to the posterior border (*b—a*). The breadth is at right angles to this, from one side of the shell to the other (*c*—*d*). The thickness is measured from one valve to the other, perpendicular to the length and breadth (*e*—*f*). In some genera (*e.g. Terebratula*) the length is greater than the breadth, in others (*e.g. Strophomena*) the breadth is greater. Between the hinge-line and the umbo there is in some brachiopods (*e.g. Cyrtia*, fig. 100) a flat or slightly concave portion of the shell, usually triangular, on which the ornamentation of the rest of the shell is absent, the surface being either smooth or with growth-lines parallel to the hinge; this is known as the *hinge-area* or *cardinal area*. It may occur on both valves (*e.g. Orthis*), but is sometimes found on the ventral valve only; and is due to the more extensive growth along the hinge margin than occurs in genera which have no hinge-area.

Fig. 100. *Cyrtia exporrecta*, Wenlock Limestone. *a*, umbo of ventral valve; *abc*, hinge-area with deltidium in the middle; *b*—*c*, hinge-line. (The pedicle opening in the deltidium is omitted.) Natural size.

Nearly all living brachiopods are fixed to a rock or other object; but some fossil forms were free, especially in old age (*e.g. Productus*). Some, like *Crania*, are attached by the close adhesion of one valve to the rock; others (*e.g. Strophalosia*) by spines given off from the surface of the shell. More commonly, however, the attachment takes place by means of a stalk or pedicle; this is a cylindrical process, in some genera long, in others short, fixed to the ventral valve, and passing out either through an opening in the ventral valve (fig. 102 A, *f*) or between the umbones (*e.g. Lingula*, fig. 106). It is composed mainly of supporting-tissue with a sheath of horny material, but in some forms there are muscular layers also. In *Lingula*, which commonly lives in burrows in the sand of the sea-floor, the contraction of the muscles of the pedicle serves to withdraw the animal from the surface into its burrow.

The opening for the passage of the pedicle varies considerably in different genera, and is a feature of importance in classification. The simplest case is that found in *Lingula* and other similar forms, in which the opening is shared by both valves. In other types we find that the pedicle-opening is confined to the ventral valve; in *Discina* the opening is completely enclosed by the shell and is often near the centre

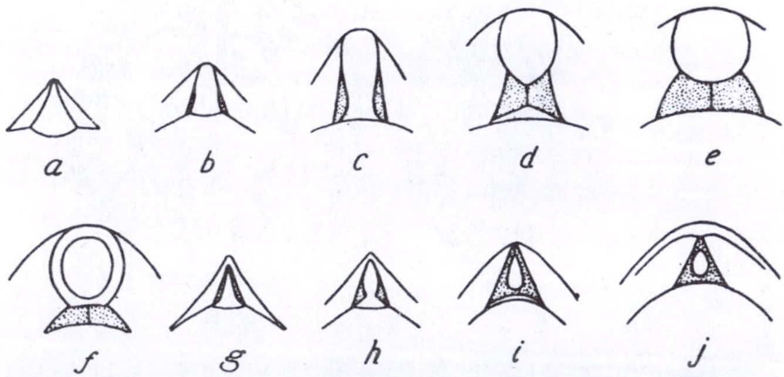

Fig. 101. Development of the deltidium in *a*—*f* a Jurassic Terebratulid; *a*, earliest stage; *f*, adult. *g*—*j*, a Jurassic Rhynchonellid; *g*, an early stage; *j*, adult. The deltidium dotted. Enlarged. (After Deslongchamps.)

of the valve, consequently the pedicle comes out at right angles to the plane of the valves. Sometimes, as in *Orthis* (fig. 111), the pedicle-opening is in the form of a triangular fissure, under the umbo, known as the *delthyrium*. In brachiopods belonging to the group Telotremata, a delthyrium is found in young individuals, but subsequently becomes partly closed by two plates, which grow inwards from the sides of the delthyrium and sometimes meet in the middle line (fig. 101). These two plates form the *deltidium* (fig. 102 A, *d*). In *Rhynchonella* the two plates usually meet, but a small circular or ovate opening (the *foramen*) is left near the centre for the pedicle. In *Magellania* (fig. 102 A, *f*) the foramen truncates the apex of the umbo, its lower boundary being formed by the deltidium (*d*); in other genera the foramen may be at any point between the apex of the umbo and the base of the delthyrium. In genera

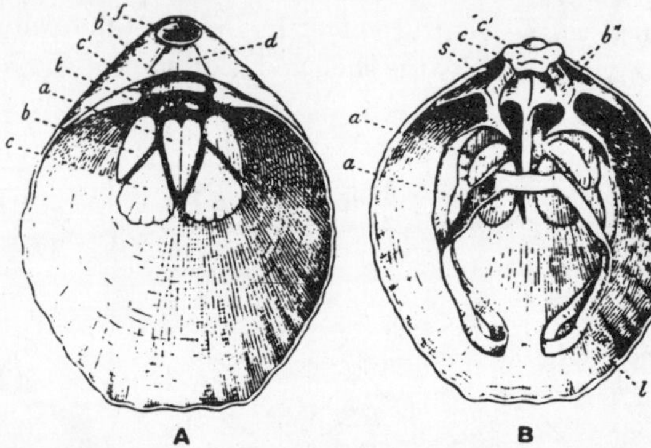

Fig. 102. *Magellania* [= *Waldheimia*] *flavescens*. Recent. A. Interior of ventral (or pedicle) valve. *f*, foramen; *d*, deltidium; *t*, teeth; *a*, impressions of adductor muscles; *c*, *c'*, impressions of divaricator muscles; *b*, *b''*, muscles of the pedicle. B. Interior of dorsal (or brachial) valve. *c*, *c'*, cardinal process; *b''*, hinge-plate; *s*, dental sockets; *l*, brachial skeleton; *a*, *a'*, adductor impressions; *c*, point of attachment of the smaller divaricator. (After Davidson.) × 1½.

belonging to the Protremata and a few of the Neotremata, the delthyrium is more or less completely closed by a single plate known as the *pseudo-deltidium* (fig. 109 B); this at first sight closely resembles the deltidium, but is really of a different nature. It originates on the dorsal surface of the body, but subsequently becomes attached to the ventral valve, and then continues to grow by secretion from the pedicle. The deltidium, on the other hand, is formed by the edge of the ventral lobe of the mantle and consists of a pair of plates which, in some cases, coalesce. The pseudodeltidium is developed at an earlier stage in the life of the individual than the deltidium, and grows from the apex of the delthyrium downwards, becoming fused to the ventral valve.

The two valves of the brachiopod can be opened and closed by means of muscles (fig. 98); those which open them are called the *divaricators* (*c*, *c'*), those which close them, the *adductors* (*a*). When the soft parts of the animal have been removed the places where the muscles were attached to the interior of the shell are indicated by a difference in the surface such as striation, or by slight depressions or elevations; these markings are termed the *muscular impressions* (fig. 102). In the articulate brachiopods there are generally five or six pairs of muscles. In the genus *Magellania* there are two pairs of divaricators (fig. 98, *c*, *c'*) and one of adductors (*a*). Both pairs of the former are attached to a process (the *cardinal process*, fig. 102 B, *c*, *c'*) on the dorsal valve between the teeth sockets, and one pair join the ventral valve near its centre (fig. 102 A, *c*), while the other pair, which are smaller, are attached nearer the posterior border (*c'*). Hence the dorsal valve forms with these two pairs of muscles a lever of the first order. The adductor muscles are united to the ventral valve near the centre (fig. 102 A, *a*) and form a single impression divided by a median line; these muscles bifurcate before reaching the dorsal valve and there form four impressions (fig. 102 B, *a*, *a'*). There are also muscles belonging to the pedicle which serve to retract it, one pair of these being united to the dorsal valve (fig. 102 B, *b''*), the others to the ventral (A, *b*, *b''*). In the Inarticulata the muscles are usually more complicated; thus, in *Lingula* (fig. 106) we find, in addition to the adductors and divaricators, muscles for moving one valve backward or forward in relation to the other, and others for giving a slight rotary motion.

In many of the Protremata there is a concave or spoon-shaped plate, or trough, inside the ventral valve next to the middle part of the hinge margin. This is known as the *spondylium* and serves for the attachment of muscles. It is sometimes supported on a median septum (fig. 112), but may be joined directly to the valve, and appears to be due to the convergence and union of the dental plates.

The arms, already mentioned as occupying in most genera the main part of the mantle-cavity, are generally coiled up. In some forms they can be protruded a greater or shorter distance. Sometimes they are supported on a calcareous ribbon—the *brachial skeleton*—which is attached to the posterior part of the dorsal valve at the sides of the cardinal process. In *Rhynchonella* (fig. 115 B, *c*) the brachial skeleton consists of two short curved processes known as the *crura*. In *Terebratula* (fig. 117) a ribbon-like band comes off from the crura and forms a short loop. In *Stringocephalus* (fig. 119) the loop is more extensive and runs parallel to and near the margin of the valves. In *Magellania* (fig. 102 B, *l*) the loop extends nearly to the anterior margin of the shell and is then bent back upon itself. In many Palæozoic and a few Mesozoic genera the brachial skeleton is in the form of two spiral ribbons which come off from the crura; in *Spirifer* (fig. 113 A) the apices of the spirals are directed towards the lateral margins of the shell, in *Glassia* they point inwards, in *Atrypa* (fig. 114 A) upwards to the centre of the dorsal surface. The brachial skeleton is absent in all the inarticulate genera, as well as in some of the articulate forms belonging to the Protremata, such as *Productus* and *Chonetes*.

The development of the brachial skeleton has been studied in some living species of *Terebratulina*, *Magellania*, and *Terebratella*. In *Terebratulina* the adult form is reached almost directly; but in *Magellania* the brachial skeleton passes through various stages before the adult condition is attained; and it is noteworthy that these stages are similar to the adult forms of certain other genera. Thus in *Magellania venosa* the brachial skeleton passes through stages which, in succession, resemble the brachial skeletons of the genera *Gwynia*, *Cistella*, *Bouchardia*, *Megerlina*, *Magas*, *Magasella*, and *Terebratella*, after which the adult condition is reached. Another striking fact is that some species, which have hitherto been referred to the genus *Magellania*, have a development differing from this; thus *M. cranium* passes through stages distinctive of the genera *Gwynia*, *Cistella*, *Platidia*, *Ismenia*, *Mühlfeldtia*, and *Terebratella*. If the stages through which an individual passes in its development be taken to indicate its ancestry, then it follows that in *Magellania* there are two groups of species having different ancestors, and these two groups must therefore be regarded as constituting two distinct genera (see page 15).

The largest brachiopod known is *Productus giganteus*, from the Carboniferous Limestone, which has a breadth of twelve inches; the size of the shell in different genera varies from this down to about a quarter of an inch. Generally the shell is thin, but in some forms, such as *Daviesiella llangollensis*, it is thick and massive. The external form varies considerably; it may be globular, ovoid, hemispherical, quadrilateral, or triangular. Usually both valves are convex, but in some genera, one is plane the other convex, or one may be concave and the other convex; in the last case the space in the interior is often small. Sometimes there is a median depression or sulcus on the anterior part of one valve (generally the ventral) and a corresponding ridge on the other valve, or there may be two sulci and two ridges (biplication). The surface of the shell is sometimes quite smooth, but is often ornamented with striæ or ribs, which generally radiate from the umbones but are occasionally concentric. In a few forms the shell is covered with spines.

In the Articulata the shell is mainly calcareous. In the genus *Magellania* it is formed of three layers (fig. 103);

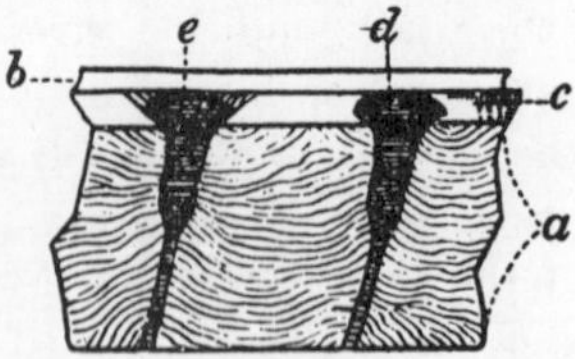

Fig. 103.

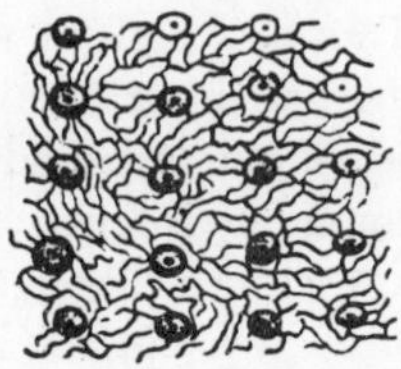

Fig. 104.

Fig. 103. Vertical section of shell of *Magellania flavescens*, Recent. *a*, prismatic layer; *b*, chitinous layer; *c*, outer calcareous layer; *e*, *d*, canals traversing the calcareous layers. (After King.) Magnified.

Fig. 104. Horizontal section through the prismatic layer of *Terebratula maxillata*, from the Great Oolite, showing prisms and canals. Magnified.

the inner (*a*), next the mantle, is the thickest and most important, and consists of flattened prisms of calcite arranged obliquely to the surface of the shell, each prism being encased in a membrane, which of course has disappeared in the fossil examples. The middle layer (*c*) is lamellated and also calcareous. The outer (*b*) consists of chitinous material. The inner and middle layers are traversed by canals (figs. 103, *e*, *d*; 104) running at right angles to the surface of the shell, and containing prolongations of the mantle; in fossil specimens, in which the chitinous layer is not preserved, the openings of these canals can be seen on the surface of the shell, giving it a punctate appearance. The shell is secreted by the mantle, its outermost border producing the chitinous layer, a zone just within this forming the lamellated layer, and the remainder giving rise to the prismatic layer which gradually encroaches on the preceding; hence the last layer is the only one which can subsequently increase in thickness. In many forms the lamellated layer is absent, and in some (*e.g. Rhynchonella*) there are no canals traversing the calcareous layers.

The shell of the Inarticulata has a different structure. In *Lingula* it consists of alternating calcareous and chitinous layers, the calcareous material being largely phosphate of lime; the canals which traverse these layers are more numerous and much smaller than those found in the articulate forms. In *Crania* the shell is calcareous and the canals branch near the surface.

In old age the valves of the brachiopod become thickened, and their margins truncated. The vertical diameter of the shell also increases, and the ornamentation tends to disappear on the marginal part of the valves. The umbones and adjacent parts may be resorbed.

In the earliest or embryonic stage of development the shell is similar in character in all the genera which have been examined. This embryonic shell has been termed the *protegulum*, and may sometimes be found at the umbones of adult shells, but generally, owing to its delicate nature, it has been worn off; it is semicircular or semi-elliptical in form, with concentric lines of growth, and is without an area: it is composed of horny material, and varies in size from ·05 to ·6 millimetre From the constancy of the occurrence of the protegulum it has been inferred that the ancestral form of the Brachiopoda possessed throughout life a shell similar to the protegulum: but, at present, no brachiopod agreeing entirely with the protegulum has been found; for although *Paterina* (fig. 105), from the Lower Cambrian, is in many respects similar, yet the possession of an area distinguishes

Fig. 105. *Micromitra* (*Paterina*) *labradorica*, from the Lower Cambrian (Olenellus Beds). A. Ventral valve. B. Dorsal valve. Enlarged.

it from a protegulum. *Rustella*, also from the Lower Cambrian, is now regarded as the most primitive brachiopod known.

The Brachiopoda can be divided into two Classes, (1) the Inarticulata, (2) the Articulata,[1] each of which may be divided into two Orders.

CLASS I. INARTICULATA

The valves are not provided with teeth, but are held together by the muscles and mantle. The intestine is long and ends in an anus. There is no brachial skeleton. The shell is usually formed to a considerable extent of chitinous material.

ORDER I. ATREMATA

The pedicle passes out between the umbones, the opening being shared by both valves. Lower Cambrian to present day.

[1] These classes have received other names, the Inarticulata being known by some authors as the *Lyopomata*, the *Ecardines*, the *Pleuropygia*, or the *Tretenterata*; and the Articulata as the *Arthropomata*, the *Testicardines* the *Apygia*, or the *Clistenterata*.

Lingula (fig. 106). Shell thin, nearly equivalve, compressed, elongate-ovate or quadrilateral, tapering towards the umbones, slightly gaping at the extremities. Dorsal valve a little shorter than the ventral. Hinge-line slightly thickened. Twelve muscular impressions in each valve, but usually indistinctly marked. Surface of shell smooth, or concentrically or radially striated. Pedicle long, passing out between the umbones. Shell composed of alternating layers of calcareous and chitinous material. Ordovician to present day. Ex. *L. anatina*, Recent; *L. ovalis*, Kimeridge Clay.

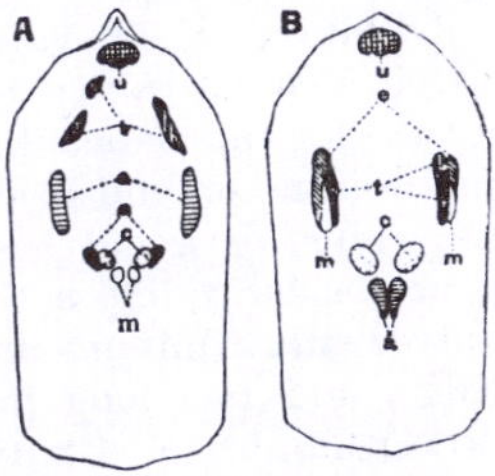

Fig. 106. *Lingula anatina*, Recent. Interior of valves showing muscular impressions. A, ventral valve. B, dorsal valve. *u*, umbonal muscle; *t*, transmedians; *c*, centrals; *a*, anterior laterals; *m*, middle laterals; *e*, external laterals. × ½.

Lingulella. External form similar to *Lingula*; in the ventral valve a distinct hinge-area and a groove for the pedicle. Lower Cambrian to Ordovician. Ex. *L. davisi*, Lingula Flags.

Kutorgina. Shell calcareous, usually broader than long, with a long, straight hinge-line; surface with concentric striæ. Ventral valve very convex, with an elevated umbo; four pairs of muscular impressions. Dorsal valve flat or slightly convex, with a small umbo and two pairs of muscular impressions. Area of ventral valve narrow, with a wide fissure; dorsal area only slightly developed. A rudimentary hinge. Lower (? also Middle) Cambrian. Ex. *K. cingulata.*

ORDER II. NEOTREMATA

The pedicle-opening is confined to the ventral valve. In the lower types (*e.g. Trematis*) the opening is in the form of a triangular slit at the margin of the valve; but in the higher forms, owing to shell-growth occurring all round the margin, it is completely surrounded by shell, and is often near the centre of the valve, in which case the pedicle passes out at right angles to the plane of junction of the two valves. The valves are commonly more or less conical. Lower Cambrian to present day.

Obolella. Shell ovate or sub-circular, lenticular, nearly equivalve. Ventral valve with a solid umbo, and a small area with a tube for the pedicle in the middle; one pair of long muscular impressions extend from near the hinge-line to the middle of the valve, between these are a pair of small impressions, and near the hinge-line a third pair of small impressions. Dorsal valve with a minute umbo, a small area, an internal median ridge, and two long muscular impressions diverging widely. Lower Cambrian. Ex. *O. crassa.*

Siphonotreta. Shell elongate-oval, biconvex, inequivalve, with spines on the surface. Hinge-line curved; no area. Ventral valve the more convex, with a prominent, straight umbo, having a small foramen at its apex continued as a tube to the interior of the valve. Dorsal valve less convex, with umbo at the margin. Muscular impressions near the hinge-line in both valves. Ordovician and Silurian. Ex. *S. micula*, Llandeilo.

Discina (group). Shell composed partly of chitinous material; sub-orbicular or sub-elliptical, surface smooth or covered with striæ of growth. Valves more or less conical, the summits of both sub-central or sub-posterior. Pedicle-opening placed either near the summit of the ventral valve or a little behind it. Four adductor impressions. Ordovician to present day. *Discina*, in the wide sense, as defined above, includes the three genera *Discina* (restricted), *Discinisca*, and *Orbiculoidea*.

Discina (restricted). Both valves convex. Pedicle-opening small, near the middle of the valve externally, passing through the shell obliquely forwards. The only species definitely known is *D. striata*, Recent.

Discinisca. Ventral valve flattened; behind the apex is a disc, which is depressed externally and interrupts the continuity of the lines of growth. The disc is perforated for the pedicle by a fissure which passes directly, not obliquely, through it. Lias to Recent; perhaps Carboniferous. Ex. *D. lamellosa*, Recent.

Orbiculoidea. Ventral valve flattened; on the surface just behind the apex is a narrow furrow, which is perforated at the point farthest from the apex, the perforation passing through the shell obliquely backwards. Ordovician (perhaps Upper Cambrian) to Oligocene. Ex. *O. morrisi*, Wenlock Limestone.

Crania (fig. 107). Shell calcareous, traversed by vertical canals which branch near the outer surface; quadrangular or sub-circular, smooth or with radiating ribs, fixed by the ventral

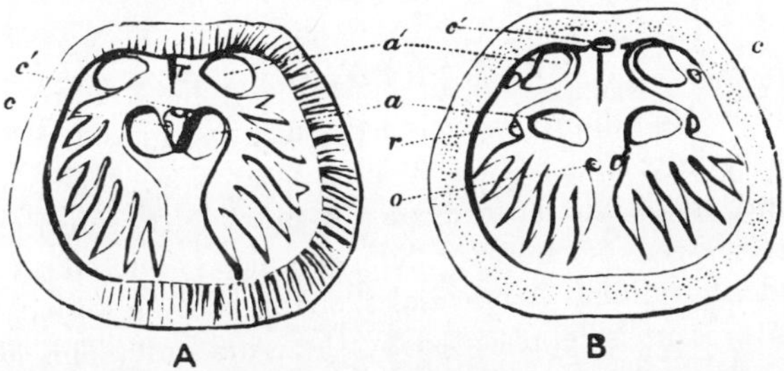

Fig. 107. *Crania anomala*, Recent. A. Interior of ventral valve. B. Dorsal valve. *a*, anterior adductors; *a'*, posterior adductors; *c*, posterior adjustors; *c'*, cardinal muscle; *r*, *o*, central and external adjustors. (From Woodward.) × 2.

valve; without pedicle-opening. Ventral valve depressed-conical: dorsal larger than the ventral, conical with a sub-central apex. Interior of each valve with a border covered with granulations. Two pairs of well-marked adductor impressions in each valve (*a*, *a'*): the posterior pair near the margin, the anterior near the centres of the valves and close together, especially so in the ventral valve; also other smaller muscular impressions. A triangular protuberance near the centre of the ventral valve. Ordovician to present day. Ex. *C. ignabergensis*, Chalk.

CLASS II. ARTICULATA

The valves articulate by means of two teeth on the ventral valve which fit into sockets on the dorsal. The intestine is short and ends blindly. A brachial skeleton may or may not be present. The shell is calcareous.

ORDER I. PROTREMATA

A pseudo-deltidium is developed, but sometimes disappears in the adult. The pedicle-opening is at the margin of the ventral valve, in the form of a fissure (delthyrium) either entirely open or more or less completely closed by the pseudo-deltidium. A brachial skeleton is often absent, and when present is represented by the crura only. This group is found mainly in the Palæozoic formations; in the Mesozoic it is represented by *Thecidea* and other allied genera; the only living forms are *Lacazella* and *Thecidellina*.

1. STROPHOMENACEA

Spondylium absent. Pseudo-deltidium nearly always present throughout life. Cardinal process well developed. Pedicle opening small, at the apex of the umbo, but closed in fixed forms. Ordovician to present day.

Productus (fig. 108). Shell free, or fixed by spines, generally transverse (*i.e.* broader than long) but sometimes elongated, often produced into 'ears' at the sides. Dorsal valve concave. Ventral valve very convex, often sharply bent, sometimes with a median sinus; umbo large, incurved, not perforated. Hinge-line straight, teeth absent or rudimentary. Area linear or absent. Surface ornamented with radiating ribs, crossed by concentric folds, especially in the umbonal region. Tubular spines, especially in the region of the umbo and ears. Muscular impressions strongly marked; in the ventral valve the adductors (*a*) are near the umbo, and in front of them are the divaricators (r). A prominent cardinal process (j) on the dorsal valve is continued as a median ridge in the interior. No brachial skeleton. Carboniferous and Permian. Ex. *P. semireticulatus*, Carboniferous Limestone. *Productella*, Devonian, is an allied form.

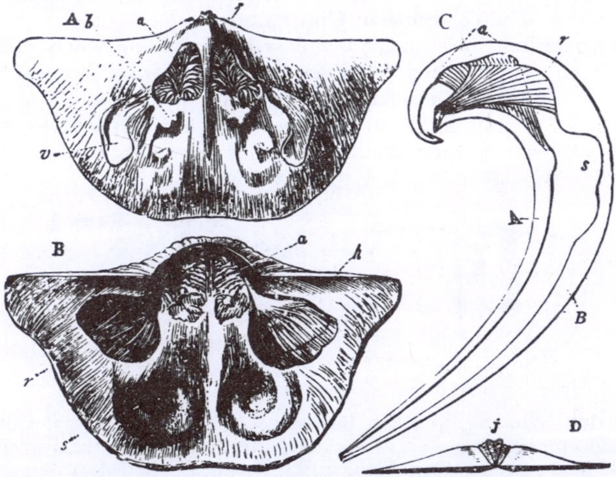

Fig. 108. *Productus giganteus*, Carboniferous Limestone. A. Interior of dorsal valve. B. Interior of ventral valve. C. Ideal section of both valves. D. Dorsal hinge-line. *j*, cardinal process; *a*, adductor; *r*, divaricator; *h*, ventral area; *b*, brachial prominence (?); *s*, hollows occupied by the spiral arms; *v*, reniform impressions. (From Woodward.) $\times \frac{1}{3}$.

Strophalosia. Shell similar to *Productus* in form; attached by umbo of ventral valve. A distinct area on each valve; the ventral area larger than the dorsal, with a pseudo-deltidium. Ventral valve with two prominent teeth. Dorsal valve with a prominent, bifid cardinal process. Surface of ventral (and sometimes also the dorsal) valve covered with spines. Middle Devonian to Permian. Ex. *S. excavata*, Permian.

Chonetes (fig. 109). Shell transverse, semicircular, concavo-convex or sometimes plano-convex. Hinge-line straight, forming the greatest width of the shell. Teeth strong. An area on each valve; dorsal area very narrow. Upper margin of area of ventral valve with a row of hollow, diverging spines, which increase in length towards the ends of the hinge-line. Delthyrium more or less completely closed by a pseudo-deltidium. Muscular impressions faintly marked. Cardinal process divided. Surface usually ornamented with radial striæ. Silurian to Permian. Ex. *C. striatella*, Upper Ludlow.

Leptæna. Shell concavo-convex, semi-oval or nearly quadrangular, ornamented with small radiating ribs, crossed by concentric folds on the flatter parts; anterior part bent sharply, often at a right angle to the posterior part. Space between the two valves very small. Hinge-line straight, forming the greatest width of the shell. A narrow area on each valve; the delthyrium is covered by a convex pseudo-deltidium. Umbo of ventral valve perforated by a small foramen except in old individuals. Two strong diverging teeth in the ventral valve supported by lamellæ which are continued round the muscular area. Muscular impressions: in the ventral valve, two narrow adductors surrounded by two large divaricators; in the dorsal, two small adductors near the centre of the valve, behind which are two larger adductors. Cardinal process divided. Ordovician to Carboniferous. Ex. *L. rhomboidalis*, Bala Beds, etc.

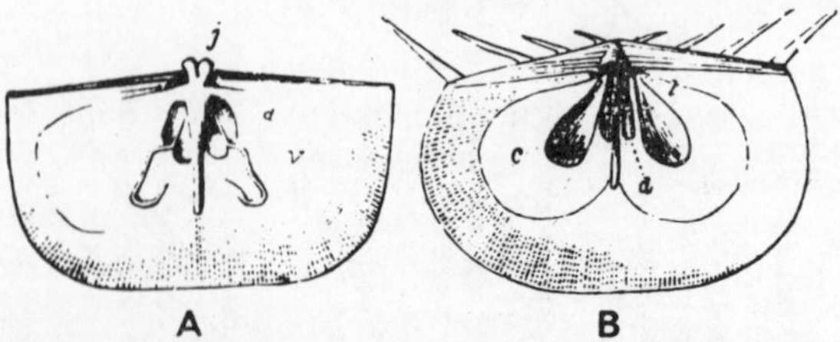

Fig. 109. *Chonetes*, from the Devonian. A, dorsal; B, ventral valve. *d*, adductor impressions; *c*, divaricators; *t*, teeth; *v*, vascular impressions; *j*, cardinal process. (From Woodward.) Enlarged.

Rafinesquina. Outline similar to *Strophomena*. Ventral valve convex; dorsal valve concave. Ornamented with radiating striæ alternating in size, crossed by concentric growth-lines. Muscular area of the ventral valve faintly limited, consisting of two broad divaricator impressions enclosing a long, narrow adductor. In the dorsal valve the posterior adductors have aborescent markings, and the anterior adductor impressions are indistinct. Cardinal process bilobed, low. Ordovician and Lower Silurian. Ex. *R. alternata*, Ordovician.

Strophonella. Shell semicircular or semi-elliptical; ventral valve concave, dorsal valve convex. Hinge-line long, straight. Dorsal area narrower than the ventral; inner margins of areas crenulate. Muscular area of ventral valve limited by a prominent border. Silurian and Devonian. Ex. *S. euglypha*, Wenlock Limestone.

Strophomena. Shell semicircular or semi-elliptical, ornamented with fine radiating ribs; hinge-line straight, forming the greatest width; dorsal valve convex; ventral valve convex near the umbo, but concave in the middle. Ventral area conspicuous, with a pseudo-deltidium; apex perforated except in old age; dorsal area narrow. Teeth diverging widely, supported by plates, which are produced into ridges nearly surrounding the muscular area; the latter is divided by a median ridge. Dorsal valve with a ridge separating two large adductor impressions, in front of which are two narrow impressions. Ordovician and Silurian. Ex. *S. antiquata*, Wenlock Limestone.

Schellwienella. Ventral valve flat or slightly concave, with a slight convexity around the umbo; dorsal valve convex. Valves ornamented with fine radiating ribs; hinge-line usually shorter than the width of the shell. Without a median septum. Ventral area prominent, often high, the two sides sometimes unequal; delthyrium closed by a pseudo-deltidium; muscular impressions fan-shaped; dental plates short, diverging. Dorsal area rudimentary or absent; the cardinal process fairly strong. Silurian to Carboniferous. Ex. *S. crenistria*, Carboniferous Limestone.

2. ORTHACEA

Spondylium and pseudo-deltidium absent except in the older genera. Delthyrium open. Cardinal process usually more or less well developed. Cambrian to Permian.

Orthis (group) (figs. 110, 111). Outline sub-circular or quadrate. Both valves more or less convex. Surface radially ribbed or striated. Hinge-line straight, sometimes equal to the width of the shell, but generally shorter. An area on each valve; usually with an open delthyrium in the ventral valve, and a similar opening in the dorsal valve. In the ventral valve two large teeth, supported by dental plates. Four muscular impressions in the dorsal valve. Two long, narrow impressions (*d*) with two smaller ones (*a*) between them in the ventral. Cambrian to Carboniferous. *Orthis*, as defined above, includes a large number of species which have been divided into numerous groups now regarded as genera, some of which are *Orthis* (restricted), *Platystrophia*, *Dalmanella*, *Schizophoria*, *Rhipidomella*, *Bilobites*; four of these are briefly described below:

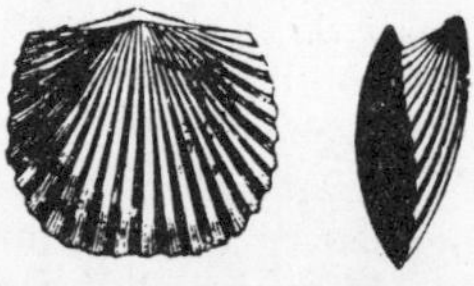

Fig. 110. *Orthis calligramma* var. *Davidsoni*, Ordovician. (From Nicholson.)

Orthis (restricted) (fig. 110). Shell plano-convex, with semicircular to semi-oval outline; with few strong sharp ribs, rarely bifurcating. Hinge-line wide. Area of the ventral valve elevated. Cardinal process in the form of a thin vertical plate. A small flat plate sometimes found in the apex of the delthyrium. Shell not punctate. Ordovician to Silurian. Ex. *O. callactis*, *O. calligramma*, Ordovician.

Platystrophia. Shell spiriferoid in form, with long hinge-line, and sharp radial folds; both valves very convex, with the two areas of nearly equal size. Ventral valve with a strong median fold, dorsal valve with a corresponding sinus. Cardinal process a simple linear ridge. Shell not punctate; surface finely granular. Ordovician and Silurian. Ex. *P. lynx*, Ordovician.

Bilobites. Shell small, bilobed, coarsely punctate; ventral valve more convex than the dorsal, both with a deep median sulcus and ornamented with fine radial ribs. Hinge-line shorter than the width of the shell. Ventral area larger than the dorsal, delthyrium open; teeth strong, dental plates thick; muscle area bilobed, divided by a median ridge. Dorsal valve with thick cardinal process and a long blade-like plate coming off from each side of the hinge. Upper Ordovician to Middle Devonian. Ex. *B. bilobus*, Silurian.

Schizophoria (fig. 111). Shell punctate, ornamented with fine hollow striæ bearing short spines. Dorsal valve more convex than the ventral. Hinge-line shorter than the width of the shell. Cardinal process with accessory ridges in old individuals. Dorsal valve with 4 to 6 deep pallial sinuses (fig. 111 A). Silurian to Upper Carboniferous. Ex. *S. resupinata*, Carboniferous.

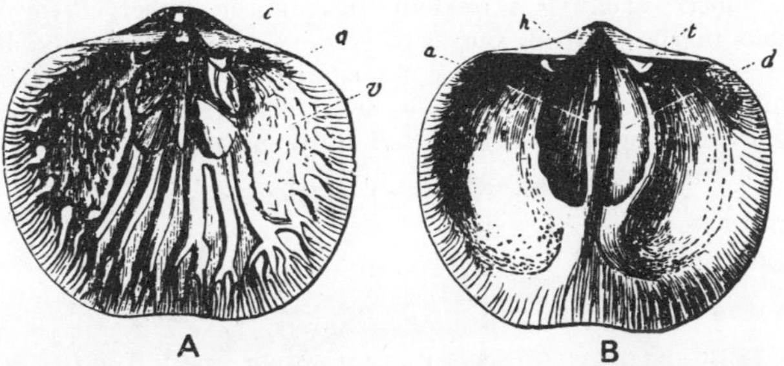

Fig. 111. *Orthis* (*Schizophoria*) *striatula*, Devonian. A. Interior of dorsal valve. B. Ventral valve. *c*, curved brachial processes (crura); *v*, genital impressions; *h*, area with delthyrium; *t*, teeth; *a*, adductors; *d*, divaricators. (From Woodward.) Natural size.

3. PENTAMERACEA

Spondylium well developed. Pseudo-deltidium present in primitive forms, but absent in later types. Brachial skeleton represented by crura. Valves tend to be elongate instead of transverse. Cambrian to Permian.

Conchidium (fig. 112). Shell **large, oval or subtrigonal,** biconvex, with strong radial ribs; not punctate. Ventral valve the more convex, with prominent incurved umbo usually touching the dorsal valve and concealing the delthyrium. Hinge-line curved. Ventral area narrow. Delthyrium wide, covered. Median dorsal fold and ventral sinus slightly developed or absent. Dental plates unite to form a long, narrow, deep spondylium supported on a long median double septum. **Dorsal valve with two long septa. Silurian.** Ex. *C. knighti*, Aymestry Limestone.

Pentamerus. Internal structure similar to *Conchidium.* Surface smooth or with faint undulations in front. Outline oval or approaching pentagonal or hexagonal; trilobed in front.

Hinge-line gently curved, no areas. Ventral umbo less prominent and valves less convex than in *Conchidium*. Delthyrium open. Silurian. Ex. *P. oblongus*, Llandovery.

Gypidula. Surface smooth or with rounded ribs. Hinge-line short, straight. A median sinus in the dorsal valve and a fold in the ventral, but usually on the anterior part only. Ventral valve usually very convex, with inflated and strongly incurved umbo, and a narrow area; delthyrium large, open. A narrow spondylium supported by a septum for part of its

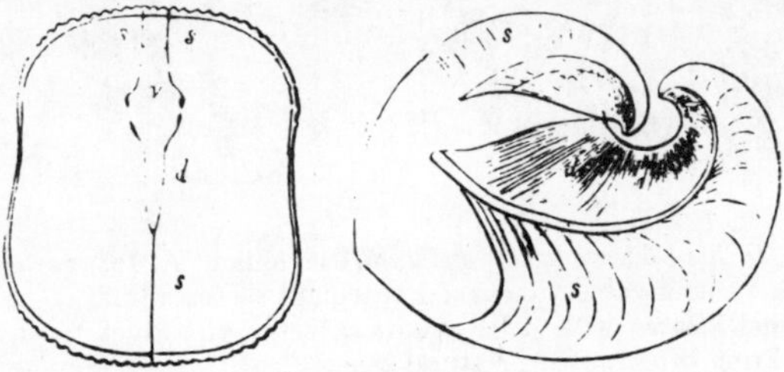

Fig. 112. *Conchidium knighti*, Aymestry Limestone. A. Transverse section. B. Longitudinal section. *s*, septa; *d*, spondylium; *v*, space between dorsal septa. (From Woodward.) × ½.

length. No area in the dorsal valve. Silurian and Devonian. Ex. *G. gypidula*, Wenlock Limestone. *Sieberella* (Silurian and Devonian) is similar to *Gypidula* externally.

Stricklandia. Shell large, oval or sub-circular, smooth or with ribs; convexity of valves nearly equal, sometimes with a fold and a sinus. Umbo of ventral valve not prominent. Hinge-line straight; an area on each valve, the dorsal being small. Delthyrium open. Spondylium and its supporting septum very short. Dorsal valve with short crural processes. Silurian. Ex. *S. lens*, Llandovery Beds.

Camerophoria. External form similar to *Rhynchonella*, with radial folds; ventral umbo sharp, incurved. In the ventral valve the dental plates converge to form a short trough (spondylium) supported by a long medium septum. In the dorsal valve a trough-like plate is supported by a median septum. Carboniferous and Permian. Ex. *C. schlotheimi*, Permian.

ORDER II. TELOTREMATA

The pedicle-opening is confined to the ventral valve in the adult, and is either at the umbo or beneath it. A deltidium is developed, and a brachial skeleton is nearly always present. There are three main divisions.

1. SPIRIFERACEA

Brachial skeleton spiral. Apex of umbo generally not perforated. Punctation of shell generally absent. Ordovician to Jurassic.

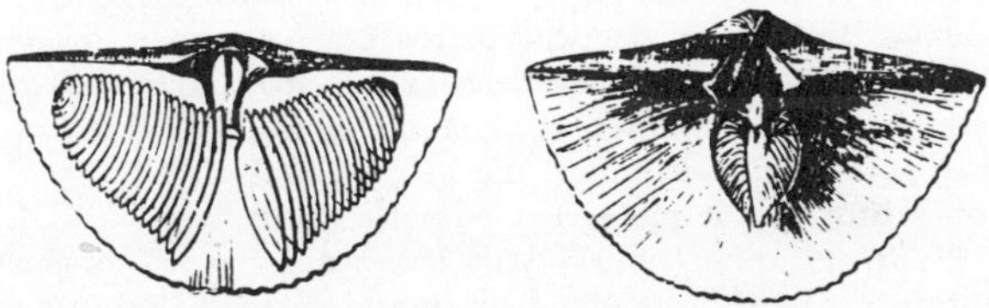

Fig. 113. *Spirifer striatus*, Carboniferous. A. Interior of dorsal valve, showing brachial skeleton. B. Interior of ventral valve, showing muscular impressions, area, and delthyrium. (From Woodward.) × ½.

Spirifer (fig. 113). Shell transverse, more or less triangular, usually alate, biconvex, ornamented with radial ribs. Often with a sinus on the ventral valve and a ridge on the dorsal. Hinge-line straight, long. An area on each valve, the ventral one triangular, often transversely striated, with a delthyrium which is partly closed by a deltidium; dorsal area small. Teeth supported by short dental plates. Brachial skeleton often filling a great part of the interior of the shell, formed mainly of two spirals, with their apices directed laterally. Silurian to Permian. Ex. *S. striatus*, Carboniferous Limestone. *Martinia* includes 'Spirifers' with a short hinge, usually smooth surface, and without dental plates (*e.g. M. glaber*, Carboniferous).

Spiriferina. Similar to *Spirifer*, with a high median septum in the ventral valve, and a punctate shell. Carboniferous to Lias. Ex. *S. walcotti*, Lias.

Syringothyris. Similar to *Spirifer*, but with a high ventral area and an internal split tube in the delthyrium. Deltidium convex, not perforated; a semi-oval pedicle-opening at its base. Carboniferous. Ex. *S. cuspidata.*

Cyrtia (fig. 100). Area on the ventral valve very large; deltidium narrow, convex, with a perforation in the middle for the pedicle. Dental plates well developed but not joining. Brachial skeleton as in *Spirifer*, but the apices of the spires are nearer the hinge-line. Silurian and Devonian. Ex. *C. exporrecta,* Wenlock Limestone.

Uncites. Shell elongate-oval, biconvex, striated. Hinge-line curved, no area. Umbo of ventral valve prominent and incurved, often distorted; pedicle-opening closed in the adult by a concave deltidium. Dental plates strong. Apex of dorsal valve incurved and partly hidden in the ventral valve; cardinal process prominent. Brachial skeleton spiral, apices of spires directed laterally. Devonian. Ex. *U. gryphus.*

Meristina. Shell biconvex, smooth; hinge-line curved, no area. Ventral umbo incurved in the adult, so as to conceal the foramen. Teeth supported by dental plates which reach to near the middle of the valve. Spires of brachial skeleton pointing laterally, joined by a band bearing a median stem which is forked at its end. Silurian. Ex. *M. tumida.*

Athyris. Shell with transversely elliptical or sub-circular outline and a median sinus; the two valves nearly equally convex. Surface often with concentric growth-lines produced into lamellæ. Hinge-line curved. Ventral umbo small, incurved, usually concealing the pedicle-opening and deltidium; with prominent teeth supported by dental plates; four muscular impressions. Dorsal valve with a tube from the interior of the valve opening at the hinge. Brachial skeleton consisting of two spires joined by a band; the apices of the spires pointing laterally. Devonian and Carboniferous. Ex. *A. concentrica,* Devonian.

Atrypa (fig. 114). Shell sub-circular or oval, ornamented with radiating ribs, often crossed by well-marked growth-rings. Ventral valve convex near the umbo, depressed in front; dorsal valve often much inflated. Hinge-line short, slightly curved; no area. A small circular foramen near the apex with a small deltidium below. Two strong crenulate teeth; muscular impressions grouped at the centre of the valve. Brachial skeleton formed of two spirals with their apices directed towards the centre of the dorsal valve; the two spires joined by a band near the umbo. Ordovician to Devonian; abundant in Silurian and Devonian. Ex. *A. reticularis*, Wenlock Limestone.

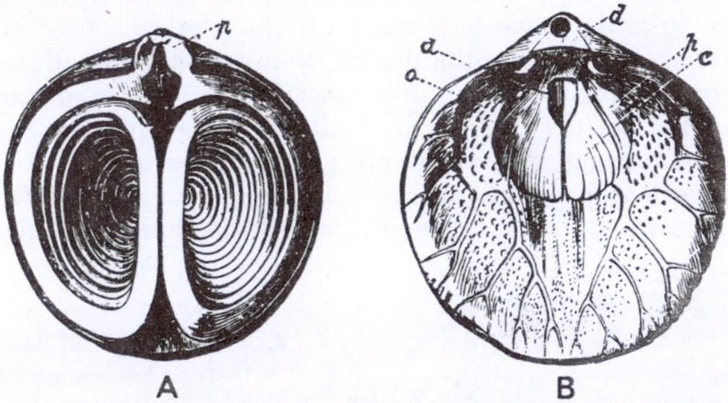

Fig. 114. *Atrypa reticularis*, Wenlock Limestone. A. Dorsal valve, showing brachial skeleton. B. Interior of ventral valve. *a*, impressions of adductor muscles; *c*, divaricator muscles; *p*, muscles of pedicle; *o*, genital impression; *d*, deltidium. (From Woodward.) Natural size.

2. RHYNCHONELLACEA

Brachial skeleton represented by crura only, but sometimes absent. Apex of umbo seldom perforated, since the pedicle opening is nearly always below it. The absence of punctation in the shell is almost constant. Ordovician to present day.

Rhynchonella. Shell triangular, sub-pyramidal owing to the sinus in the ventral valve which is broad and deep in front where it produces a tongue-shaped extension; the dorsal valve with a corresponding fold. Surface with radial striæ. Ventral valve: umbo small, sharp, slightly curved, with the foramen below it and surrounded by the deltidium. Teeth large, crenulate; dental plates short, vertical. Adductor impressions oval, surrounded in front and at the sides by the elongate divaricator impressions. Posterior divaricator and pedicle muscle impressions small, situated in the umbonal part. Dorsal valve: cardinal process small. Crura narrow, slightly curved towards the ventral valve. Dorsal septum reaches to the centre of the valve. Anterior adductors large, oval, on each side of the septum; posterior adductors a little smaller and narrower. Upper Jurassic. Ex.' *R. loxia.*

Cyclothyris. Shell oval or triangular, not perforated by canals, ornamented with numerous radial ribs. Both valves convex; usually a median sinus on the ventral valve and a corresponding ridge on the dorsal. Ventral umbo small, acute, more or less incurved; foramen just below the umbo, almost surrounded by the deltidium. Ventral valve with two strong teeth; dental plates short, diverging; muscular area oval—two large divaricator impressions enclosing small adductors. Brachial skeleton consists of short crura; no cardinal process; median

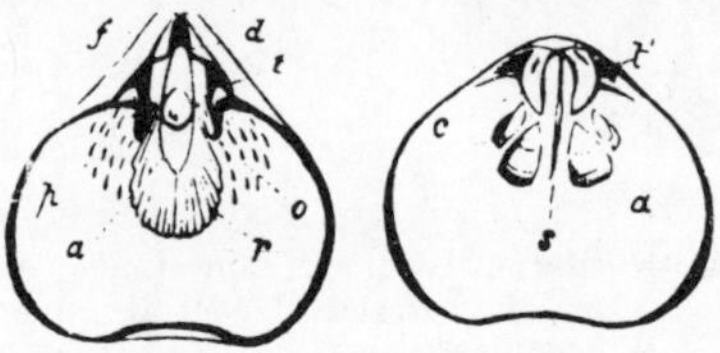

A B

Fig. 115. *Hemithyris psittacea*, Recent. A, interior of ventral; B, interior of dorsal valve. *f*, foramen; *d*, deltidium; *t*, teeth; *a*, adductor impressions; *r*, divaricator impressions; *p*, peduncular impressions; *o*, genital impressions; *t'*, dental sockets; *c*, crura; *s*, septum. (From Woodward.) Natural size.

septum in the dorsal valve feeble or absent. Cretaceous. Ex. *C. latissima*, Lower Greensand. Numerous species found in the Mesozoic are similar in form to *Rhynchonella* and *Cyclothyris*, but differ in internal structures and are regarded as distinct genera.

Acanthothyris. Differs from *Cyclothyris* mainly by the development of numerous spines all over the surface of the shell. Ventral sinus and dorsal fold usually little developed. Jurassic. Ex. *A. spinosa*, Inferior Oolite.

Hemithyris (fig. 115). Shell oval to sub-triangular, with a single fold in front; smooth or faintly ribbed. Ventral umbo high, with the pedicle opening below it; the two plates of the deltidium triangular and separate. Teeth prominent, with dental plates. Crura short, curved. Pliocene to present day. Ex. *H. psittacea.*

Rhynchotreta. Shell elongate-triangular, with strong radial folds; anterior sinus and fold small. Ventral valve: umbo erect, foramen at the apex, nearly surrounded by the large, high deltidium. Dental plates vertical. Divaricator impressions elongate; adductor impression central, small. Dorsal valve: crura long, slightly curved; cardinal process small; a median septum extends to half the length of the shell and divides at the posterior end. Silurian. Ex. *R. cuneata*, Wenlock Limestone.

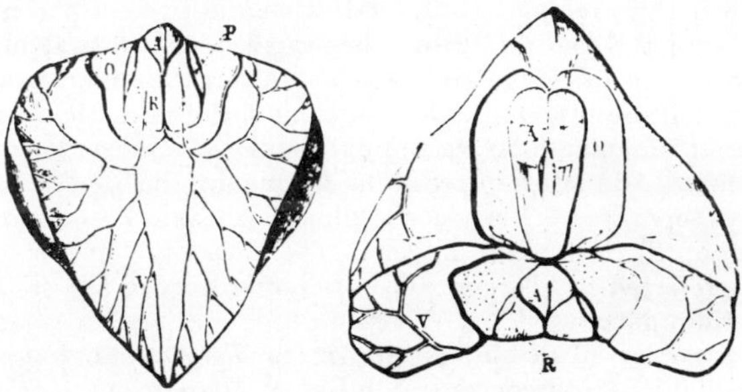

Fig. 116. *Pugnax acuminatus*, Carboniferous Limestone. Internal casts. A. Ventral valve. B. Dorsal valve and posterior part of ventral. *V*, 'vascular' impressions; *O*, genital impressions; *A*, adductors; *R*, divaricators: *P*, muscles of the pedicle. (From Woodward.) Natural size.

Wilsonia. Shell sub-cuboidal or sub-pentahedral; with small radial ribs, each with a fine median line in front. Margins of valves sharply serrated. Fold and sinus not sharply developed except at the anterior margin; anterior slope abrupt. Dental plates united to the lateral walls of valve. Divaricator impressions large and deep in the ventral valve. Cardinal process absent. A median septum in the dorsal valve. Silurian. Ex. *W. wilsoni*, Wenlock Limestone.

Pugnax (fig. 116). Ventral valve shallow, dorsal valve deep. Median sinus and fold very prominent, causing the front margin to be elevated and often acuminate. Some radial ribs present. Dental plates short. No median septum in the dorsal valve. Devonian and Carboniferous. Ex. *P. acuminatus*, Carboniferous Limestone.

3. TEREBRATULACEA

Brachial skeleton in the form of a loop. Apex of the umbo generally truncated for the passage of the pedicle. Shell punctate. Devonian to present day.

Terebratula (figs. 99, 117). Shell biconvex; oval, elongate or rounded; surface nearly always smooth, finely punctate; often with two folds on the dorsal valve and two corresponding sinuses on the ventral Hinge-line curved. No dental plates. Umbo of ventral valve truncated by a large circular foramen with a deltidium at its base. Brachial skeleton in the form of a short triangular loop extending about a third the length of the shell. Adductor impressions strong in the dorsal valve, widely separated. Eocene to Pliocene. Ex. *T. terebratula*, Pliocene; *T. bisinuata*, Eocene. The Mesozoic species commonly referred to 'Terebratula' are now regarded as belonging to distinct genera.

Dictyothyris (Jurassic) is similar to *Terebratula*, but with fine radial ribs and concentric lines; ex. *D. coarctata*. *Dielasma* (Devonian to Permian) and *Cœnothyris* (Trias) are distinguished from the Terebratulæ of Jurassic and later formations mainly by the possession of well-developed dental plates. Ex. *Dielasma hastatum*, Carboniferous; *Cœnothyris vulgaris*, Trias.

Terebratulina (fig. 118). Form similar to *Terebratula*. Ornamented with fine radiating ribs. Umbo short, foramen large; the two plates of the deltidium small, separate. Two ear-like processes at the sides of the dorsal umbo. Brachial loop short, with a ring formed by a band joining the crural points. No septum in the dorsal valve. Jurassic to present day. Ex. *T. striata*, Chalk; *T. caput-serpentis*, Recent.

Magellania (= *Waldheimia*) (fig. 102). Shell biconvex, ovate; smooth or with radial folds on the later parts of the valves. Hinge-line curved. Umbo prominent, truncated by a large circular foramen; plates of deltidium united. Brachial skeleton extending to near the front of the shell and then bent back on itself; a median septum in the dorsal valve, but not joined to the brachial skeleton. Muscular impressions as in fig. 102. Tertiary and living. Ex. *M. flavescens*, Recent.

Magellania in the restricted sense includes species of the type of the recent *M. flavescens.* A large number of species found in the Mesozoic rocks resemble *Magellania,* but show some differences, mainly in internal characters; they are now regarded as constituting a number of distinct genera, some of the more important being: *Eudesia* (ex. *E. cardium,* Great Oolite); *Zeilleria* (ex. *Z. cornuta,* Lias); *Ornithella* (= *Microthyris*) (ex. *O. ornithocephala,* Cornbrash); *Aulacothyris* (ex. *A. resupinata,* Lias); *Obovothyris* (ex. *O. obovata,* Cornbrash); *Digonella* (ex. *D. digona,* Bradford Clay).

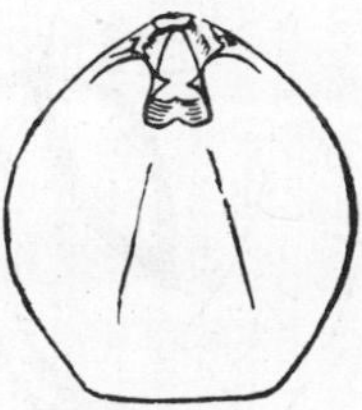

Fig. 117.

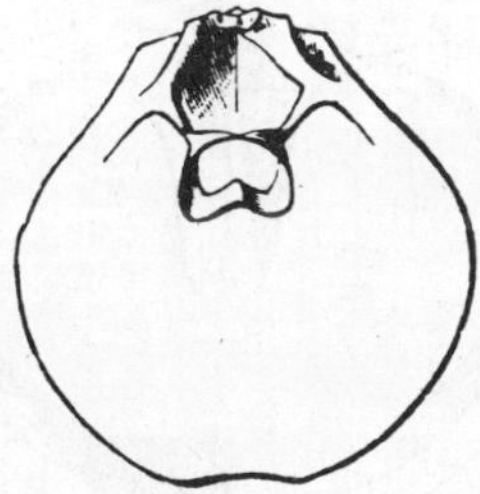

Fig. 118.

Fig. 117. *Terebratula (Gryphus) vitrea,* Recent. Interior of dorsal valve, showing the brachial skeleton. (From Woodward.) × ⅔.

Fig. 118. *Terebratulina caput-serpentis.* Interior of dorsal valve. Recent. (From Woodward.) × 2.

Terebratella. Shell oval, usually with radiating ribs. Ventral valve very convex; dorsal more or less flattened. Hinge-line broad, straight or slightly curved; an area present. Umbo with a large foramen, below which are the two plates of the deltidium, either touching or nearly touching. Brachial skeleton similar to *Magellania,* but descending branches joined by a band to a septum in the middle of the dorsal valve. Lias to present day. Ex. *T. pectita,* Upper Greensand; *T. dorsata,* Recent.

Stringocephalus (fig. 119). Shell smooth, circular or oval in outline. Ventral valve with a sharp, prominent, incurved umbo; area present. Pedicle opening large in young individuals, but smaller and oval in adults on account of the development of the deltidium. Ventral valve with a median septum (*vs*), which extends from the umbo almost to the front of the valve, and increases in height towards the latter. Dorsal valve less convex, with a small septum (*s*), and a long slightly curved cardinal process (*j*), divided at its extremity to embrace the ventral septum. The brachial skeleton consists of the crura (*c*), arising from the hinge-plate (*p*), which pass to the middle of the shell; from the crura the descending branches (*l*) come off and form a ring parallel and near to the margin of the valve. Devonian. Ex. *S. burtini.*

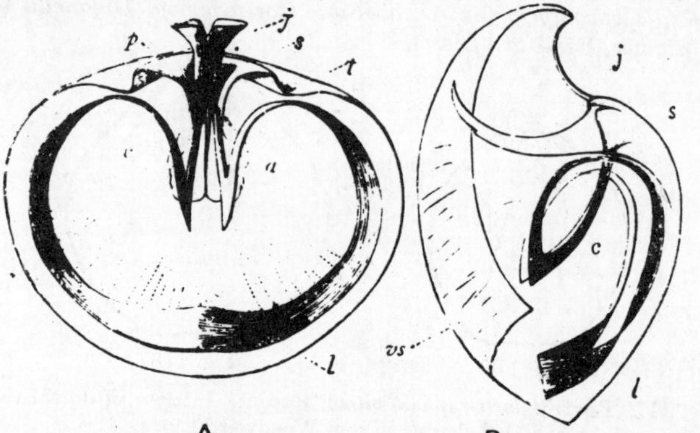

Fig. 119. ***Stringocephalus burtini*, Devonian. A. Dorsal valve. B. Profile. *a*, adductor; *c*, crura; *l*, descending branch; *j*, cardinal process; *p*, hinge-plate; *s*, dorsal septum; *vs*, ventral septum; *t*, dental sockets. (From Woodward.) × ½.**

Distribution of the Brachiopoda

The Brachiopods are all marine, and are found in all parts of the world. At the present time they are much less numerous than in former periods of the earth's history, there being only about 175 living species belonging to 60 genera; of these species 38 are Inarticulates, 137 Articulates. Of the latter group five species belong to the Protremata and 132 to the Telotremata—this being the predominating group of Brachiopods at the present day and represented by 16

Rhynchonelliids and 116 Terebratulids. Many forms occur more abundantly where the sea-bottom is rocky, or stony, or formed of corals, than where it is soft and muddy: frequently they are much localized, being found in enormous numbers at one spot, whilst, in the adjoining areas, they are sparsely distributed. Over 70 per cent. of the existing species are found between the shore-line and a depth of 100 fathoms, and several of these do not extend beyond this limit. Brachiopods are most abundant between 15 and 100 fathoms: their relative scarcity in the Littoral zone (p. 295) is probably due to the fact that most of them are attached by the pedicle and would easily become displaced in the rough waters of the shallow sea. As a whole the Brachiopoda are characteristic of shallow water. Below 150 fathoms they soon become comparatively rare; but some species occur in deep water and in abyssal regions down to 2900 fathoms and are characterised by their thin shells.

The majority of the Inarticulata are found between low-water mark and a depth of 15 fathoms; of the remainder, all but one occur between 15 and 100 fathoms. The principal littoral genera are *Lingula*, *Discina* and *Discinisca* which extend from the shore-line to a depth of about 20 fathoms. The littoral and shallow water species characterise warm seas, and are more numerous and possess thicker and often larger shells than those found in deep water and abyssal regions. *Crania* ranges from 2 to 800 fathoms and is the only Inarticulate genus living in the shallow water of cooler regions, mainly those of the Northern Hemisphere.

The Articulata, although represented by 15 species in water of less than 15 fathoms deep, are mainly characteristic of depths between 15 and 100 fathoms. The Rhynchonellids do not live at depths of less than 15 fathoms and are found mainly in deep water; they occur in nearly all parts of the oceans from the Arctic to the Antarctic regions; some of the species are found in warm seas but the majority live in cool waters; some species have a restricted geographical range, others occur in several provinces, and one (*Hemithyris psittacea*) is found throughout the greater part of the Northern Hemisphere. In depth *Hemithyris* ranges from 15 to 2084 fathoms. The Terebratulids are most abundant, both in individuals and species, between the shore-line and a depth of 100 fathoms, where 67 per cent. of the species are found.

Geographically, the Brachiopoda which live in comparatively shallow water are distributed in provinces, agreeing generally with the Molluscan provinces (p. 297), and these can be grouped into larger regions. Each province is characterised by the presence or abundance of certain species, the ranges of which are determined mainly by climate. A few species, as for example *Terebratulina caput-serpentis*, have a very wide geographical distribution, extending from polar to tropical regions, and also have a great range in depth, the form mentioned being found from the shore-line down to 1180 fathoms.

The species found in deep water have generally a much wider geographical range than those confined to shallow water; and the polar or boreal species have a wider range than those found in warmer regions, since, in lower latitudes, they can find a suitable temperature at greater depths. Since Brachiopods are fixed animals it is only in the free-swimming larval stage that the range of a species can be extended. In most genera this stage is of short duration, so that migration can take place only in shallow water, where there is some foundation to which the Brachiopod can attach itself at the end of the free-swimming stage.

Brachiopods are very abundant in the Palæozoic and Mesozoic formations, and are usually well preserved on account of the fact that their shell generally consists of calcite. The majority of the living genera are represented by species in the Tertiary formations, and a few by species in earlier deposits; of these *Lingula* and *Crania* have existed since the Ordovician period, and others (*e.g. Lacazella, Megathyris, Terebratella, Terebratulina*) since Jurassic times. In connection with the remarkably long range in time of *Lingula* it is interesting to note the habitat of the living species. *Lingula* lives in tubes which it burrows in the sediment on the sea-floor, and is attached to the tube by means of the pedicle; it survives when left uncovered by the sea for several hours, and can live in places which have become putrid owing to the decomposition of organic matter; further, when buried by a rapid deposit of sediment which kills molluscs and other brachiopods, *Lingula* survives by tunnelling to the surface.

Cambrian. The earliest Brachiopods occur in the Lower Cambrian (*Olenellus* Beds), where more than 20 genera are represented. The majority of the Cambrian species belong to the Inarticulata; the Protremata are also represented, but do not become important until the Upper Cambrian. Atremata: *Rustella, Micromitra, Obolus, Lingulella, Kutorgina*. Neotremata: *Obolella, Acrothele, Acrotreta*. Protremata: *Eoorthis, Billingsella*.

Ordovician. Brachiopods are much more numerous than in the Cambrian, especially the Articulate forms belonging to the Protremata, of which the Orthids and Strophomenids show a great development. The Telotremata appear first in the Middle Ordovician. Atremata: *Lingula, Lingulella*. Neotremata: *Siphonotreta, Trematis*. Protremata: *Orthis, Platystrophia, Strophomena, Rafinesquina, Leptæna, Clitambonites*. Telotremata: *Protorhyncha*.

Silurian. Brachiopods have attained their maximum development. The Inarticulates now form a relatively small proportion of the total number. Most of the genera found in the Ordovician survive into the Silurian, but *Pentamerus*, *Conchidium*, *Gypidula* and *Chonetes* now appear, and the Telotremata have become important. Atremata: *Lingula*, *Trimerella*, *Dinobolus*. Neotremata: *Orbiculoidea*. Protremata: *Orthis*, *Dalmanella*, *Bilobites*, *Pentamerus*, *Gypidula*, *Stricklandia*, *Leptœna*, *Strophonella*, *Strophomena*, *Chonetes*. Telotremata: *Meristina*, *Eospirifer*, *Atrypa*, *Cyrtia*, *Rhynchotreta*, *Camarotœchia*, *Wilsonia*.

Devonian. Although showing some decline Brachiopods form a very important part of the Devonian faunas. In the main the genera are similar to those of the Silurian, but *Stringocephalus*, *Uncites*, *Megalanteris* and others are restricted to the Devonian, and the Productids and Terebratulids are now represented by *Productella* and *Dielasma* respectively. Protremata: *Productella*, *Dalmemella*, *Schizophoria*, *Rhipidomella*, *Leptœna*, *Sieberella*. Telotremata: *Spirifer*, *Cyrtina*, *Uncites*, *Athyris*, *Atrypa*, *Uncinulus*, *Hypothyridina*, *Centronella*.

Carboniferous. Orthids, Strophomenids, Productids, Spiriferids and Rhynchonellids are abundant. Under favourable conditions Inarticulates are common. Atremata: *Lingula*. Neotremata: *Orbiculoidea*, *Crania*. Protremata: *Productus*, *Chonetes*, *Schizophoria*, *Schellwienella*. Telotremata: *Spirifer*, *Martinia*, *Syringothyris* (Carboniferous only), *Athyris*, *Pugnax*, *Dielasma*.

Permian. In England the chief genera are *Productus*, *Strophalosia*, *Camerophoria*, *Spirifer*, *Dielasma*. In the Salt Range of India, Mongolia, etc. many others are found: *Schizophoria*, *Rhipidomella*, *Streptorhynchus*, *Lyttonia*, *Aulosteges*, *Richthofenia*, *Spiriferella*.

Trias. Most of the Palæozoic genera have died out, but *Spiriferina* and *Cyrtina* persist into the Trias. *Koninckina*, belonging to the Spiriferacea is confined to the Trias. The chief forms are Rhynchonellids and Terebratulids (*Cœnothyris*).

Jurassic. The Inarticulates are represented by *Lingula*, *Orbiculoidea*, and *Crania* which are sometimes abundant. Protremata: represented only *Lacazella* and its allies and *Cadomella*. Telotremata: *Spiriferina* survives into the Lias. The abundance of species commonly referred to '*Terebratula*,' '*Magellania*,' and '*Rhynchonella*' is the striking feature of the Jurassic Brachiopoda. *Terebratella* and *Terebratulina* also occur.

Cretaceous. Neotremata: *Crania* is common in the Chalk. Protremata; *Thecidia* and allied forms. Telotremata: Terebratulids and Rhynchonellids are still abundant. *Terebratulina* and *Terebratella* are more important than in the Jurassic. *Magas, Kingena, Trigonosemus* and *Terebrirostra* are confined to the Cretaceous.

Tertiary. Brachiopods are poorly represented and form an insignificant part of the Tertiary faunas. Nearly all belong to genera which are still living, *e.g. Lingula, Terebratula, Terebratulina,* and *Magellania*. In England Brachiopods are found chiefly in the London Clay and the Pliocene deposits.

PHYLUM POLYZOA

Classes	*Orders*	*Sub-Orders*
1. Ectoprocta ...	1. Phylactolæma	
	2. Gymnolæma	1. Ctenostomata
		2. Cyclostomata
		3. Trepostomata
		4. Cryptostomata
		5. Cheilostomata
2. Entoprocta		

With the exception of the genus *Loxosoma* all the Polyzoa[1] are colonial animals, numerous individuals living in association. The colony is nearly always fixed, and may be arborescent, laminar, almost massive, or encrusting shells, stones, or plants. The entire colony is known as the *zoarium*; each individual (fig. 120 A) has a sac-like form; at the upper end there is a platform or disc, the *lophophore*, on which tentacles (*t*) are placed, arranged either in a circle or in the form of a horse-shoe. In most forms the tentacles are not contractile, but are provided with cilia, which produce à current of water that conveys food to the mouth (*o*). The anal aperture (*a*) is near the mouth, generally below the lophophore, but in some forms within the circle of tentacles. On account of this approximation of the mouth and anus the alimentary canal is bent into a U-shape; in it may be distinguished œsophagus (*oes*), stomach (*st*), and intestine (*int*). Between the alimentary canal and the body-wall is a spacious body-cavity. The nervous system consists of a single ganglion (*g*) placed on the side of the œsophagus facing the intestine. The polyzoa multiply by

[1] The name *Bryozoa* is used for this Phylum by many authors.

budding and sexually, and are generally hermaphrodite Heart and blood-vessels are absent.

The structures described above form together what is known as the *polypide*; this is contained in the body-wall or *zoœcium*. The outer layer of the zoœcium, known as

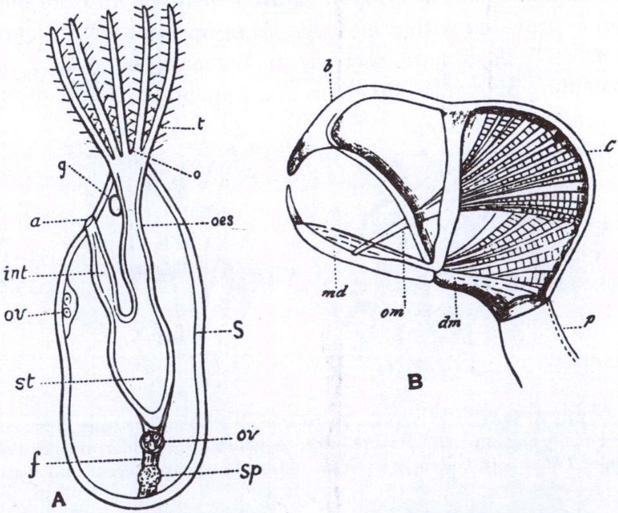

Fig. 120. A. Diagram of the structure of a single Polyzoan individual *S*, body-wall; *t*, tentacles; *o*, mouth; *oes*, œsophagus; *st*, stomach; *int*, intestine; *a*, anus; *g*, ganglion; *f*, funiculus; *ov*, ovary; *sp*, testis. B. Avicularium of *Bugula*, enlarged. *b*, beak; *md*, mandible; *C*, chamber; *p*, peduncle; *om*, occlusor muscles; *dm*, divaricator muscles. (After Hincks.)

the *ectocyst*, generally becomes hardened by calcareous or chitinous matter, and after the death of the animal this alone remains; its surface is usually ornamented with ribs, etc. The anterior part of the polypide can be withdrawn by means of longitudinal muscles into the zoœcium, just as the finger of a glove can be pulled into the hand. In some Polyzoa (the Cyclostomata, etc., fig. 121 B) the zoœcium is tube-like, the aperture is at the end and is of the same diameter, or nearly so, as the rest of the tube. In others (the Cheilostomata, fig. 121 A) the zoœcium is more or less box-shaped, the aperture (*m*) is contracted and is not terminal, but is situated in front near the anterior end, and is provided with a movable lid or operculum. In many of the Cheilostomata there is at the anterior end of the zoœcium, above the aperture, a projecting chamber (*o*),

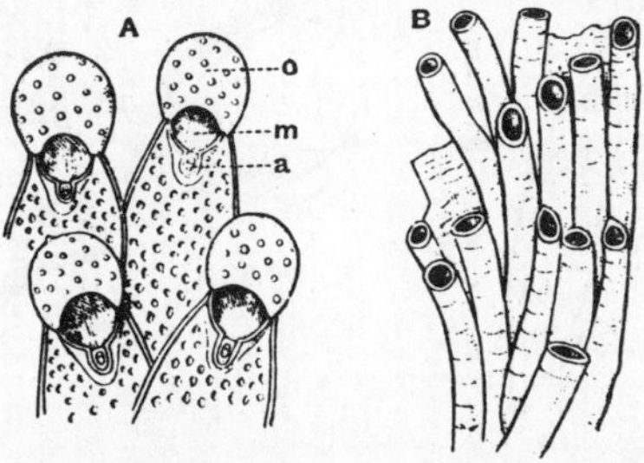

Fig. 121. A. Portion of *Smittia landsborovi*, a Cheilostomatous Polyzoan, Recent. *o*, oœcium; *m*, aperture of the zoœcium; *a*, avicularium. B. Portion of *Tubulipora fimbria*, a Cyclostomatous Polyzoan, Recent. Enlarged.

termed the *ovicell*, into which the ova pass. In many forms of Cheilostomata some of the individuals are modified so as to form appendages termed *avicularia* and *vibracula*. The avicularium (fig. 120 B) may be sessile or placed on a peduncle (*p*), and in the more specialized forms has somewhat the appearance of a bird's head, consisting of a chamber (*C*) produced into a beak and provided with a mandible (*md*) which is kept constantly snapping by means of muscles in the chamber. The vibraculum consists of a long seta kept in motion by means of muscles at its base. The individuals of a colony may communicate with one another, either directly, or by means of *communication-plates*; these are portions of the zoœcium which are thinner and perforated. The surface of the zoœcium may be smooth or punctate, or ornamented with spines, granules, or ribs.

The Polyzoa are divided into two classes, (1) the Ectoprocta, (2) the Entoprocta. The Ectoprocta only are found fossil

CLASS I. ECTOPROCTA

The anal aperture is not situated within the area of the lophophore. There are two orders, (1) the Phylactolæma, (2) the Gymnolæma.

ORDER I. PHYLACTOLÆMA

The lophophore is horse-shoe-shaped. There is a tongue-shaped lip in front of the mouth, known as the epistome. The forms included in this order are found only in fresh water and do not occur fossil.

ORDER II. GYMNOLÆMA

The lophophore is circular, and there is no epistome. There are five sub-orders, (1) Ctenostomata, (2) Cyclostomata, (3) Trepostomata, (4) Cryptostomata, (5) Cheilostomata. The third and fourth are extinct.

SUB-ORDER I. *CTENOSTOMATA*

The zoarium is horny or membranous. The zoœcia are usually isolated, and develop by budding from a tubular stolon; their orifices are terminal and can be closed by setæ. This group begins in the Ordovician and is represented by *Rhopalonaria*, *Vinella* and a few other genera in the Palæozoic.

SUB-ORDER 2. *CYCLOSTOMATA*

The zoœcia are calcareous and tubular, and seldom divided by transverse partitions; as a rule all are of one size, since mesopores, acanthopores, avicularia, and vibracula, are generally absent; the apertures are round and terminal, not constricted and not provided with an operculum. There may be a brood-pouch, formed of one or more modified zoœcia. Such a brood-pouch is called an *oœcium*, and is a *gonœcium* if composed of one, or a *gonocyst* if of more than one zoœcium. But ovicells, such as are characteristic of the Cheilostomata, and are not modified individuals, are never present.

Stomatopora. Zoarium encrusting, of branching rows of zoœcia in single file. Ordovician to present day; common in Jurassic and Cretaceous. Ex. *S. granulata*, Cretaceous.

Berenicea. Zoarium a thin, flat, encrusting sheet—discoid, fan-shaped, or irregular. Zoœcia simple, tubular, arranged in irregularly alternating lines. Ordovician to present day—common in the Jurassic and Cretaceous. Ex. *B. diluviana*, Lias to Oxfordian.

Idmonea. Zoarium encrusting or erect. Zoœcia arranged in alternating transverse rows on one face only of the zoarium. Jurassic to present day. Ex. *I. hagenowi*, Lower Greensand.

Entalophora. Zoarium of erect cylindrical branches, with the zoœcia opening on all sides of the branch and arranged irregularly or quincuncially. Jurassic to present day. Ex. *E. virgula*, Cretaceous.

Theonoa (=*Fascicularia*). Zoarium large, generally massive and globose. Zoœcia in the form of long tubes, with horizontal tabulæ, in contact laterally, and forming bundles which are either distinct and radiate from the base to the periphery, or fuse into laminæ which intersect. Jurassic to Pliocene. Ex. *T. aurantium*, Coralline Crag.

Fistulipora. Zoarium varying from encrusting to massive or rarely branching. Zoœcia more or less rounded or pyriform, with hood-like projections to the zoœcial aperture called *lunaria*; walls thin, diaphragms few, complete. Spaces between the zoœcia smooth or granular on the surface, occupied internally by one or more series of vesicles. Ordovician to Permian, chiefly Silurian to Lower Carboniferous. Ex. *F. cornavica*, Wenlock; *F. minor*, Carboniferous.

Ceramopora. Zoarium discoidal, free, attached by the centre of the base or encrusting (often on smooth-shelled brachiopods); under surface with one or more layers of small irregular cells. Openings of zoœcia radiating outwards from a depressed centre on the upper surface; apertures large and oblique, with lunaria; diaphragms absent. Mesopores irregular, short, numerous. Pores in walls of zoœcia and mesopores. Ordovician and Silurian. Ex. *C. imbricata*, Silurian.

SUB-ORDER 3. *TREPOSTOMATA*

The Zoaria are massive or branching, and composed of calcareous tubes of two sizes, the larger ones being the zoœcia, the smaller being known as *mesopores*. Both mesopores and zoœcia are crossed by horizontal partitions (*diaphragms*); these are more closely spaced in the mesopores. Curved partitions known as *cystiphragms* are sometimes present in the zoœcia in addition to the diaphragms. Tubular spines (*acanthopores*) occur in some genera at the angles between the zoœcia and the mesopores. Both mesopores and acanthopores probably contained modified zooids (polyzoan individuals), and are therefore comparable with the vibracula and avicularia of other groups. The zoœcial apertures are round, polygonal or irregular, and usually without opercula. The surface of the zoarium is typically marked by regularly spaced elevations (*monticules*) consisting of enlarged zoœcia, or of large clusters of mesopores, sometimes forming slightly depressed areas, constituting *maculæ*.

The Trepostomata are known to have been subject to periodic degeneration and regeneration of the polypides, in the same way that recent forms are. In fact the diaphragms in a single zoœcium probably represent the 'floors' of successively superimposed individuals.

The Trepostomata are probably confined to the Palæozoic, although the living Heteroporidæ have an essentially similar structure.

It should be noted that many of the diagnostic features of Trepostome genera only appear in the mature or peripheral zone of the zoarium.

Diplotrypa. Zoarium hemispherical, discoidal or irregularly massive, generally free. Zoœcia large, prismatic, with thin walls, and with diaphragms. Mesopores few or many, variable in size, with closely spaced diaphragms. Ordovician and Silurian. Ex. *D. petropolitana*, Ordovician.

Stenopora. Zoarium branching, massive or laminar. Walls of zoœcia usually periodically thickened, giving a beaded appearance in longitudinal sections. Acanthopores present. True mesopores apparently absent, but zoœcia smaller than the average are occasionally present. Diaphragms complete, few or many. Carboniferous (Permian?). Ex. *S. redesdalensis*, Carboniferous.

SUB-ORDER 4. *CRYPTOSTOMATA*

The zoœcia are calcareous and tubular, often with transverse partitions, and often of two sizes. Avicularia and vibracula are absent. The external orifices of the zoœcia are round, but these are not the true apertures; the latter are situated at the bottom of a tubular vestibule, the round orifice of which is seen on the surface of the zoarium. Probably a chitinous operculum covered the true aperture, but is never found in the fossils. Oœcia are absent. The Cryptostomata range from the Ordovician to the Permian.

Fenestrellina (= *Fenestella*). Zoarium funnel-shaped or fan-shaped. Branches straight, united by cross-bars, so as to form a network. The cross-bars do not bear zoœcia. On each branch there is a median ridge or carina, on the sides of which the zoœcia occur. Openings of zoœcia round. Ordovician to Permian. Ex. *F. plebeia*, Carboniferous.

Rhabdomeson. Zoarium of cylindrical branches with an axial tube to which the proximal ends of the zoœcia are attached; the surface is divided into rhombic areas, arranged regularly, in the middle of which are the round orifices. Carboniferous and Permian. Ex. *R. rhombiferum.*

SUB-ORDER 5. *CHEILOSTOMATA*

The zoœcia are sometimes calcareous, sometimes horny, often both; they are more or less box-shaped, never tubular; and not divided by transverse partitions. Zoœcia, differing from the normal forms in size and shape, and modified for protective purposes, are often present, and are called avicularia and vibracula—according to whether their function is to pinch or to sweep away foreign bodies which would settle on the zoarium. The apertures of the zoœcia are contracted and not terminal, of varying outline, and provided with a movable operculum, which being horny is not found in fossil specimens. Globular ovicells are often present; these are not modified individuals, but outgrowths in front of the distal end of each zoœcium. The Cheilostomata range from the Jurassic to the present day, but are rare in deposits earlier than the Chalk.

Membranipora. Zoarium encrusting or erect; the top of each zoœcium is covered with a chitinous membrane in which is situated the aperture; consequently in fossil specimens each zoœcium has a rim enclosing an unroofed space; the rim may have spines around it. Jurassic to present day. Ex. *M. elliptica*, Chalk.

Cribrilina. Zoarium usually encrusting. Zoœcia as in *Membranipora*, but the spines of the rim meet and fuse with their neighbouring and with their opposite fellows, and form an incomplete roof over the zoœcium. Tertiary and present day. Ex. *C. punctata*, Coralline Crag to Recent.

Pelmatopora. Like *Cribrilina*, but the costæ, or spines that form the front wall, are very coarse, and their broken upturned ends form a row or rows of hob-nail-like markings on each side of the mid-line of the front wall. Upper Cretaceous. Ex. *P. solearis*, Chalk.

Micropora. Zoarium encrusting. Zoœcia with an encircling rim as in *Membranipora*, but the chitinous roof is replaced by one of carbonate of lime; and this roof is perforated by two holes, one on each side, near the rim and proximally to the orifice. Cretaceous to present day. Ex. *M. cribriformis*, Barton Beds.

Cellepora. Zoœcia heaped irregularly upon an irregularly encrusting or erect zoarium; the front wall entirely calcareous and very convex; the aperture terminal, more or less round, always accompanied by one or more small avicularia; in addition larger avicularia are often present between the normal zoœcia. Tertiary to present day. Ex. *C. tubigera*, Coralline Crag.

Distribution of the Polyzoa

By far the larger number of the Polyzoa are marine; they occur both in shallow and deep water. The deep-water forms belong mainly to the Cheilostomata; a few Ctenostomata occur at considerable depths, but the group is characteristic of shallow water. The Cyclostomata are comparatively rare at the present day, except in the Northern seas. The extinct Trepostomata and Cryptostomata are usually associated with reef conditions.

The earliest Polyzoa occur in the Ordovician rocks. Nearly all the Palæozoic genera are extinct; they belong mainly to the Trepostomata and Cryptostomata. The Cyclostomata are represented by a few genera in the Palæozoic rocks, and become increasingly abundant in the Mesozoic, attaining their maximum in the Upper Cretaceous. A few Cheilostomata have been recorded from the Jurassic rocks, but the group does not become abundant until the Cretaceous period; in the Tertiary it is better represented than the Cyclostomata. Very many of the Pliocene forms belong to species which are still living.

The chief genera found in the different systems are:

Palæozoic. *Archimedes, Ceramopora, Diplotrypa, Fenestrellina* (=*Fenestella*), *Fistulipora, Hallopora, Hemitrypa, Pinnatopora, Polypora, Ptilodictya, Rhabdomeson, Thamniscus.*

Jurassic. *Berenicea, Ceriopora, Diastopora, Entalophora, Haplocœcia, Idmonea, Proboscina, Spiropora, Stomatopora.*

Cretaceous. *Crisina, Diastopora, Entalophora, Heteropora, Lunulites, Membranipora, Onychocella, Pelmatopora, Proboscina, Stomatopora.*

Eocene. *Hornera, Idmonea, Membranipora.*

Pliocene. *Alveolaria, Cellepora, Cribrilina, Hornera, Levralia Membranipora, Theonoa.*

PHYLUM MOLLUSCA

Classes	*Orders*	*Sub-Orders*
1. Lamellibranchia	1. Taxodonta	
	2. Anisomyaria	
	3. Eulamellibranchia	
2. Amphineura	1. Polyplacophora	
	2. Aplacophora	
3. Gasteropoda	1. Prosobranchiata	1. Aspidobranchia
		2. Pectinibranchia
	2. Opisthobranchiata	1. Nudibranchia
		2. Tectibranchia
	3. Pulmonata	
4. Scaphopoda		
5. Cephalopoda	1. Nautiloidea	
	2. Ammonoidea	
	3. Dibranchia	1. Decapoda
		2. Octopoda

The majority of the molluscs (oysters, whelks, cuttlefish, etc.) are marine, but some live on land, others in fresh water. Unlike the worms and arthropods, they are unsegmented animals, and they bear no serially repeated appendages. Typically the body is bilaterally symmetrical, and there is consequently a repetition of the same organs on each side; but in most gasteropods this symmetry is more or less completely lost. From the dorsal surface arises a fold of the skin forming what is known as the *mantle*; this generally secretes a calcareous shell, consisting of one or two (occasionally more) pieces. On the ventral surface of the body is the *foot*—a muscular organ used in locomotion. In most cases respiration takes place by means of gills, which are placed in the cavity enclosed by the mantle. A heart is present, and is on the dorsal surface; it consists usually of a ventricle and two auricles. The mouth is situated anteriorly, and, except in the lamellibranchs, is provided with a rasping organ, the *odontophore*; the anus, in typical forms, is placed posteriorly. Renal organs (nephridia) are present and place part of the body-cavity in communication with the exterior. The nervous system consists of a ring round the œsophagus, and usually of three main groups of ganglia, from which nerves are given off. Only sexual reproduction occurs; most forms are unisexual, a few hermaphrodite.

The Mollusca are divided into five classes: (1) Lamellibranchia, (2) Amphineura, (3) Gasteropoda, (4) Scaphopoda, (5) Cephalopoda.

CLASS I. LAMELLIBRANCHIA

In the lamellibranch, as in the brachiopod, the shell is generally calcareous and consists of two valves, but these instead of being dorsal and ventral as in the latter, are placed one on the right, the other on the left side of the body, and the two are joined together by means of a hinge and a ligament at the dorsal margin. The interior of the shell is lined by a fold of the skin, the *mantle* (fig. 122, *m*), which is divided into two lobes, one being placed in each valve. In the middle of the space enclosed by the mantle (the mantle-cavity) and projecting from the ventral surface of the visceral mass, is the foot (*f*). This is a laterally flattened muscular organ, frequently hatchet-[1] or ploughshare-shaped, and is used for crawling, or for burrowing in sand or mud. Sometimes, as in the case of *Trigonia*, by means of a rapid movement it enables the animal to jump to a considerable distance. In the genus *Mytilus* the foot is very much reduced: in others which have lost the power

[1] Hence the name *Pelecypoda* used by some authors for this class.

of locomotion (*e.g. Ostrea*) it is absent altogether. On the posterior part of the foot there is in some genera (*e.g. Mytilus, Pinna, Arca*) a gland which secretes threads of a viscous substance which gradually harden, and then form a bundle of horny fibres, known as the *byssus*, by means of which the animal moors itself to foreign objects. On each side of the foot, between it and the mantle, and attached to the body dorsally, are the gills or branchiæ (fig. 122, *g*); these consist of filaments which usually become connected so as to form leaf- or plate-like bodies, whence the name Lamellibranchia.

In some forms the margins of the two mantle-lobes although in contact are not united, and when this is the case there are usually at the posterior margin two openings leading from the exterior to the mantle-cavity; these are produced by adjoining excavations or notches in the two lobes of the mantle. A current of water, caused by the cilia on the gills and mantle, flows in through the ventral opening, and provides the animal with food and oxygen; another current flows out through the dorsal opening, carrying with it fæcal matters. In many cases, however, the two lobes of the mantle are fused at one or more points; this union occurs between the exhalent and inhalent openings, and also, in many forms, below the latter opening. In this way the mantle becomes a kind of bag, having three openings, a ventral for the protrusion of the foot, and two posterior for the inhalent and exhalent currents of water. Frequently, at the posterior openings, the mantle is greatly produced so as to form two complete tubes, known as *siphons* (fig. 122, *s*, *s'*); these are sometimes free, sometimes united, and may be as much as four times the length of the shell. The ventral is generally the longer; it is furnished with tactile papillæ, and is known as the *branchial siphon* (*s*),

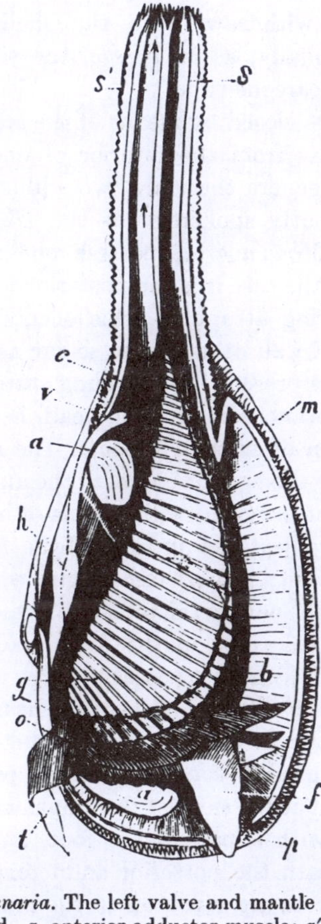

Fig. 122. *Mya arenaria*. The left valve and mantle and half the siphons have been removed. *a*, anterior adductor muscle; *a′*, posterior adductor; *b*, visceral mass; *c*, cloacal chamber into which the anus opens; *f*, foot; *g*, branchiæ; *h*, heart; *m*, cut edge of the mantle; *o*, mouth; *p*, edge of mantle; *s*, branchial siphon; *s′*, anal siphon; *t*, labial palps; *v*, anus. (From Woodward.) Natural size.

the dorsal being the *anal siphon* (s'). In many forms the siphons can be withdrawn into the shell by means of muscles. Occasionally, as in *Teredo*, the siphons are surrounded by a calcareous tube.

The shell can be closed by means of the adductor muscles (a, a'), which pass from the interior of one valve to the other. In many genera there are two adductors. and these forms are frequently spoken of as the *Dimyaria*; others, known as the *Monomyaria*, possess one adductor only. and when this is the case it is the posterior which is present, the anterior having atrophied; this occurs in the oyster, but in this, and in all other forms so far as is known, the anterior muscle is present in the young state.

In the lamellibranchs there is no head, hence the class is sometimes spoken of as the *Acephala*. The mouth (o) is in the middle line of the body, ventral to the anterior adductor muscle, and is not provided with organs of mastication. At each side are two leaf-like processes, the *labial palps* (t). The mouth leads into a short œsophagus, which passes into a globular stomach; next is the intestine, which, after undergoing many convolutions in the foot, reaches the dorsal surface of the body, where it passes through the pericardium and is surrounded by the ventricle of the heart. The anus (v) is situated dorsally to the posterior adductor muscle. The nervous system usually consists of three pairs of ganglia. One pair is placed at the sides of the mouth and is connected by nerve-cords with a pair in the foot, and with a third pair placed beneath the posterior adductor muscle. From these ganglia nerves are given off to the muscles, gills, etc. Tactile organs are present on the margin of the mantle and especially on the ventral siphon. In some forms eyes occur at the ventral margin of the mantle-lobes; they are especially well-developed in the genus *Pecten*. The heart (h) is placed dorsally, just below the hinge, and is surrounded by a large pericardial cavity; it consists of two auricles and a ventricle, which, as already mentioned, extends round the intestine. The renal organs consist of a pair of glandular tubes underneath the pericardium. In almost all cases the sexes are separate, but a few forms are hermaphrodite.

As already mentioned, the two valves of the shell are on the sides of the animal. The margin near the hinge

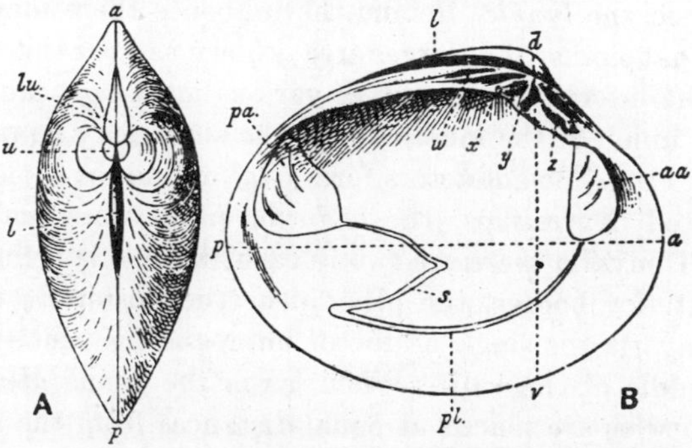

Fig. 123. *Meretrix* (*Macrocallista*) *chione*, Recent. A. Dorsal view of the two valves. B. Interior of left valve. × ½. *a*, anterior border; *p*, posterior; *d*, dorsal; *v*, ventral; *lu*, lunule; *u*, umbo; *l*, ligament; *aa*, anterior adductor impression; *pa*, posterior adductor: *pl*, pallial line; *s*, pallial sinus: *w*, *x*, *y*, cardinal teeth; *z*, anterior lateral tooth.

(fig. 123 B, *d*– -*l*) is dorsal, the opposite (*v*), where the valves open, is ventral; that near the mouth is anterior (*a*), that near the anus and siphons posterior (*p*) In the majority of cases the two valves are equal or almost equal in size, and each valve is generally inequilateral. But in some (*e.g. Glycimeris*) the shell is nearly equilateral and in others (*e.g. Ostrea*) it is inequivalve. When the shell is equilateral the direction of greatest growth is perpendicular to the hinge-line: when inequilateral the direction is oblique to the hinge-line. Each valve may be regarded as a greatly depressed hollow cone, the apex of which forms the *umbo* (fig. 123 A, *u*); these umbones are sometimes straight (*e.g. Pecten*), but generally curved towards the anterior margin; in a few genera (*e.g. Nucula, Trigonia, Exogyra*) they are directed posteriorly; in *Diceras* they are spiral. Sometimes there is in front of the umbones, and bounded by a groove, an oval depressed area (*lu*), half being on each valve; this is termed the *lunule*. Behind the umbones there may be a somewhat similar, but larger area, known as the *escutcheon*.

In the interior of the valves various markings, produced by the union of the muscles with the shell, may be noticed (fig. 123 B). The adductors form oval, round, or sometimes elongated depressions (the *adductor impressions, aa, pa*); in the Dimyaria there are two in each valve, one being near the anterior border, the other near the posterior; in the Monomyaria the single adductor impression is usually near the middle of the valve. When, as in the genus *Mya*, the two muscles are placed at equal distances from the hinge-margin, they are of nearly the same size, since on account of their position they are equally efficient in closing the valves; but in forms like *Mytilus*, where the shell is very inequilateral and the anterior muscle is close to the umbo but the posterior at a considerable distance from it, the latter is much larger than the former, since it is placed in a more advantageous position for closing the valves. For the same reason the single muscle of the Monomyaria is attached near the centre of the valves. Less important than the adductor impressions are those produced by the muscles for the movement of the foot (protractors and retractors); these occur close to the anterior and posterior adductors. Passing from one adductor impression to the other in each valve is a linear depression, caused by the attachment of the muscles of the mantle to the shell, and known as the *pallial line* (*pl*). In some forms this line runs evenly between the two adductor impressions and parallel with the margin of the valve; it is then said to be *simple* or *entire*. But in those genera which possess retractile siphons the pallial line bends inward just before reaching the posterior adductor; this indentation is known as the *pallial sinus* (*s*), and is caused by bending inwards of the part of the pallial muscles which serve for the retraction of the siphons.

The hinge is formed by projections known as teeth, which alternate in the two valves, the teeth of one valve fitting into the depressions between those of the other. Its function is to ensure that the valves should close perfectly. The margin of the valve on which the teeth occur is known as the *hinge-line*; generally it is curved, but in some genera it is straight (*e.g. Arca*). Several types of hinge may be recognised: (1) *Taxodont:* the teeth are numerous and more or less similar in form and size, *e.g. Nucula* (fig. 124 A). (2) *Dysodont:* the teeth are of a simple type, and are developed from internal ribs at the margin of the valve; the hinge-margin may be simple or somewhat thickened, *e.g. Mytilus*. The dysodont hinge appears to have been derived from a taxodont form in which the teeth radiate outwards from the umbo. (3) *Isodont:* there are strong teeth of equal size, which fit into corresponding sockets in the other valve; the teeth are placed symmetrically on each side of the median ligament-pit, *e.g. Spondylus* (fig. 124 D, E). (4) *Schizodont:* the teeth are few in number, thick, and sometimes grooved; a typical form is *Trigonia* (fig. 124 B, C) in which the teeth diverge from below the umbo, and the middle tooth of the left valve is bifid. In others (*e.g. Unio*) the teeth are less definite in shape and position. (5) *Heterodont:* the teeth are few in number and not all of uniform shape and size. They are divisible into those (usually two or three) which are placed immediately under the umbo and are known as the *cardinal* teeth, and others, termed *laterals*,

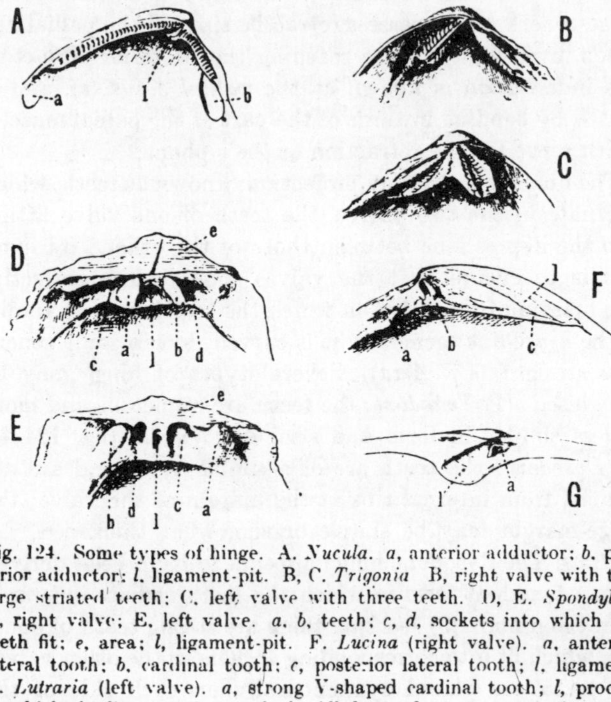

Fig. 124. Some types of hinge. A. *Nucula*. *a*, anterior adductor; *b*. posterior adductor; *l*. ligament-pit. B, C. *Trigonia* B, right valve with two large striated teeth; C. left valve with three teeth. D, E. *Spondylus*. D, right valve; E. left valve. *a*. *b*, teeth; *c*, *d*, sockets into which the teeth fit; *e*, area; *l*, ligament-pit. F. *Lucina* (right valve). *a*, anterior lateral tooth; *b*, cardinal tooth; *c*, posterior lateral tooth; *l*, ligament. G. *Lutraria* (left valve). *a*, strong V-shaped cardinal tooth; *l*, process to which the ligament is attached. All drawn from recent specimens.

which are in front of and behind the umbo, forming the *anterior* and *posterior laterals* respectively; some or all of the cardinals or of the laterals may be absent; the hinge-margin is extended as a vertical lamina or flange known as the *hinge-plate* (fig. 124 F) on which the teeth are borne, *e.g. Meretrix.* (6) *Desmodont:* true teeth and a hinge-plate are absent, but one or more laminæ or ridges are developed at the hinge-margin, *e.g. Pleuromya.*

In some genera in which the hinge-line is straight (*e.g. Arca*) there is, between the hinge-line and the umbo of each valve, a flattened triangular part of the shell, known as the *area* (fig. 124 D, *e*); when this is present the umbones of the two valves are of course widely separated. The area is due to the more extensive growth at the hinge-margin than occurs in genera in which the umbones are close together. The lunule and escutcheon (p. 240) appear to represent the anterior and posterior parts of the area. Some lamellibranchs with a straight hinge-line (*e.g. Pecten*) have, on each side of the umbo, triangular or wing-like extensions of the shell, known as *ears.*

In the brachiopods the valves are opened by divaricator muscles, but in the lamellibranchs the work of these muscles is performed by the ligament. This consists of two parts, the external (fig. 123, *l*), and the internal (sometimes erroneously termed the cartilage) (fig. 124 G, *l*). One or other may be absent. The *external ligament* is composed of horny material; it is placed at the hinge-margin, usually posterior to the umbones, and is frequently attached to more or less prominent ridges; in some genera (*Glycimeris*) the external ligament extends both in front of and behind the umbones. The *internal ligament* consists of parallel elastic fibres, and is placed in a hollow or pit on the hinge-plate (fig. 124 G, *l*), so that when the valves are closed it is compressed, and, being elastic, tends to force the valves apart—its action is similar to that of a piece of indiarubber placed in the hinge-line of a door. The external ligament acts like a C-spring, and is bent when the valves are closed. Consequently, in order to open the shell, the animal has merely to relax its adductor muscles. Occasionally the ligament is preserved in fossil specimens.

The *length* of a lamellibranch shell is measured from the anterior to the posterior margin (fig. 123 B, *a—p*), the *breadth* or *height* from the umbo to the ventral margin (*d—v*), the *thickness* from one valve to the other at right angles to the lines of length and breadth.

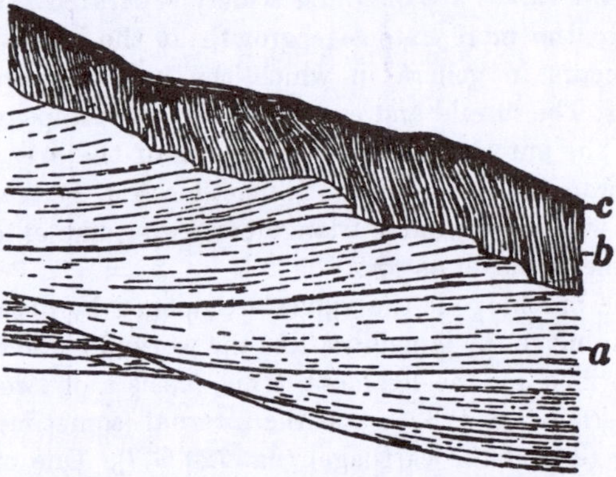

Fig. 125. Vertical section of the shell of a recent *Unio*, cut in a radial direction from the umbo; the right-hand side of the section is near the ventral margin. *a*, pearly or nacreous layer, in which the later lamellæ overlap the earlier and extend on to (*b*) the prismatic layer; *c*, periostracum. ×10.

The shell is secreted by the mantle; its structure varies in different groups. Three layers may be distinguished. (1) On the external surface (fig. 125, *c*) is a green or brownish layer formed of horny material (conchiolin) and known as the *periostracum* (frequently referred to as the 'epidermis'). This layer is not usually preserved in fossils; (2) in the middle is the *prismatic* layer (*b*) (fig. 126), consisting of prisms, usually of calcite, each being encased in a sheath of organic material. The prisms are often arranged more or less perpendicularly to the surface of the shell; (3) the inner layer (*a*) consists of aragonite and may be formed of very thin lamellæ separated by thin layers of organic material when it is *pearly* or *nacreous*, or it may be formed of thicker lamellæ when it becomes *porcellanous*. The chief genera in which this layer is nacreous are *Nucula*, *Pteria*, *Perna*, *Trigonia*, *Unio*. The prismatic layer is formed by the margin of the mantle only; the pearly layer by the general surface of the mantle, and this layer gradually encroaches on the former, which consequently cannot afterwards increase in thickness, whereas the pearly layer may do so throughout the life of the animal.

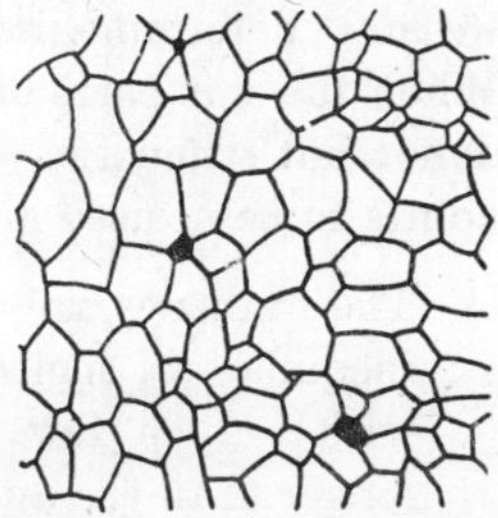

Fig. 126. Section of prismatic layer of recent *Pinna*, parallel to the surface of the shell and at right angles to the prisms. Magnified.

The surface of the shell may be smooth, or may be ornamented with radiating or concentric ribs and striæ, or with tubercles, or spines. Often the exterior shows concentric lamellæ, which represent periods of growth. The part of the shell at the umbo is that which was first formed, and often differs in ornamentation and form from the other parts. The margins of the valves may be smooth or crenulated; sometimes, as in some species of *Pecten*, the entire shell is corrugated, thus increasing its strength without materially adding to the weight. In many genera the two valves can be completely closed, in others they are always open at some part, and are then said to be *gaping*; this gape occurs most frequently at the posterior end and serves for the passage of the siphons; sometimes there is also an anterior gape through which the foot or byssus passes to the exterior. Sometimes the small embryonic shell, known as the *prodisso-*

conch, is found at the umbo of the adult shell; this represents the protegulum of the Brachiopods (p. 199) and the protoconch of the Gasteropods and Cephalopods.

In order to be able to distinguish the right and left valves we must determine first the anterior and posterior margins. When the soft parts of the animal are present this is easily done; but when the shell only is before the observer the points to be noticed are the following:

(1) The umbones are generally directed anteriorly; and in inequilateral shells, the posterior part of the valves is, with only a few exceptions (*Nucula*, *Lima*, *Donax*), longer than the anterior part.
(2) The lunule is anterior to the umbones.
(3) The external ligament is commonly posterior to the umbones, and is never entirely in front of them.
(4) The pallial sinus is posterior.
(5) When one adductor impression only is present, it is the posterior.
(6) When one adductor impression is distinctly larger than the other, the larger is the posterior.

Having found the anterior and posterior margins, the shell should be placed with the dorsal surface uppermost and the anterior margin pointing away from the observer, then the right and left valves will be on his right- and left-hand sides respectively.

Most of the lamellibranchs are free and can move by means of the foot. Since these live with the median plane of the body in a vertical position, the two valves are of equal size. A few genera (*Pecten*, *Lima*) move by the rapid closing of the valves, which causes water to be forcibly expelled from the mantle-cavity. Some forms (*Pecten*) rest on one valve, which then becomes more convex than the upper valve. Others, such as the oyster, are permanently attached by one valve, which has been secreted by the mantle directly on to a rock or some other object to which it adheres firmly. In some cases the right valve is fixed, in others the left. The shell in these forms becomes irregular and the fixed valve is larger, more convex and thicker than the free valve. Other genera, *e.g. Pteria*, are attached by means of a byssus (p. 236), which often passes out through a notch or sinus in the margin of one or both valves—such forms tend to become inequilateral. Many genera burrow in the sediment on the sea floor and live with the posterior end uppermost. The valves are elongated posteriorly, so that the shell becomes distinctly inequilateral, and the line joining the two adductor muscles is parallel with the dorsal margin of the valves. Since these lamellibranchs are sufficiently protected in their burrows there is no necessity for the complete closing of the valves, and there is a gape at the posterior end for the siphons and often another anteriorly for the foot. As frequent opening and closing of the valves is not needed the hinge tends to become degenerate. In order to facilitate movement through the sediment these burrowing forms often have laterally compressed valves with a smooth surface. A few genera are capable of making borings into various substances; in these the shell tends to become cylindrical in form. Thus *Teredo*, the ship-worm, bores into wood, *Lithophaga* and *Saxicava*, into limestone, and *Pholas* into various materials, such as sandstone, limestone, gneiss, peat, and amber. Wood perforated by *Teredo* has been found fossil in various formations of Eocene and Oligocene age.

The features which more especially characterise the lamellibranchs as a class are: the absence of a head and of organs of mastication, the bilateral symmetry, the division of the mantle into two lobes, the bivalve shell and the lamellar gills. Although at first sight the shell appears to resemble closely that of the brachiopods, it differs in several important respects: (1) the valves are right and left, instead of dorsal and ventral, (2) they are generally inequilateral and equivalve, (3) teeth occur on both valves, (4) a ligament is present, (5) the umbones are never perforated for a peduncle, (6) the microscopic structure of the shell is different.

The classification adopted here is based primarily on the character of the hinge, but with other features taken into account. The three main divisions are: (1) Taxodonta, (2) Anisomyaria, (3) Eulamellibranchia.

ORDER I. TAXODONTA

Hinge taxodont. Two nearly equal adductor muscles. Siphons usually wanting. Lower Ordovician to present day.

Nucula[1] (fig. 124 A). Shell equivalve, trigonal or oval, closed, posterior side very short; umbones directed posteriorly. Surface smooth or with fine radial lines. Interior nacreous. Margins of valves smooth or crenulated. Hinge-line angular, with a median internal triangular ligament-pit, and numerous sharp teeth. Adductor impressions nearly equal. Pallial line simple. The character of the hinge, the simple type of gill structure and other anatomical features indicate that *Nucula* is one of the most archaic of living lamellibranchs. Silurian to present day. Ex. *N. hammeri*, Lias; *N. dixoni*, Bracklesham Beds.

Nuculana (=*Leda*). Similar to *Nucula*. Posteriorly the shell is produced and pointed, and provided with a ridge or carina. Pallial line with a small sinus. Margins smooth. Escutcheon lanceolate. Silurian to present day. Ex. *N. lachryma*, Inferior Oolite to Cornbrash; *N. caudata*, Pliocene to present day.

[1] All the genera of Mollusca described are marine unless otherwise stated.

Ctenodonta. Shell oval or elongated, nearly equilateral, smooth or with concentric striæ. Ligament external. No area. Hinge curved or angular, with numerous small teeth. No internal ligament-pit. Pallial line simple. Cambrian to Carboniferous. Ex. *C. pectunculoides*, Ordovician.

Arca. Shell equivalve, sub-quadrangular, ventricose, with a carina from the umbo to the postero-ventral angle. Surface with radiating ribs and concentric striæ: margins smooth or dentate; gaping ventrally. Hinge straight, with numerous, small, similar, transverse teeth. Umbones prominent, separated by the large areas, which have numerous ligamental grooves converging from the hinge-margins to below the umbones. Adductor impressions sub-equal, the anterior rounded, the posterior divided. Pallial line simple. Jurassic to present day. Ex. *A. biangula*, Eocene; *A. noæ*, Miocene to present day. Sub-genus *Barbatia*, with very narrow area, and the end teeth oblique. Jurassic to present day. Ex. *B. barbata*, Miocene to Recent. *Anadara*, with thicker shell, regular radial ribs, closed valves, less inequilateral than *Arca*. Miocene to present day. Ex. *A. diluvii*, Miocene.

Cucullæa. Shell similar to *Arca*; ventricose, sub-equilateral, valves closed. Hinge with short central transverse teeth, and two to five lateral teeth nearly parallel to the hinge-margin. Posterior adductor fixed to a thin raised plate. Jurassic to present day. Widespread in the Mesozoic. Living in the Indian Ocean and China Sea. Ex. *C. fibrosa*, Upper Greensand; *C. crassatina*, Eocene.

Glycimeris (=*Pectunculus*). Shell thick, solid, sub-orbicular, equivalve, almost equilateral. Surface smooth or radially striated. Ligament external, on the area. Umbones central, slightly curved posteriorly; a small triangular area provided with diverging grooves for the ligament. Hinge arched or semicircular, with a row of numerous, small, strong, transverse teeth, obliterated at the centre in the older forms by the growth of the area; towards the ends the teeth tend to become horizontal. Margins crenulate inside; adductor impressions sub-equal—the anterior sub-triangular, the posterior oval or rounded. Pallial line with a very small sinus. Cretaceous to present day. Ex. *G. glycimeris*, Pliocene to present day.

ORDER II. ANISOMYARIA

Usually inequivalve. Anterior adductor small or absent. Hinge dysodont or isodont, or without teeth. Fixed by a byssus or by cementation. Often with ears and a byssal notch. No siphons. Pallial line entire.

1. MYTILACEA

Equivalve, very inequilateral, obliquely elongated. Umbo at or near the anterior end; no ears, no byssal notch but an anterior gape for the byssus. Hinge dysodont or without teeth. Ligament long, nearly always external, behind the umbones. No area. Anterior adductor small. Interior nacreous or porcellanous. Ordovician to present day

Mytilus. Shell thin, equivalve, very inequilateral, elongated, sub-triangular, posterior border rounded; with a small gape for the well-developed byssus. Umbones sharp, terminal, anterior. A few small teeth near the umbo, sometimes absent. Ligament linear, marginal, sub-internal. Anterior adductor impression small, placed near the umbo; posterior large; pallial line simple. Trias to present day. Ex. *M. edulis*. Pliocene to present day.

Modiola (= *Modiolus*). Shell similar to *Mytilus*, but oblong, inflated in front. Umbones obtuse, anterior, but not terminal. No teeth. Devonian to present day. Ex. *M. modiola*, Recent; *M. imbricata*, Inferior Oolite.

Lithophaga (*Lithodomus*). Shell similar to *Modiola*; sub-cylindrical, rounded in front, wedge-like behind. *Lithophaga* bores into limestone, etc. Carboniferous to present day. Ex. *L. inclusa*, Inferior Oolite to Corallian; *L. lithophaga*, Recent.

Modiolopsis. Shell thin, smooth, elongate-oval, very inequilateral, anterior part small, posterior part enlarged. Umbones nearly terminal, close together; a depression crosses the valves obliquely from the umbo. No teeth. Anterior adductor impression deep; posterior adductor large, faintly marked. Ordovician to Permian. Ex. *M. complanata*, Silurian.

Myoconcha. Similar to *Modiolopsis*, but usually with a long cardinal and a long slender posterior lateral tooth in the right valve. Carboniferous to Chalk. Ex. *M. crassa*, Inferior Oolite; *M. cretacea*, Chalk.

2. PTERIACEA

Often inequivalve and with ears. Inequilateral. Anterior adductor small or absent. Usually fixed by a byssus; frequently with a byssal notch below the right anterior ear. Hinge-line straight; often with an area. Hinge dysodont or without teeth. Interior nacreous. Ordovician to present day.

Pinna. Shell generally thin, outer layer with coarse prismatic structure (fig. 126), inner layer nacreous; equivalve, inequilateral, triangular, without ears. Umbones sharp, anterior, terminal. Valves truncated and gaping posteriorly. Hinge-line straight, long. No teeth. Ligament long, narrow, lodged in a groove. Posterior adductor large, sub-central; anterior adductor close to the umbo. Carboniferous to present day. Ex. *P. hartmanni*, Lias; *P. affinis*, London Clay.

Gervillia (fig. 127). Shell obliquely elongated, very inequilateral, slightly inequivalve, the left valve a little more convex than the right; umbones almost terminal. Hinge straight, with an area on which are numerous perpendicular, widely-separated ligament-pits; with two or more oblique ridge-like teeth. Ears indistinctly limited from the rest of the shell, the anterior very short, the posterior long. Posterior adductor impression large, sub-central. Trias to Eocene. Ex. *G. forbesiana*, Gault; *G. sublanceolata*, Lower Greensand. Sub-genus *Hœrnesia*. Left valve convex. Right valve more or less flattened. One strong tooth under the umbo in each valve, and several small teeth on the posterior side in the left valve. Trias. Ex. *H. socialis*, Muschelkalk.

Inoceramus. Shell variable in form, circular, oval, or oblong; inequilateral, inequivalve, ventricose or compressed, with ears indistinctly limited. Umbones prominent, rather anterior. No teeth. Surface with concentric (or rarely [illegible]) furrows. Hinge-line straight, usually long, with numerous parallel, close-set, transverse ligament-pits. Adductor impression rarely visible. Inner layer of shell thin and nacreous; outer layer thick, formed of large prisms. Lias to Chalk; common in Upper Cretaceous. Ex. *I. concentricus*, Gault; *I. brongniarti*, Chalk.

Perna (=*Isognomon*). Shell nearly equivalve, inequilateral, compressed, sub-quadrate or sub-circular. Umbones at the anterior end. Hinge-line straight, without teeth; hinge-area broad, with numerous transverse, elongated ligament-pits placed close together and parallel with one another. Right valve with a byssal sinus. Adductor impression large, sub-

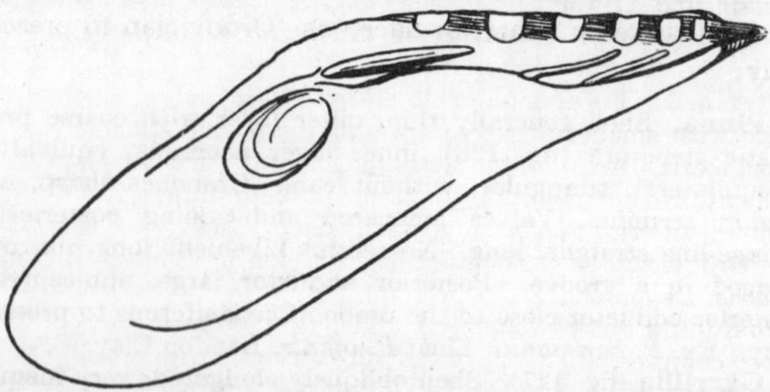

Fig. 127. *Gervillia sublanceolata*, Lower Greensand. Left valve. Showing ligament-pits, teeth, and posterior adductor impression. × $\frac{2}{3}$.

central, double; pallial line simple. Posterior ear often large, not distinctly limited. Trias to present day. Ex. *P. mytiloides*, Upper Jurassic; *P. ephippium*, Recent.

Pteria (=*Avicula*). Shell oblique, inequilateral, inequivalve, left valve more convex than the right. Interior nacreous. Hinge long, straight, with one small tooth near the umbo in each valve and a long lamellar posterior tooth in the right valve. Posterior ear wing-like and longer than the anterior. A byssal sinus under the right anterior ear. Area narrow. Ligament long, partly internal, partly external, in a groove. Posterior adductor impression large, sub-central. Trias to present day. Ex. *P. media*, Barton Beds; *P. hirundo*, Recent. Sub-genera or closely allied genera, are *Actinopteria*, *Leiopteria*, *Pteronites* (all Upper Palæozoic), and *Oxytoma* (Mesozoic).

Pseudomonotis. Similar to *Pteria*, but the shell is oval, the left valve large and very convex, and the right valve flattened; the anterior ear small or rudimentary. Carboniferous to Cretaceous. Ex. *P. echinata*, Cornbrash.

Aucella (=*Buchia*). Shell thin, obliquely elongate, inequilateral, inequivalve, with concentric folds or ribs. Left valve convex, with prominent incurved umbo; ears indistinctly limited. Right valve flattened, anterior ear triangular, with a deep byssal sinus; posterior ear indistinctly limited. Hinge-line straight, short, without teeth. Ligament external. Upper Jurassic and Lower Cretaceous. Ex. *A. keyserlingi*, Speeton Series. *Aucellina* is closely allied to *Aucella*, Gault and Lower Chalk; Ex. *A. gryphæoides*.

Pterinea. Form similar to *Pteria*; left valve flattened. Hinge with small transverse anterior teeth, and laminar posterior teeth. Area large, with longitudinal grooves for the ligament. No ligament pit. Posterior adductor impression large, shallow; anterior impression small, deep, below the anterior ear. Ordovician to Carboniferous; common in Devonian. Ex. *P. lævis*, Devonian.

Posidonia (=*Posidonomya*). Shell thin, oblique, oval, equivalve, compressed, with concentric furrows. Umbones small, sub-central. Hinge-line straight, short, without teeth; posterior ear compressed, indistinctly limited. Silurian to Jurassic. Ex. *P. becheri*, Carboniferous.

Vulsella. Sub-equivalve, irregular, vertically elongated, gaping in front and behind. Without teeth, ears and byssus. An area on each valve, with a triangular ligament pit. Umbones directed posteriorly. Posterior adductor only, sub-central. Interior nacreous. Eocene to present day. Ex. *V. deperdita*, Eocene; *V. lingulata*, Recent.

Conocardium. Shell more or less trigonal, very inequilateral, with radiating ribs; posterior side short, truncated, forming a cordate posterior end, produced into a long tube; anterior side oblique, compressed, wing-like, gaping. Umbones small, pointed, incurved. Hinge-line long, straight. Ligament partly external, partly internal, attached to a plate behind the umbones. Anterior adductor impression large, deep; posterior impression shallow. Inner margins of valves toothed. The truncated end bearing the tube is regarded by some authors as anterior, and the wing-like end as posterior. The affinities of this genus have not yet been determined. Ordovician to Carboniferous. Ex. *C. hibernicum*, *C. alæforme*, Carboniferous Limestone.

Myalina. Shell thick, trigonal, oblique, very inequilateral, with pointed umbones at the anterior extremity. Anterior marginal part of valves sharply bent. Posterior part compressed, wing-like. Hinge-line straight, long; teeth absent. Hinge-margin broad with longitudinal striations. Anterior adductor near the ventral edge of the anterior end of the hinge-plate. Posterior adductor large, oval. Pallial line simple. Surface with growth-lines, often lamellar. Silurian to Permian; common in Carboniferous. Ex. *M. verneuili*, Carboniferous.

3. PECTINACEA

Usually inequivalve and nearly equilateral, tending to be vertically elongated. Posterior adductor only. Ears and usually a byssal notch present. A narrow external ligament, and a triangular internal ligament. Hinge isodont. Often with radial ribs. Interior not nacreous. Silurian to present day.

Spondylus (fig. 124 D, E). Shell irregular, with ears and straight hinge-line, attached by the right valve; surface with radiating ribs which are spiny or foliaceous. Right valve larger and more convex than the left, with a triangular area. Ligament internal, in a deep triangular pit; the strong teeth in each valve fit into corresponding sockets in the other valve; in the left valve one, and in the right valve two teeth on each side of the ligament pit. Adductor impression large, subcentral. Jurassic to present day. Living in warm seas. Ex. *S. spinosus*, Chalk; *S. rarispina*, Bracklesham; *S. gœderopus*, Recent.

Plicatula. Similar to *Spondylus*. Surface smooth, folded or scaly. Ears absent or indistinct. Area very small. Ligament internal. Adductor impression excentric. Trias to present day. Living mainly in warm seas. Ex. *P. spinosa*, Lias; *P. inflata*, Chalk; *P. cristata*, Recent.

Pecten. Shell sub-circular. ovate or trigonal, closed, almost equilateral, inequivalve or nearly equivalve. Surface frequently with radiating ribs or striæ, sometimes smooth or with concentric ridges. Hinge-line straight; with well-developed ears, with or without a byssal sinus. A central, triangular pit for the internal ligament. Adductor impressions large, a little excentric. Carboniferous to present day. *Pecten* includes a very large number of species, which are grouped into sub genera and sections, of which the more important are: *Æquipecten* (ex. *Pecten asper*, Upper Greensand, *P. opercularis*, Pliocene); *Amusium* (ex. *P. pleuronectes*, Recent); *Camptonectes* (ex. *P. lens*, Jurassic); *Chlamys*, *Hinnites*, *Neithea* (see below) *Syncyclonema* (ex. *P. orbicularis*, Chalk).

Chlamys: shell ovate or trigonal, nearly equivalve, surface with radial ribs. Ears large—the anterior larger than the posterior and with a deep sinus for the byssus on the right valve. Trias to present day. Ex. *P.* (*C.*) *islandicus*, Pleistocene and Recent.

Hinnites: the young shell is like *Chlamys*; the adult is irregular like *Ostrea*, and is attached by the right valve. Cretaceous to present day. Ex. *H. crispa*, Pliocene.

Pecten (restricted): nearly equilateral; right valve very convex, left flattened. Ears nearly equal. No byssal sinus. Three or four lamellar teeth on each side of the ligament pit. Cretaceous to present day. Ex. *P. maximus*. Pliocene to present day.

Neithea: similar to the last; with numerous small denticles on the hinge. Cretaceous. Ex. *P.* (*Neithea*) *quadricostatus*, Upper Greensand.

Lima. Shell obliquely oval, anterior part larger than the posterior part, equivalve, compressed, with radiating striæ or ribs. Valves usually gaping anteriorly and sometimes posteriorly. Umbones distant, sharp. Hinge-line straight without teeth, with unequal ears. On each valve a triangular area, with a central ligament-pit. Adductor impression large. Two small pedal impressions. Trias to present day; maximum in the Mesozoic. Ex. *L. lima*, Recent; *L. elongata*, Chalk. Sub-genera *Plagiostoma* (*L. gigantea*, Lias; *L. cardiiformis*, Middle Jurassic); *Limatula* (*L. fittoni*, Upper Greensand); *Limatulina* and *Palæolima*, Carboniferous, are allied to *Lima*.

Aviculopecten. Shell ovate, slightly inequilateral; right valve less convex than the left. Umbones distinct; hinge-line straight, long; ears distinctly limited, the posterior larger than the anterior and often wing-like; a byssal sinus beneath the anterior ear in the right valve. Hinge-margin with narrow, nearly parallel grooves. A median pit for the internal ligament. Adductor impression large, sub-central. Surface usually with radial ribs, and concentric lines, the ornamentation different on the two valves. Devonian to Permian. Ex. *A. tabulatus*, Carboniferous.

Pterinopecten. Similar to *Aviculopecten*; posterior ear not distinctly limited; both valves with the same kind of ornamentation. Devonian and Carboniferous. Ex. *P. papyraceus*, Carboniferous.

4. ANOMIACEA

Inequivalve. Posterior adductor only. No ears. Byssus passes out through a rounded opening in the right valve. Hinge short, without teeth. Ligament more or less internal. Trias to present day.

Anomia. Shell thin, irregular or sub-circular, attached by a calcified byssus, which passes through a rounded sinus near the umbo of the right valve. Ligament short. Right valve flattened, with a central adductor impression; left valve larger, convex, with three impressions of the byssal muscles and one of the adductor. Hinge short. Teeth absent. Jurassic to present day. Ex. *A. ephippium*, Pliocene to present day.

5. OSTREACEA

Fixed by the left valve, which is the larger. Posterior adductor only. Hinge-line short; no teeth. Internal ligament in a triangular pit. Shell lamellar in structure, irregular in form. Trias to present day.

Ostrea. Shell with lamellar structure, irregular, inequivalve, slightly inequilateral, fixed by the left (larger) valve. Left valve convex, often with radiating ribs or striæ; umbo prominent, sometimes directed anteriorly, sometimes posteriorly. Right valve flat or concave, often smooth. Ligament-pit triangular or elongated. Hinge short, without teeth. Adductor impression sub-central; pallial line indistinct. No foot. Trias to present day. Ex. *O. delta*, Kimeridgian; *O. bellovacina*, Eocene; *O. edulis*, Pliocene and Recent.

Lopha (=*Alectryonia*). Includes the forms of *Ostrea* in which both valves have coarse angular folds; edges of valves toothed. The forms included in *Alectryonia* are polyphyletic. Trias to present day. Ex. *A. gregaria*, Corallian; *A. diluviana* (=*frons*), Chalk.

Gryphæa. Shell similar to *Ostrea*, but fixed in the young stage only, free in the adult; left valve large and convex, with a prominent incurved umbo. Right valve flattened or concave. Lias. Ex. *G. arcuata* (=*incurva*), Lias. (In later formations *Gryphæa*-like forms have originated independently from more than one species of *Ostrea*.)

Exogyra. Similar to *Ostrea*. Shell fixed by the left (larger) valve. Right valve flat, resembling an operculum. Umbones more or less spiral, directed posteriorly. Upper Jurassic to Chalk. Ex. *E. columba*, Upper Greensand.

ORDER III. EULAMELLIBRANCHIA

Two equal or nearly equal adductor muscles. Often with siphons. Hinge schizodont, heterodont or desmodont.

1. SCHIZODONTA

Hinge schizodont. No hinge-plate. Adductors equal or nearly equal. Ligament external, behind the umbones. Pallial line entire. Nearly all equivalve. Interior nacreous. Devonian to present day.

Trigonia (fig. 124 B, C). Shell thick, usually ornamented with concentric rows of tubercles or with concentric (sometimes radiating) ribs; trigonal, very inequilateral, anterior margin rounded, posterior produced and angular. Generally with a ridge extending from the umbones to the posterior border, cutting off a portion which has a different ornamentation. Umbones anterior, directed posteriorly. Teeth strong, grooved; in the right valve two teeth diverge from below the umbo; in the left three teeth, the central tooth being bifid, the posterior (next the hinge-margin) very thin. Ligament marginal, thick. Adductor impressions deep, the anterior smaller than the posterior, and placed near the umbones. A pedal impression in front of the posterior adductor of each valve and also one in the umbo of the left valve. Pallial line simple. Interior of shell nacreous. Rhætic to present day; abundant and widely distributed in the Jurassic and Cretaceous; found in the Australian region in the Tertiary and at the present day. Ex. *T. costata*, Inferior Oolite to Cornbrash; *T. clavellata*, Corallian; *T.* (*Neotrigonia*) *margaritacea*, Recent.

Schizodus. Similar in form to *Trigonia*; shell thin and smooth, umbones placed anteriorly. Three teeth in each valve; the anterior inconspicuous in the right valve. Adductor impressions distinct, but shallow. Carboniferous and Permian. Ex. *S. obscurus*, Permian.

Myophoria. Allied to *Schizodus*. Shell oval, triangular, or trapezoidal. Umbones anterior, often with a ridge extending to the lower part of the posterior border. Surface nearly smooth or with radial ribs. Right valve with one median and one anterior tooth, and ridge-like posterior tooth. Left valve with a triangular, sometimes bifid median tooth, and one anterior and one posterior tooth. Adductor impressions with a ridge passing to the hinge. Trias and Rhætic. Ex. *M. lævigata*, Trias.

Unio. Shell thick oval or elongated, with a thin periostracum. Surface smooth, tuberculate, striated, or folded; interior nacreous. Umbones more or less anterior, often corroded. Ligament external, elongated. In the right valve one or two thick, irregular teeth in front of the umbo, and a long lamellar posterior tooth; in the left valve, two thick irregular teeth near the umbo, and two long lamellar posterior lateral teeth. Anterior adductor impression very deep, the posterior shallow. Pallial line simple. Trias to present day. Lives in fresh water. Ex. *U. pictorum*, Pleistocene and Recent.

Anodonta. Allied to *Unio*; shell relatively thin, without teeth. Fresh water. Miocene (perhaps Purbeck) to present day.

Carbonicola (= *Anthracosia*). Similar in form to *Unio*, but the anterior part of shell is high and tumid, the posterior part low and compressed; usually a constriction at the ventral border. Hinge-margin triangular, with or without median teeth, no laterals. Ligament external. Adductors large, the anterior near the margin. Pedal impression above the anterior adductor. Pallial line simple. Carboniferous and Permian. Probably fresh water. Ex. *C. robusta, C. ovalis*, Coal Measures.

Anthracomya. Differs from *Carbonicola* chiefly in having the posterior part of the shell broad and expanded. Hinge-margin narrow, with a median and one posterior lateral tooth. Carboniferous. Probably fresh water. Ex. *A. modiolaris.*

Cardinia. Shell trigonal, oval, or oblong, very inequilateral, compressed, thick, marked by lines of growth. Interior not nacreous. Umbones small, sharp, close together. Ligament external. Cardinal teeth small or obsolete; in the right valve one anterior lateral tooth; in the left, one posterior lateral. Impression of anterior adductor very deep. Pallial line simple. Trias to Middle Jurassic (chiefly Lias). Ex. *C. listeri*, Lias.

Hippopodium. Shell very thick, very convex, oblong; surface with lines of growth. Umbones large, anterior. Hinge thick, with one oblique tooth which may disappear in old specimens. Adductor impressions deep. Pallial line simple. Lias to Great Oolite. Ex. *H. ponderosum*, Lower and Middle Lias. (Perhaps a gerontic form of *Cardinia.*)

2 HETERODONTA

Hinge heterodont. Hinge-plate usually well-developed. Two equal adductors. Ligament usually external, behind the umbones; sometimes internal. Pallial line entire or with sinus. Interior porcellanous, Trias (perhaps Silurian) to present day.

(a) Pallial line usually simple.

Cyprina. Shell orbicular or oval, convex, with concentric striæ and a thick periostracum. Umbones prominent, incurved. Ligament external, prominent. Lunule seldom present. Right valve with a very small anterior lateral tooth, two triangular cardinals, an oblique posterior cardinal, and a posterior lateral. Left valve with a small anterior lateral, a vertical cardinal, and a long oblique posterior cardinal. Adductor impressions oval. Pallial line entire. Margins of valves smooth. Lias to present day. Ex. *C. islandica*, Coralline Crag to present day.

Isocardia. Similar to *Cyprina*. Umbones inflated, curved anteriorly or spirally inrolled. In each valve two nearly parallel cardinal teeth and one posterior lateral. Cretaceous to present day. Ex. *I. humanus* (*cor*), Coralline Crag to present day.

Astarte. Shell thick, inequilateral, more or less trigonal or sub-orbicular, compressed, closed. Surface usually with concentric furrows or striæ. A thick periostracum is present. Umbones prominent, pointed. Lunule distinct. Escutcheon elongated. Ligament external. Right valve with a stout vertical cardinal, and a very small cardinal on each side; anterior lateral small. Left valve with two diverging cardinals, and a small posterior lateral. Adductor impressions strongly marked; above the anterior one is a pedal impression. Pallial line simple. Trias to present day. Ex. *A. omalii*, Coralline Crag.

Opis. Shell trigonal, cordiform, convex, with an oblique keel extending from the umbo to the postero-ventral angle. Umbones prominent, incurved or sub-spiral. Ligament external. Lunule large and very deep. Surface generally with concentric furrows. One cardinal tooth in the right valve, two in the left. Pallial line simple. Trias to Chalk. Ex. *O. lunulatus*, Inferior Oolite.

Crassatella (= *Crassatellites*). Shell solid, oblong or sub-trigonal, attenuated behind. Surface smooth or concentrically furrowed. Margins of valves smooth or crenulated. Umbones small, close together. Lunule distinct. Ligament internal, placed in a pit under the umbo. Hinge with two (sometimes three) cardinal teeth, and some small laterals. Adductor impressions deep. Pallial line simple. Cretaceous to present day. Ex. *C. sulcata*, Barton Beds.

Cyrena. Shell cordiform, oval, or trigonal, usually with concentric ridges; umbones often corroded. Hinge with three cardinal teeth; one anterior and one posterior lateral in the left valve, and two of each in the right valve. Ligament prominent, external. Pallial line usually entire. Margins of valves smooth. Lias to present day. Lives in fresh and brackish water. Ex. *C. ceylanica*, Recent.

Corbicula. Similar to *Cyrena*, but with the lateral teeth lamellar and transversely striated. Eocene to present day. Fresh water. Ex. *C. fluminalis*, Pliocene to present day.

Cardita. Shell ovoid or oblong, elongated, inequilateral, with broad radial ribs, narrower on the posterior part, usually scaly; often a little gaping and sinuous at its ventral margin. Umbones prominent, anterior. Lunule small. Ligament external. In the right valve two long, parallel cardinal teeth, which are nearly horizontal, and a very small anterior lateral tooth. In the left valve one short anterior cardinal, and one long posterior cardinal tooth. Adductor impressions large. Pallial line simple. Margins of valves coarsely crenulate. Trias to present day. Ex. *C. antiquata*, Miocene to Recent.

Venericardia. Shell oval, triangular, or heart shaped, inequilateral, with radiating ribs of uniform character. Umbones prominent. Ventral margin crenulated internally, not sinuous. Ligament external. Hinge-plate thick and high; in the right valve two oblique cardinal teeth and one small or rudimentary anterior lateral· in the left two diverging cardinal teeth. Adductor impressions unequal. Pallial line simple. Margins crenulate. Cretaceous to present day. Ex. *V. pectuncularis*, Eocene; *V.* (*Venericor*) *planicosta*, Bracklesham Beds.

Chama. Shell irregular, thick, inequivalve, fixed by the umbo of the larger valve (generally the left, sometimes the right). Umbones spiral or sub-spiral, directed anteriorly, that of the fixed valve longer than the other. Surface with concentric lamellæ or spines. The fixed valve larger and much deeper than the other. In the fixed valve a strong anterior cardinal tooth, and a narrow posterior cardinal below the ligament; in the free valve two cardinals. Ligament external, in a deep groove, prolonged towards the umbones. Adductor impressions large, the anterior commencing near the hinge-line. Pallial line simple. Upper Cretaceous to present day. Living in warm seas. Ex. *C. squamosa*, Barton Beds.

Diceras. Shell thick, inequivalve, fixed by umbo of larger (usually the left) valve. Umbones large, inrolled, directed forwards. Ligament external, in a curved groove at the posterior margin of the hinge. Hinge-plate very thick; right valve with two large teeth separated by a pit which receives the large ear-shaped tooth of the left valve. Adductor impressions distinct, the posterior on a raised elongated plate. Pallial line simple. Upper Jurassic. Ex. *D. arietinum*. *Requienia* and *Caucasia* (Lower Cretaceous) are fixed by the left valve and are related to *Diceras*.

Hippurites (figs. 128, 129). Shell very large and massive, conical or sub-cylindrical, not spiral, very inequivalve, fixed by

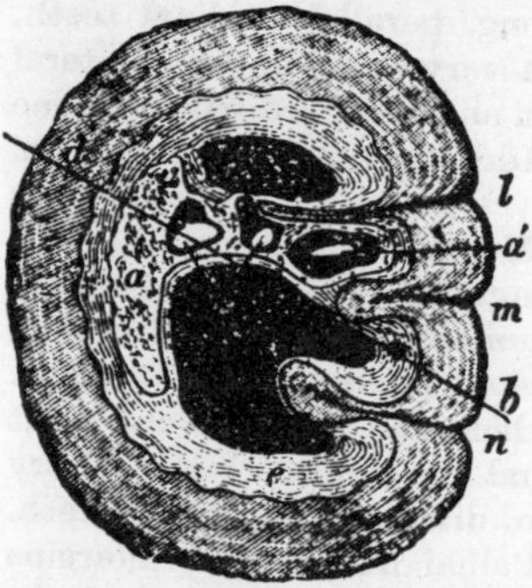

Fig. 128.

Fig. 129.

Fig. 128. Transverse section of the large valve of *Hippurites cornu-vaccinum*. *r*, umbonal cavity; *e*, internal layer of shell; *d*, external layer; *l*, *m*, *n*, folds; *t*, cardinal teeth; *a*, anterior adductor; *a′*, posterior adductor; *c*, cavity; *c′*, cardinal fossa. Cretaceous. (From Woodward.) × ⅓.

Fig. 129. Longitudinal section of the small valve and part of the large valve of *Hippurites cornu-vaccinum*, along the line *d-b* of fig. 128. *u*, umbonal cavity of small valve; *d*, external layer of shell; *r*, internal layer; *i*, part of cavity between the valves; *a*, anterior adductor; *a′*, posterior adductor; *t*, *t′*, anterior and posterior cardinal teeth of small valve; *l*, cardinal tooth of large valve. (From Woodward.) × ½.

the apex of the larger valve. The *large* (*lower*) *valve* elongate-conical, striated or smooth, and with three parallel furrows extending from the apex to the cardinal margin, due to folds of the shell-wall which give rise to three corresponding ridges in the interior. Hinge consists of a small cardinal tooth and of cardinal pits; anterior adductor impression large and divided into two separate parts; posterior adductor in a depression.

Small (upper) valve flattened or slightly convex, operculiform, porous, the pores leading into canals; with a central umbo and two prominent teeth; the anterior tooth very large with two surfaces at its base for the attachment of the adductors; the posterior tooth smaller with a tooth-like process for the posterior adductor. The small valve is formed of two layers; the outer is thin and prismatic, the inner is porcellanous and traversed by numerous canals. The outer layer (*d*) of the large valve is compact and formed of small prisms arranged in parallel layers obliquely to the surface of the shell; the inner (*e*) is porcellanous and formed of thin leaflets. Upper Cretaceous. Ex. *H. cornu-vaccinum*.

Radiolites. Shell large, thick, valves very unequal. The *large (lower) valve* conical or sub-cylindrical, generally straight, fixed by its apex (umbo); surface with vertical ribs, and thick, horizontal projecting layers which are more or less regularly folded; with a ligamental fold extending from the apex to the margin, and two vertical undulations corresponding to the positions of the anal and branchial orifices; outer layer of shell very thick, formed of polygonal or prismatic cells; inner layer thin, porcellanous, often not preserved; an elongate median tooth; two adductor impressions widely separated. The *small (upper) valve* generally convex or conical, sometimes flat, with central umbo; two straight, elongate, grooved teeth; the two adductor muscles were attached to plates on either side of the teeth; shell structure similar to that of the larger valve, but with the external layer thinner. Upper Cretaceous. Ex. *R. angeiodes*. A Radiolitid (*Durania mortoni*) is found in the Chalk of England.

Mactromya (= *Unicardium*). Shell oval or rounded, inflated; surface with concentric lines or ridges. Umbones prominent, curved inwards. In each valve a small cardinal tooth which is often obsolete, and a posterior ridge separated from the margin by a furrow in which is the external ligament. Adductor impressions elliptical. Trias to Cretaceous. Ex. *M. cardioides*, Lias.

Lucina (fig. 124 F). Shell orbicular or oval, slightly inequilateral, usually ornamented with concentric lines or ridges. Lunule usually distinct. An oblique furrow extends from the umbo to the posterior border. Hinge usually with two cardinal and one or two lateral teeth in each valve; the lateral, or the cardinal, may be absent. Ligament elongated, external, sometimes sunk in a groove. Adductor impressions well marked, the anterior elongated and placed mainly within the pallial line, the posterior oval. Pallial line entire. Margins of valves smooth or finely crenulated. Trias to present day. Ex. *L. columbella*, Miocene.

Megalodon. Shell thick, equivalve, smooth or with concentric lines, convex, inequilateral, oval or rounded triangular. Umbones prominent, curved forward. Ligament external, long. Hinge-plate very large and thick; teeth thick; in the right valve two cardinals separated by a pit; in the left valve one cardinal under the umbo and a small anterior cardinal; no laterals. Anterior adductor impression small, semilunar; posterior adductor long, shallow, on a ridge extending from the hinge to the posterior border. Devonian to Lias. Ex. *M. cucullatus*, Devonian. *Pachyrisma* (Trias and Jurassic) is allied to *Megalodon*.

Cardium. Shell convex, slightly inequilateral, cordate or oval, generally closed. Umbones prominent, incurved, turned slightly to the anterior end. Surface with radiating ribs, which are often spiny. Margins of valves crenulated. No distinct hinge-plate. Right valve with one or two cardinal teeth, two anterior laterals, and one or two posterior laterals; left valve with two cardinals, one anterior lateral and one posterior lateral. Ligament external. Adductor impressions shallow. Pallial line entire. Trias to present day. Ex. *C. costatum*, Recent; *C.* (*Acanthocardia*) *aculeatum*, Pleistocene and Recent; *C.* (*Cerastoderma*) *edule*, Pliocene to present day.

Protocardia. Similar to *Cardium*, but with radiating ribs on the posterior part of the shell only, the remainder with concentric ribs. Jurassic to present day. Ex. *P. hillana*, Upper Greensand.

Thetironia (= *Thetis*). Shell thin, oval, rounded, very convex, slightly or moderately inequilateral. Umbones prominent, curved inward and slightly forward. No lunule. Ligament external. Two small conical or tubercular cardinal teeth under the umbo in each valve; no laterals. Adductor impressions near the anterior and posterior margins. Pallial line simple. Two internal ribs meet at an acute angle near the umbo and extend ventrally to the level of the adductors. Surface of shell nearly smooth, with concentric lines and radial rows of small pits which are more distinct on the posterior part than elsewhere. Cretaceous. Ex. *T. minor*, Lower Greensand.

(*b*) Pallial line usually with a sinus, but sometimes sinuous only.

Venus. Shell thick, oval, convex, ornamented with concentric lamellæ, sometimes with radial ribs; lunule distinct. Margins of valves finely crenulate. Hinge-plate wide; in each valve three thick cardinal teeth, no lateral teeth. Ligament external, prominent. Pallial sinus short, angular. Miocene to present day. Ex. *V. casina*, Pliocene to present day; *V. verrucosa*, Recent.

Meretrix (fig. 123). Shell thick, ovate, sub-trigonal, convex, smooth or with concentric ornament. Margins of valves smooth. Lunule present. Ligament external. Hinge-plate thick, with three cardinal teeth in each valve, two anterior laterals in the right, and one in the left valve. Pallial sinus angular or rounded. Cretaceous to present day. Ex. *M. meretrix*, Recent; *M.* (*Macrocallista*) *planus*, Upper Greensand; *M.* (*Cordiopsis*) *incrassata*, Oligocene; *M.* (*Macrocallista*) *chione*, Recent. *Meretrix* is here used in a wide sense, and includes *Macrocallista* (= *Callista*), *Cytherea*, *Tivela*, *Pitaria*, etc.

Dosinia (= *Artemis*). Shell orbicular, compressed, with concentric ridges or striæ. Lunule depressed. Escutcheon narrow. Ligament sunk. Three cardinal teeth in each valve, one anterior lateral in the left valve and two (rudimentary or absent) in the right. Margins smooth. Pallial sinus very deep, pointed. Oligocene to present day. Ex. *D. exoleta*, Coralline Crag to present day.

Tellina. Shell elongate-oval, slightly inequivalve, compressed, rounded in front, attenuated behind, and furnished with an oblique fold from the umbo to the posterior border. Umbones small, turned slightly to the posterior. No lunule. Margins of valves smooth. Two cardinal teeth in each valve (one being bifid), and one anterior and one posterior lateral. Ligament external, prominent. Pallial sinus very deep. Jurassic to present day. Ex. *T. virgata*, Recent; *T. rostralis*, Eocene.

Macoma is shorter and more oval in outline than *Tellina*, and without lateral teeth. Eocene to present day. Ex. *M. balthica*, Pliocene to Recent.

Psammobia (=*Gari*). Shell thin, elongate, sub-equilateral, gaping at the ends, anterior side rounded, posterior side more or less truncate and angular; with a fold from the umbo to the postero-ventral angle. Surface smooth or with striæ. Ligament external, thick, joined to prominent ridges. Usually two cardinal teeth in each valve, some being bifid. Adductor impressions near the dorsal border. Pallial sinus very deep. Cretaceous to present day. Ex. *P. ferroensis*, Coralline Crag to present day.

Donax. Shell trigonal or oval, inequilateral, anterior side longer than the posterior. Umbones small, directed posteriorly. Surface smooth or with radial grooves and concentric striæ. Right valve with two cardinal teeth, the posterior sometimes bifid, a lamellar anterior lateral and a short posterior lateral. Left valve with two diverging cardinal teeth and one posterior lateral. Ligament very short, external or partly internal. Pallial sinus deep, rounded. Margins of valves usually crenulate. Eocene to present day. Ex. *D. vittatus*, Pliocene and Recent.

Solen. Shell very long, sub-cylindrical, straight, smooth or finely striated, the dorsal and ventral margins parallel; gaping at both extremities. Margins of valves smooth. Umbones at the anterior end. Hinge terminal, with one cardinal tooth in each valve. Ligament long, external. Anterior adductor impression elongated, parallel to the dorsal margin. Pallial sinus short. Eocene (perhaps earlier) to present day. Ex. *S. obliquus*, Bracklesham Beds; *S. vagina*, Recent.

Mactra. Shell oval or trigonal, nearly equilateral, smooth or with concentric striæ. Internal ligament in a large triangular pit. External ligament in a groove. In front of the internal ligament-pit is a bifid cardinal tooth (in the form of an inverted V); anterior and posterior lateral teeth long and lamellar; two of each in the right valve, one in the left. Adductor impressions semicircular. Pallial sinus round or angular. Cretaceous to present day. Ex. *M. stultorum*, Pliocene to present day.

3. DESMODONTA

Hinge desmodont. External ligament behind the umbones; internal ligament sometimes present. Two equal adductors. Pallial sinus usually present. Valves often somewhat unequal. A few genera in the Upper Palæozoic, mainly Trias to present day.

Mya (fig. 122). Shell oblong, gaping at both ends, particularly at the posterior; the left valve a little smaller than the right. Surface with concentric ridges. In the left valve a large spoon-like process (chondrophore) to which the internal ligament is fixed, and a corresponding pit under the umbo of the right valve. External ligament thin. Anterior adductor impression elongated. Pallial sinus large and rounded. Eocene to present day. Ex. *M. truncata*, Pliocene to present day.

Corbula (= *Aloidis*). Shell convex, oval, inequivalve, closed, rounded in front, somewhat angular and contracted behind, with a ridge passing from the umbo to the posterior angle. Surface generally with concentric grooves. Umbones prominent. Right valve larger and more convex than the left, and with a strong cardinal tooth in front of the ligament-pit; left valve with a spoon-like process for the internal ligament. External ligament present. Adductor impressions well marked. Pallial line slightly sinuous posteriorly. Trias to present day. Ex. *C. pisum*, Eocene; *C. gibba*, Miocene to Recent.

Panopea. Shell equivalve, inequilateral, oblong, rounded in front, thick, concentrically striated, gaping at each end, especially at the posterior. Ligament external, on prominent ridges. One cardinal tooth in each valve. Pallial sinus deep. Cretaceous to present day. Ex. *P. faujasi*, Coralline Crag to present day.

Saxicava (= *Hiatella*). Shell small, more or less oblong, gaping; umbones anterior. Ligament external. Teeth absent in the adult, one or two cardinals present in the young. Pallial line not continuous, sinuous. *Saxicava* bores into rocks, etc. Jurassic to present day. Ex. *S. rugosa*, Coralline Crag to present day.

Pholas. Shell elongate, very inequilateral, cylindrical, gaping at both ends. Surface with spiny ridges, best marked in front. On the dorsal region are one or more accessory calcareous plates. No teeth; no ligament. Hinge-margin reflected over the umbonal region. In the interior, under the umbones, is a process for the insertion of the muscle of the foot. Pallial sinus very deep. *Pholas* bores into rocks, etc. Eocene (perhaps Mesozoic) to present day. Ex. *P. dactylus*, Pleistocene and Recent.

Teredo. Shell more or less globular, gaping at the ends, valves tri-lobed, with concentric ridges; without teeth; adductors unequal. In the interior, under the umbones, is a long narrow plate for the insertion of the pedal muscle. Posterior part covered by a long, calcareous tube, which is sub-cylindrical, straight or curved, and often with partitions. *Teredo* perforates wood. Jurassic to present day. Ex. *T. norvegica*, Coralline Crag to present day.

Pleuromya. Shell elongated, anterior side short and rounded, posterior long and generally compressed, sometimes gaping; surface with concentric folds. Hinge without teeth, but each valve with a tooth-like boss near the umbo. Ligament external. Adductor impressions faintly marked; pallial sinus deep. Trias to Lower Cretaceous. Ex. *P. donacina*, Corallian and Kimeridgian.

Gresslya. Shell oval, elongate, very inequilateral, smooth or with concentric furrows; anterior side high and inflated, posterior side narrowing and somewhat compressed. Umbones anterior, close together; lunule sometimes well marked. Ligament internal. Right valve a little higher and larger than the left. Adductor impressions shallow; pallial sinus deep. Behind the umbo of the right valve is a tooth-like projection and an internal plate—the latter appears as a furrow in casts of the shell. Jurassic. Ex. *G. gregaria*, *G. abducta*, Inferior Oolite.

Ceratomya (= *Ceromya*). Shell heart-shaped, inflated, inequilateral, finely granular, with concentric grooves. Left valve not quite so convex as the right. Anterior side short, posterior longer and compressed. Umbones prominent, anterior, curved forward. Hinge thickened, with a ridge behind the umbones; teeth absent. Pallial line sinuous. Jurassic. Ex. *C. concentrica*, Inferior Oolite to Cornbrash.

Pholadomya. Shell thin, translucent, oblong or oval, ventricose, equivalve, gaping posteriorly and sometimes anteriorly.

Anterior side short and rounded. Surface with radiating ribs (corrugations) crossed by concentric folds or striæ. Umbones prominent. Ligament external, short. Hinge without teeth or with a small transverse tubercle. Adductor impressions very faint. Pallial sinus moderately deep. Lias to present day; abundant and widespread in the Mesozoic; two species living, one in the Antilles, one in Japan. Ex. *P. margaritacea*, London Clay.

Homomya. Similar to *Pholadomya*. Without radial ribs; surface smooth or ornamented with fine granules. Trias to Cretaceous. Ex. *H. gibbosa*, Inferior and Great Oolite.

Goniomya. Similar to the last two, but with V-shaped ribs pointing ventrally. Lias to Cretaceous. Ex. *G. literata*, Great Oolite to Kimeridge Clay.

Thracia. Shell rather thin, oblong, compressed, attenuated and gaping posteriorly: surface smooth or concentrically striated. Umbones nearly central, turned a little to the posterior side. Right valve usually larger than the left. External ligament short, prominent. Hinge without teeth. Behind the umbo is a stout process or ossicle in each valve to which the internal ligament is fixed. Adductor impressions small. Pallial sinus not deep. Trias to present day. Ex. *T. pubescens*, Coralline Crag to present day.

4. PALÆOCONCHA

Shell thin. Hinge either without teeth or with only imperfectly developed teeth, and without hinge-plate. Ligament external. Two nearly equal adductors. Pallial line simple. This group is a provisional one and includes primitive Palæozoic genera, the affinities of which have not yet been determined.

Grammysia. Shell elongate-ovate, very inequilateral, ornamented with concentric furrows, and one or more radial folds passing from the umbo to the postero-ventral border. Umbones placed anteriorly. Lunule very deep. Hinge-margin thick, without teeth. Anterior adductor very small, posterior large. Pallial line simple. Silurian and Devonian. Ex. *G. cingulata*, Silurian; *G. hamiltonensis*, Devonian.

Cardiola. Shell thin, convex, oval, generally inequilateral; umbones prominent, incurved. Surface with well-marked radiating and concentric grooves. Hinge-line straight, probably with very small teeth; ligamental area large, horizontally grooved. Muscular impressions unknown. Silurian and Devonian. Ex. *C. interrupta*, Lower Ludlow, etc.

Cardiomorpha. Shell thin, smooth or with concentric lines; sub-quadrate or rounded, inequilateral, very convex. Umbones prominent, curved forwards; no lunule. Hinge toothless. External ligament small. Adductor impressions shallow; pallial line simple. Principally Carboniferous. Ex. *C. oblonga*, Carboniferous Limestone.

Edmondia. Shell sub-quadrate or ovate, convex, inequilateral; surface with concentric lines or ridges. Umbones anterior; no lunule, no escutcheon. Hinge toothless, with a thick ridge posterior to the umbones and separated from the edge of the valve by a groove. Posterior to the hinge is an internal, elongated 'ossicle'. External ligament small. Pallial line simple. Devonian and Carboniferous. Ex. *E. unioniformis*, Carboniferous Limestone.

Sanguinolites. Shell elongate, very inequilateral, with rounded ends, the posterior part usually higher than the anterior parts; surface with concentric ribs or lines. Umbones near the anterior end, with a ridge passing to the lower part of the posterior end; lunule and escutcheon distinct. Anterior adductor impression large, deep, limited posteriorly by a ridge; posterior adductor shallow, near the hinge. Pallial line entire. Hinge toothless. Carboniferous. Ex. *S. angustatus*, Carboniferous Limestone.

Distribution of the Lamellibranchia

All the Lamellibranchs are aquatic animals, and by far the larger number are marine. The marine forms range from the shore-line down to a depth of 2900 fathoms; they are most abundant in shallow water, and are scarce at depths greater than 500 fathoms, but the following, and a few other genera, have been found below 1500 fathoms: *Nucula*, *Nuculana*,

Arca, *Limopsis*, *Malletia*, *Verticordia*, *Cuspidaria* (= *Neæra*) (see p. 295).

Two genera, which may be Lamellibranchs, have been recorded from the Lower Cambrian of North America. In England the earliest forms appear in the Lower Ordovician. Lamellibranchs are not usually common in the Ordovician, but the Taxodonta, Mytilacea, Pteriacea and Palæoconcha are represented. These groups continue throughout the Palæozoic, the Palæoconcha being especially characteristic of the Ordovician, Silurian and Devonian. The Schizodonts are well represented in the Carboniferous and Permian by *Schizodus*, *Carbonicola*, etc. The Pectinacea begin in the Silurian and become numerous in the Upper Palæozoic. In the Carboniferous the Palæoconcha are represented by *Edmondia* and *Sanguinolites*. *Allorisma* (Carboniferous and Permian) is one of the few Palæozoic genera with a pallial sinus.

Lamellibranchs form an important part of the faunas of the Mesozoic and Tertiary formations. Most of the Palæozoic genera die out before the beginning of the Mesozoic. In the Trias the Pteriacea, Pectinacea and Mytilacea are abundant. The Ostracea and Anomiacea now appear. The Schizodonta, Heterodonta and Desmodonta are represented by a few genera. In the Jurassic and Cretaceous all these groups increase in importance. The Taxodonta are represented by numerous forms allied to *Arca* and *Cucullæa*; the Schizodonts by many species of *Trigonia*. The Cretaceous period is distinguished by the abundance of *Inoceramus*, and by the presence of the Rudistids (*Hippurites*, *Radiolites* and others). In the Tertiary period the Heterodont group attains the greatest importance.

Fresh water lamellibranchs are generally rare in the Palæozoic and Mesozoic formations. Probably the earliest form is *Archanodon* (=*Amnigenia*) *jukesi* from the Old Red Sandstone. In the Coal Measures several species of *Carbonicola*, *Anthracomya*, and *Naiadites* occur. The living type *Unio* has been found in the Inferior Oolite of Yorkshire, and is fairly common in the Purbeckian and Wealden of the south of England, where it is associated with *Cyrena*. Fresh water lamellibranchs also occur in the Woolwich Beds, the Oligocene deposits, and in the Pleistocene river-gravels.

The principal genera of Lamellibranchs found in the different systems are as follows:

Ordovician. *Ctenodonta, Cyrtodonta, Heikia, Modiolopsis, Ambonychinia, Mytilarca, Actinopterinia, Shaninopsis.*

Silurian. *Ctenodonta, Cardiola, Pterinea, Palæopecten, Ambonychia, Modiolopsis, Grammysia.*

Devonian. *Nucula, Ctenodonta, Cardiola, Pterinea, Aviculopecten, Actinopteria, Megalodon, Conocardium, Grammysia.*

Carboniferous. *Nucula, Parallelodon, Posidonia, Pinna, Conocardium, Leiopteria,* subgenera of *Pecten, Aviculopecten, Pterinopecten, Schizodus, Protoschizodus, Carbonicola* (=*Ancosia*), *Anthracomya, Edmondia, Sanguinolites, Cardiomorpha.*

Permian. *Bakevellia, Schizodus, Pseudomonotis* (*Eumicrotis*).

Trias. *Nucula, Nuculana, Palæoneilo, Gervillia, Hœrnesia, Pteria, Monotis, Cassianella, Halobia, Ostrea, Pecten* (subgenera), *Lima, Myophoria, Megalodon, Cardium, Palæocardita.*

Jurassic. *Nucula, Nuculana, Arca, Grammatodon, Myoconcha, Modiola, Hippopodium, Pteria, Pseudomonotis, Gervillia, Perna, Pinna, Ostrea, Gryphæa, Lopha* (=*Alectryonia*), *Pecten* (various subgenera of), *Lima, Cardinia, Trigonia, Diceras, Cardium, Mactromya* (=*Unicardium*), *Astarte, Opis, Anisocardia, Pachyrisma, Pleuromya, Ceratomya, Gresslya, Pholadomya, Homomya, Goniomya, Thracia.*

Cretaceous. *Nucula, Arca, Cucullæa, Modiola, Myoconcha, Gervillia, Inoceramus, Perna, Pteria, Buchia* (=*Aucella*), *Aucellina, Ostrea, Exogyra, Lopha* (=*Alectryonia*), *Pecten* (the subgenera *Chlamys, Syncyclonema, Neithea,* etc.), *Lima, Spondylus,*

Plicatula, Unio, Trigonia, Hippurites, Radiolites, Sphærulites, Cardium, Protocardia, Thetironia, Cyprina, Cyrena, Macrocallista, Panopea, Pholadomya.

Eocene. *Nucula, Arca, Glycimeris* (*Pectunculus*), *Pinna, Pecten* (*Chlamys*, etc.), *Ostrea, Chama, Cardium, Venericardia, Cardita, Astarte, Crassatella, Cyprina, Lucina, Cyrena, Corbicula, Meretrix, Psammobia, Tellina, Corbula, Panopea, Pholadomya.*

Oligocene. *Mytilus, Dreissensia, Ostrea, Cyrena, Corbula, Erodona* (= *Potamomya*), *Lucina, Meretrix, Venus, Psammobia.*

Miocene. *Barbatia, Anadara, Pecten, Anomia, Ostrea, Cardita. Lucina, Venus, Meretrix, Dosinia.*

Pliocene. *Nucula, Glycimeris* (= *Pectunculus*), *Mytilus, Pecten, Chlamys, Cardium, Cardita, Astarte, Cyprina, Isocardia, Lucina, Venus, Dosinia, Tellina, Mya, Pholas, Thracia.*

CLASS II. AMPHINEURA

The Amphineura include *Chiton* and its allies. The body is bilaterally symmetrical and more or less elongated, with the mouth in front and the anus at the posterior end. The head is without tentacles and eyes. The mantle covers the dorsal surface and the sides of the body. A nerve-ring surrounds the œsophagus and from it two nerves come off on each side and extend to the posterior end of the body; ganglia are poorly or not at all developed. A radula is sometimes present. All the Amphineura are marine. There are two Orders.

(1) *Polyplacophora*. The foot is large and flat, and forms the whole of the ventral surface of the body. There are numerous (6 to 80) pairs of gills, which are placed in a groove between the foot and the mantle. The shell consists of eight transverse plates placed in a longitudinal row on the dorsal surface of the body; each plate usually overlaps the one behind it, and a flexible band or girdle encircles the whole series of plates.

Polyplacophora live chiefly in shallow water, but a few examples have been found at great depths. Although of great antiquity and represented by a large number of living species, they are rarely found fossil. The earliest forms (*Priscochiton*) occur in the Ordovician; *Helminthochiton* is found in the Silurian; *Gryphochiton* and other genera in the Carboniferous; *Lepidopleurus, Chiton* and others in the Tertiary.

(2) The *Aplacophora* are worm-like in form, and being without a shell are not known fossil.

CLASS III. GASTEROPODA

Well-known examples of the Gasteropoda are the snail, the whelk, and the cowry. The bilateral symmetry, so characteristic of the lamellibranchs, is generally to a large extent obliterated, owing to the twisting of the visceral mass and the atrophy of some of the organs on one side of the body. There is a distinct head, which bears one or two pairs of tentacles, and usually also eyes. On the ventral surface of the body is the foot; this is usually large and sole-like and used for crawling, but in the Heteropods it is in the form of a flattened fin, and in the Pteropods it is wing-like. The mantle is never divided into two lobes. Respiration takes place in some cases through the skin, but generally by means of a lung-cavity or by gills; the latter are placed in a sac formed by the mantle; sometimes they are present on both sides of the body, but usually the original left gill has disappeared. In some forms the mantle, at the opening of the gill-sac, is produced into a tube, known as the *siphon*, by means of which water passes to the gills. The heart is on the dorsal surface, and consists of a ventricle and usually one, but in some cases two auricles. In many forms the gills are placed in front of the heart, but in others behind it. The mouth is at the anterior end of the body, and the anus is near the opening of the gill chamber. On the floor of the cavity of the mouth is a dental apparatus, known as the *odontophore*: this consists of a cartilaginous and muscular ridge on which rests a chitinous ribbon (the *radula*); the radula bears numerous teeth placed in rows, and serves as a rasping organ. The arrangement of the teeth varies in different genera and is of considerable importance in classification, but since the radula has never been definitely recognised in fossil forms, it can only be used by the palæontologist in the case of genera which have existing representatives. In most gasteropods, except the carnivorous genera and the Heteropods, there are also one or two horny jaws in the upper part of the mouth which are used for biting. The nervous system consists of ganglia which are connected by nerve-cords. Typically there are three pairs of principal ganglia—the *cerebral* placed above the œsophagus, and the *pleural* and *pedal* placed below it; a visceral nerve-cord, which may bear ganglia, comes off from the pleural ganglia, and forms a loop ventral to the intestine. In some gasteropods this loop is simple, but in others (the Prosobranchiata) one side is bent over so that the loop forms a figure of 8. In some gasteropods the sexes are separate; others are hermaphrodite.

In the majority of the gasteropods a shell is secreted by the mantle; in a few forms, as for instance the slugs, it is internal, but usually it is external and covers the visceral mass. The shell consists of a single piece, and is hence said to be *univalve*. In the limpet (*Patella*) it has the form of a hollow cone; but in most cases it consists of a long tube, open at one end, and tapering to a point at the other. This tube is coiled into a spiral, generally screw-like, each coil being termed a *whorl*; in a few genera (*e.g. Vermetus*, *Siliquaria*) the whorls are separated, but as a rule they are in contact (fig. 130), the line between two contiguous whorls being known as the *suture* (*su*). All the whorls, except the last, together form the *spire* (*S*) of the shell, the point of which is termed the *apex* (*a*). The last whorl is nearly always larger—frequently much larger—than the one preceding, and the part of it farthest from the apex is called the *base* of the shell. The spire varies in form in different genera and species; sometimes it is composed of a large number of whorls, sometimes of few, and it may be long, short, or depressed; occasionally all the whorls are in one plane. The angle of the spire (*spiral angle*) consequently varies; this is measured by lines drawn from the apex to the base of the shell on opposite sides of the exterior of the whorls. The coiling of the shell is usually *dextral*; so that when the apex of the shell is pointed away from the observer (as in fig. 130) the aperture will be on the right-hand side; in a few cases it is *sinistral*, when the aperture will be on the left. Sinistral forms may occur as 'sports' in a dextral species, or they may be characteristic of a species or occasionally of a genus.

Frequently the inner parts of the whorls coalesce, and form an axial pillar extending from the apex to the base of the shell (fig. 130) and known as the *columella*. In other cases the inner parts do not fuse, and in the place of the columella there is left a tube-like space, extending from the base of the shell a greater or less distance towards the apex; this space, which opens at the base of the shell, is called the *umbilicus*. When there is a columella the shell is said to be *imperforate*, when instead there is an umbilicus it is *perforate*. The opening of the umbilicus sometimes becomes partly or completely filled up with *callus* (see p. 278). The animal is attached to the columella by means of a muscle, the contraction of which enables it to withdraw completely into the shell; but, when not retracted, the coiled visceral mass only is covered by the shell

Usually the cavity of the gasteropod shell is continuous from the apex to the aperture, but in a few cases partitions are thrown across the earlier parts of the shell (fig. 130), forming chambers which remain empty. The form of the aperture varies considerably in different genera and is of great importance in classification; in shape, it may be circular, oval, elongate, oblong, etc. Its margin is termed the *peristome*: the outer part forms the *outer lip* (*L*), the inner part (that next the columella) the *inner lip*. As the gasteropod crawls along, the shell is carried on the dorsal surface of its body with the apex directed backward and upward, and the aperture downward; consequently the part of the aperture farthest from the apex is *anterior*, the opposite (nearest the apex) is *posterior*. Sometimes, as in *Natica*, there is no break in the peristome, and it is then said to be entire or *holostomatous*; in other cases the anterior border is notched or produced into a tube (*ac*) in which the incurrent siphon is placed, and these forms are said to be *siphono-*

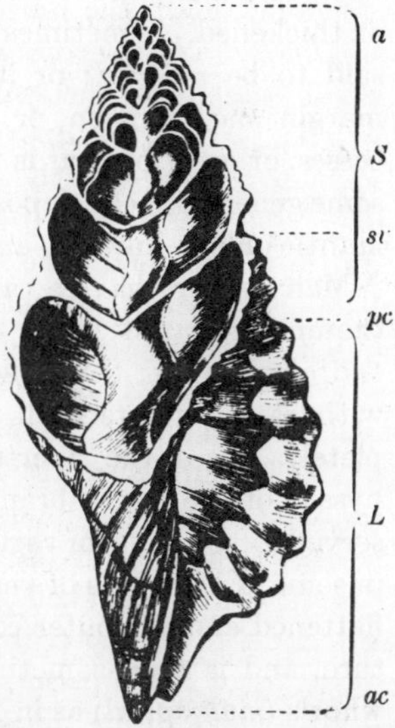

Fig. 130. Longitudinal section of *Cymatium* [*Tritonium*] *corrugatum*. The upper part of the spire has been partitioned off many times successively. *a*, apex; *su*, suture; *S*, spire; *L*, outer lip of the aperture; *ac*, anterior canal; *pc*, posterior canal. (From Woodward.)

stomatous; sometimes there is also at the posterior border another notch or a canal (*pc*), in which the excurrent or anal siphon is placed. The outer lip may be thin and sharp, or thickened. Sometimes it is curved outwards, and is then said to be *reflected*; or it is curved inwards—*inflected*. Its margin may be even, or crenulated, or produced into processes, or grow outwards to form a wing-like projection. In some genera a shelly deposit, termed *callus*, is secreted by the mantle on the inner lip and adjoining part of the last whorl.

Many genera have a calcareous or horny plate, known as the *operculum*, attached to the dorsal part of the posterior end of the foot; this is so arranged that when the animal withdraws into its shell the operculum more or less completely closes the aperture. It probably represents the byssus of the lamellibranch. The operculum is seldom preserved fossil; its form varies considerably in different genera, in some (*Turbo*) it is of very large size with the inner surface flattened and the outer convex; it may have a spiral structure, and is then sometimes formed of a large number of whorls (*multispiral*) as in *Trochus*, or of a few whorls (*paucispiral*) as in *Littorina*. When not spiral it may be *concentric*, if growth takes place equally all round; it is then marked with concentric lines, the nucleus being nearly central, as in *Viviparus*; or it may be *unguiculate* or *claw-shaped* when the nucleus is at the apex as in *Fusinus*.

The form of the shell in the spiral gasteropods varies considerably, depending on the arrangement of the whorls in one plane or in a helicoid spiral, on the spiral angle, on the number and shape of the whorls, on the size of the last whorl and whether it conceals the earlier whorls or not. The chief types are the following:

1. *Discoidal;* all the whorls are nearly or quite in one plane, and all are visible on the exterior as in *Planorbis*.

2. *Conical* or *trochiform;* conical with a moderately acute spire and a flat base. The whorls increase in diameter uniformly and have nearly flat surfaces, *e.g. Trochus.*

3. *Turbinate;* conical with a convex base, as in *Turbo.*

4. *Turreted;* spire long, very acute, formed of numerous whorls which increase in diameter slowly and uniformly, as in *Turritella.*

5. *Fusiform;* spindle-shaped, thickest in the middle and tapering to each end. The spire is elongated and the base of the last whorl is produced into a long neck, as in *Fusinus.*

6. *Cylindrical;* after the first few whorls their diameter remains constant or may decrease near the anterior end of the shell, as in *Pupilla.*

7. *Globular;* spire short, last whorl large and rounded, as in *Natica.*

8. *Convolute;* when the last whorl is very large and convex and covers all or nearly all the others and the aperture is consequently as long as the shell, as in *Cypræa.*

The surface of the shell is frequently ornamented with spines, knobs, ribs, or striæ; these are said to be *spiral* when they run parallel with the sutures, and *transverse* when they cross the whorls from suture to suture. In some genera (*e.g. Murex*) rows of spines, or lamellar processes, extend across all the whorls from the apex to the base of the shell, forming what are termed *varices.* The surface of the shell in recent gasteropods is generally coloured, often variegated; in fossil examples the colour has nearly always disappeared, but a few specimens, from various formations, even as early as the Carboniferous, have been found showing the colour more or less perfectly preserved. The shell consists of an outer chitinous layer, and of a calcareous layer, usually aragonite, which is thick and porcellanous; in some cases there is also an inner nacreous or pearly layer.

The protoconch or embryonic shell is often found at the apex of the shell, and usually consists of several whorls differing in character from those of the rest of the shell. The gradual development of the ornamentation on the gasteropod shell may be traced on the whorls which follow the protoconch.

The Gasteropoda are divided into three Orders: (1) Prosobranchiata, (2) Opisthobranchiata, (3) Pulmonata.

ORDER I. PROSOBRANCHIATA

In the Prosobranchiata (or Streptoneura) the visceral nerve-cord is twisted into a figure of 8. Usually one gill only is present, and it is generally placed in front of the heart. The sexes are separate. An operculum is found in most cases. The Prosobranchiata are divided into two sub-orders, (1) the Aspidobranchia, (2) the Pectinibranchia.

SUB-ORDER 1. *ASPIDOBRANCHIA*

The axis of the gill is attached at its base only and bears two rows of plates (*bipectinate*). This group includes the more primitive gasteropods in which signs of the original symmetry are shown by the presence of two kidneys, two auricles and, in some cases, two gills. The Aspidobranchs are nearly all marine and include the majority of the Palæozoic gasteropods.

Patella. Shell conical, outline oval or sub-circular; apex sub-central, nearer the anterior border; surface with radiating ribs or striæ, rarely smooth. Margin simple or spinose. Muscular impression horse-shoe shaped, open in front. Jurassic (perhaps Palæozoic also) to present day. Ex. *P. vulgata*, Pliocene to present day.

Pleurotomaria. Shell trochiform, conical, turbinate, or nearly discoidal; interior nacreous. Umbilicus present or absent.

Aperture sub-quadrate or oval, outer lip sharp, with a slit which, as the shell grows, becomes filled up, leaving a band on the whorls, towards which the lines of growth are directed obliquely backwards. Operculum horny. Trias to present day; common and widespread in Jurassic; four species living in the seas of the West Indies and Japan at depths of from 70 to 200 fathoms. Palæozoic representatives of the Pleurotomariidæ begin in the Silurian. Ex. *P. anglica*, Lias; *P. ornata*, Inferior Oolite.

Murchisonia. Shell turreted, with many, more or less angular whorls, provided with a band as in *Pleurotomaria*. Aperture oblong, with a slit, and a very short anterior canal. Devonian to Trias; mainly Devonian and Carboniferous. Ex. *M. verneuiliana*, Carboniferous Limestone. Allied genera (*Cyrtostropha*, *Hormotoma*, etc.) occur in the Lower Palæozoic.

Bellerophon (fig. 131). Shell globular, smooth or with growth-lines; umbilicus small or closed; whorls few, embracing, symmetrically coiled in one plane; with or without umbilicus. Aperture sub-circular or oval, with a deep median slit, which is replaced by a band or keel dividing the shell into two similar parts; columellar edge often with callus. Silurian to Permian; maximum in Carboniferous. Ex. *B. tenuifascia*, Carboniferous Limestone. Allied genera are *Bucaniopsis*, *Cymbularia*, *Euphemus*, *Sinuites*, *Waagenella*.

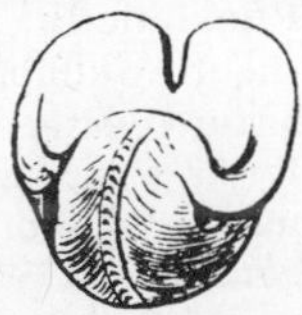

Fig. 131. *Bellerophon*, from the Carboniferous Limestone, showing the slit in the aperture. × ⅔.

Emarginula. Shell conical, surface generally ornamented with a trellis-work of longitudinal and transverse ribs; apex not perforated, curved posteriorly. Anterior border with a well-marked slit, which becomes filled up during growth, leaving a raised band. Muscular impression horse-shoe shaped; no internal septum. Jurassic (perhaps Carboniferous) to present day. Ex. *E. fissura*, Coralline Crag to present day.

Fissurella. Shell similar to *Emarginula*, but more or less depressed; apex perforated and nearer the anterior than the posterior border; no marginal slit. Muscular impression as in *Patella*. *Fissurella* is divided into several sub-genera; many of the fossil species belong to the sub-genus *Fissuridea*. Jurassic to present day. Ex. *F. crassa*, Recent; *F. græca*, Coralline Crag to present day.

Euomphalus. Shell depressed, discoidal or conical, with a wide and large umbilicus; whorls convex with a ridge on the upper surface. Aperture polygonal; outer lip with a slit on its upper surface. Silurian to Trias; maximum in Carboniferous. Ex. *E. pentangulatus*, Carboniferous Limestone.

Poleumita (=*Horiostoma* of some authors). Form similar to *Euomphalus*. Whorls ornamented with spiral keels and numerous transverse striæ or fine ribs. Aperture without a slit. Common in the Silurian. Ex. *P. discus*, Silurian.

Turbo. Shell solid, turbinate or conical, whorls convex, interior nacreous. Aperture large, circular, entire, slightly produced anteriorly; outer lip sharp. Columella curved, flattened. Imperforate, or with a small umbilicus. Operculum thick, calcareous, exterior convex, interior flat and spiral, nucleus central or sub-central. Jurassic to present day. Ex. *T. marmoratus*, Recent. There are numerous sub-genera.

Phasianella. Shell elongated, ovate, smooth, polished, without an umbilicus, interior porcellanous. Whorls rounded. Aperture ovate, entire, rounded anteriorly, angular posteriorly; outer lip thin, simple, sharp. Columella smooth, flattened, with a narrow band of callus. Operculum calcareous, with an excentric nucleus. Upper Cretaceous to present day. Ex. *P. australis*, Recent; *P. gosauica*, Upper Cretaceous.

Amberleya. Shell turbinate, elongate, without umbilicus. Whorls ornamented with several spiral keels which are usually spiny or nodular; between the keels are numerous transverse striæ or fine ribs. Base rounded. Aperture sub-oval; outer lip often crenulated. Trias to Cretaceous (chiefly Jurassic). Ex. *A. ornata*, Inferior Oolite.

Cirrus. Shell sinistral, conical or turbinate, or sometimes nearly discoidal, with a very large umbilicus. Spire acute. Whorls irregular, ornamented with strong transverse nodular ribs and finer spiral ribs; last whorl large. Aperture rounded, entire. Trias to Inferior Oolite. Ex. *C. nodosus*, Inferior Oolite.

Trochus. Shell conical, whorls numerous and flat or slightly convex, spire sharp, interior nacreous; base flat or nearly so, angular at the periphery. Aperture entire, rhomboidal; outer lip sharp, oblique. Columella twisted, with a prominent anterior tooth-like protuberance or a fold. Operculum horny, multispiral, nucleus central. Trias to present day. Ex. *T. niloticus*, Recent. There are numerous sub-genera.

Nerita. Shell thick, solid, ovoid or semi-globose, without an umbilicus; interior not nacreous. Spire very short, flat. Surface smooth, or with spiral ribs. Aperture semicircular, entire; outer lip thick, the interior generally denticulate; inner lip flattened, with callus, and a straight denticulate border. Operculum calcareous, nucleus excentric. Cretaceous to present day. Ex. *N. ustulata*, Recent; *N. globosa*, London Clay and Bracklesham Beds.

Theodoxus (=*Neritina*). Form similar to *Nerita*. Shell relatively thin, usually smooth and with colour marking. Outer lip sharp, not thickened, with interior not denticulate; inner lip flattened, with sharp or finely denticulate border. Eocene (perhaps earlier) to present day. Lives in brackish or fresh water. Ex. *T. concavus*, Headon Beds; *T. zebra*, Recent.

SUB-ORDER 2. *PECTINIBRANCHIA*

The gill is attached to the mantle throughout its length, and bears one row of plates only. There is no sign of bilateral symmetry in the circulatory, respiratory and excretory organs—only one kidney, one auricle, and one gill being present.

Macrochilina (=*Macrocheilus*). Shell elongate-oval, with sharp spire, and last whorl high. Surface smooth or with growth-line. Sutures shallow. No umbilicus. Aperture ovate, angular behind, with a shallow anterior canal; outer lip thin, inner lip with a weak anterior fold. Silurian to Trias. Ex. *M. arculata*, Devonian.

Loxonema. Shell turreted, spire very long, consisting of numerous convex whorls; ornamented with sinuous growth-lines; sutures deep. No umbilicus. Aperture long, enlarged in front, with shallow canal; outer lip sharp, sinuous. Ordovician to Carboniferous; mainly Silurian. Ex. *L. sinuosum*, Silurian. *Pseudozygopleura*, *Eoptychia*, *Microptychis* found in the Carboniferous, are allied to *Loxonema*.

Pseudomelania. Shell elongate, with many nearly flat whorls, without umbilicus, spire long, surface smooth or with growth-lines. Aperture oval, entire, rounded in front, narrowed and angular behind; outer lip sharp. Columella smooth. Trias to Cretaceous; common in Jurassic. Ex. *P. heddingtonensis*, Corallian.

Epitonium (=*Scalaria*). Shell turreted, spire elongate; whorls numerous, very convex, only just in contact or slightly separated, ornamented with regular varices, frequently with spiral ridges also. Umbilicus more or less distinct. Aperture circular, entire, margin thickened. Operculum horny, paucispiral. Jurassic to present day. Ex. *E. scalaris*, Recent; *E. grœnlandicum*, Red Crag to present day.

Architectonica (=*Solarium*). Shell conical, depressed, angular at the periphery. Aperture entire, sub-quadrate; lip sharp. Ornamented with spiral ridges. Base flat. Umbilicus wide and deep, limited by a sharp edge which is generally crenulated. Operculum horny, spiral. Jurassic to present day. Ex. *A. perspectiva*, Recent; *A. canaliculata*, Barton and Bracklesham Beds.

Purpuroidea. Shell thick, oval, spire rather short, last whorl inflated. Whorls step-like, flattened below the suture, with tubercles or spines at the angles. Aperture with a small notch anteriorly; outer lip thin. Inferior Oolite to Upper Cretaceous. Ex. *P. nodulata*, Great Oolite.

Littorina. Shell thick, without a nacreous layer, turbinate, with few whorls, without umbilicus. Aperture rounded, angular behind, outer lip sharp. Columella flattened. Operculum horny, paucispiral. Lias to present day. Ex. *L. littorea*, Red Crag to present day.

Capulus. Shell thin, conical, with apex bent considerably backward and more or less spirally inrolled; with fine radial ribs. Aperture rounded or irregular. Muscular impression horse-shoe shaped. Lower Palæozoic to present day. Ex. *C. hungaricus*, Coralline Crag to present day.

Platyceras. Allied to *Capulus*; apical part usually more extensively coiled, dextral. Surface smooth, or with concentric striæ, or radial folds or spines. Silurian to Carboniferous. Ex. *P. cornutum*, Silurian.

Calyptræa. Shell thin, conical, trochiform, spiral, apex central; interior with a spiral plate under the apex and attached at the periphery. Aperture nearly circular. Cretaceous to present day. Ex. *C. chinensis*, Miocene to present day.

Natica. Shell oval, globular, generally smooth, spire short, last whorl very large. Aperture semi-lunar or oval, entire; outer lip sharp, oblique; inner lip thickened with callus, not crenulate. Umbilicus present, partly covered by callus. Operculum of the same size as aperture, horny or calcareous, pauci-spiral, nucleus excentric. Trias to present day. Ex. *N. canrena*, Recent: *N. millepunctata*, Coralline Crag to present day. There are numerous sub-genera.

Xenophora (=*Phorus*). Shell conical, low, with flattened or concave base; periphery of last whorl sharp. Surface with fine growth ridges. Aperture large, oblique, lower part concave, outer lip sharp and oblique. Umbilicus generally small and deep, sometimes partly or completely closed by callus. Whorls flattened, covered with agglutinated foreign bodies. Cretaceous to present day. Ex. *X. agglutinans*, Barton Beds.

Viviparus (=*Paludina*). Shell thin, turbinate, with a thick periostracum; whorls convex, smooth or with faint ribs. Umbilicus very small or absent. Aperture entire, oval, slightly angular behind. Operculum horny with concentric striæ, and excentric nucleus. Inferior Oolite to present day. Lives in fresh water. Ex. *V. lentus*, Bembridge Beds.

Turritella. Shell without umbilicus, turreted with many flat or slightly convex whorls, ornamented with spiral ribs and with striæ of growth; spire very long and acute. Aperture oval or sub-quadrate, entire, outer lip thin, sinuous, slightly produced in front. Operculum horny. Cretaceous to present day. Ex. *T. communis*, Pliocene to present day; *T. imbricataria*, Barton and Bracklesham Beds.

Thiara (=*Melania*). Shell with dark periostracum, elongate, turreted, with many whorls, without umbilicus, apex sharp but usually corroded. Surface smooth, or ornamented with spiral and transverse striæ, sometimes with tubercles or spines. Aperture oval, entire, narrow behind, rounded in front; outer lip sharp, slightly sinuous behind. Columella smooth. Operculum horny, oval, sub-spiral. Wealden to present day. Lives in fresh water. Ex. *T. amarula*, Recent; *T. acuta*, Bembridge Beds.

Nerinea (fig. 132). Shell elongate, usually without an umbilicus, whorls numerous. Aperture sub-quadrangular, oval, or elongate, with a short anterior canal; outer lip thin, with a posterior slit near the suture, which becomes filled, leaving a continuous band. Columella and also the interior of the whorls furnished with folds which are continuous to the apex. Inferior Oolite to Upper Cretaceous. Ex. *N. cingenda*, Inferior Oolite.

Cerithium. Shell without an umbilicus, turreted, without periostracum. Whorls numerous, narrow, the last whorl always much shorter than the spire; with spiral ornament. Aperture oblong or semi-oval, with a short posterior canal and a well-marked recurved anterior canal; outer lip more or less thickened and often somewhat reflected; columellar edge concave. Operculum horny, oval, paucispiral, with sub-marginal nucleus. Cretaceous to present day. Ex. *C. adansoni*, Recent; *C. mutabile*, Barton and Bracklesham Beds. There are several sub-genera. *Cerithium* in the restricted sense includes forms in which there is a strong ridge on the posterior part of the inner lip forming the inner boundary of the posterior canal, the outer lip is expanded in front, and the whorls are provided with varices. Ex. *C. nodulosum*, Recent.

Fig. 132. *Nerinea trachea*, partly sliced to show the form of the interior. Great Oolite. (From Woodward.) ×⅔.

Potamides. Form similar to *Cerithium*. Shell with a brown or blackish periostracum. Aperture rounded or sub-quadrangular; either with a fold in front or a very short and not recurved anterior canal; outer lip rather thin. Operculum circular, multispiral, with central nucleus. Lives in brackish water. Cretaceous to present day. Ex. *P. lapidus*, Eocene; *P. lamarcki*, Oligocene.

Aporrhais. Shell fusiform, without umbilicus, whorls numerous, angular, spire elongate. Aperture produced in front into a straight or curved canal. Outer lip expanded, thick, with an anterior sinuosity, lobed or digitate, the posterior process being attached partly or entirely to a part of the spire forming a canal. Operculum small. Cretaceous to present day. Ex. *A. pes-pelicani*, Coralline Crag to present day.

Dicroloma (=*Alaria*). Similar to *Aporrhais*, but without a process from the outer lip attached to the spire, and without anterior sinuosity. Jurassic and Cretaceous. Ex. *D. armata*, Great Oolite.

Strombus. Shell ovoid, ventricose, tuberculate or spiny, without umbilicus. Spire with several whorls. Last whorl very large. Aperture long, narrow, with a short anterior channel; canaliculate posteriorly; outer lip expanded, wing-like, thick, often lobed behind, with a sinus near the anterior margin and at the posterior end. Inner lip with callus. Operculum small, horny, claw-shaped, with serrated edge. Eocene to present day. Ex. *S. pugilis*, Recent.

Tibia (=*Rostellaria*). Shell fusiform, spire elongate, composed of many whorls, which are smooth or faintly ribbed. Aperture oval, with a long straight or slightly curved anterior canal, and a posterior canal applied to the spire; outer lip expanded, with tooth-like processes, and an anterior bay. Operculum small, oval or claw-shaped, edge not serrated. Eocene to present day. Ex. *T. curta*, Recent; *T. lucida*, London Clay, etc.

Hippochrenes. Similar to *Tibia*, but outer lip more expanded and wing-like and extending up to, or nearly to the apex; its margin without tooth-like processes. Upper Cretaceous to Oligocene. Ex. *H. amplus*, Barton Beds.

Rimella. Similar to *Tibia*, but with cancellate ornamentation; outer lip but little expanded, reflected outwards; posterior canal reaching nearly to the apex of the spire; anterior canal shorter. Eocene to present day. Ex. *R. rimosa*, Barton Beds.

Cypræa. Shell ovoid or elongate, convex, convolute, surface covered with shining enamel. Spire almost or quite concealed by the last whorl. Aperture oblong and narrow, as long as the shell, with a short canal at each end; outer lip inflected and crenulated; inner lip crenulated. In the young form the outer lip is thin and the spire prominent. Eocene to present day. Ex. *C. mappa*, Recent; *C. oviformis*, London Clay. *Trivia* is similar but smaller and with transverse ribs. Eocene to present day. Ex. *T. europæa*, Recent.

Cymatium (fig. 130). Shell thick, oval, or fusiform. Spire elongate, with varices which are continued over a few whorls only. Aperture with a posterior notch, and a short, slightly curved anterior canal; outer lip thick, crenulate internally; inner lip with callus and usually with folds. Eocene to present day. Ex. *C. femorale*, Recent; *C.* (*Lampusia*) *corrugatum*, Recent.

Buccinum. Shell ovoid or elongate, without an umbilicus. Spire of moderate length; last whorl large. Whorls convex, with small spiral ribs crossed by transverse folds. Aperture oval, large; outer lip simple, thin; anterior canal short, truncated, a little reflected; inner lip a little sinuous, with callus. Operculum small, oval or circular. Pliocene to present day. Ex. *B. undatum*, Coralline Crag to present day.

Liomesus (=*Buccinopsis*). Spire shorter, and whorls less convex than in *Buccinum*; without transverse folds. Eocene to Recent. Ex. *L. dalei*, Pliocene and Recent.

Nassarius (=*Nassa*). Shell solid, ovate, elongate, without umbilicus, usually with transverse and spiral ornament. Aperture oval, pointed behind, with a very short reflected anterior canal; inner lip with callus, reflected on to the last whorl; outer lip thick, crenulate internally. Columella truncated, provided with an oblique fold in front. Upper Cretaceous to present day. Ex. *N. mutabilis*, Pliocene and living.

Chrysodomus (=*Neptunea*). Shell solid, fusiform, spire more or less elongate; sometimes sinistral. Whorls rounded, smooth or with spiral lines. Aperture oval; outer lip simple, inner lip smooth; anterior canal short, slightly twisted. Operculum horny, unguiculate. Eocene to present day. Ex. *C. antiquus*, Red Crag to present day.

Sipho. Shell more slender, whorls less convex, and aperture narrower than in *Chrysodomus*. Pliocene to present day. Ex. *S. gracile*, Red Crag to Recent.

Thais (=*Purpura*). Shell tuberculate, striated or lamellar, without varices. Spire rather short, last whorl large; no umbilicus. Aperture oval, large, with either an anterior notch or a short oblique anterior canal, and a posterior notch or groove. Columella flattened, with callus. Operculum lamellar, nucleus marginal. Miocene to present day. Ex. *T. persica*, Recent. Sub-genus *Nucella*, with rather longer spire, distinct anterior canal, and spiral ribs. *T.* (*Nucella*) *lapillus*, Red Crag to present day; *T.* (*N.*) *tetragona*, Pliocene.

Murex. Shell thick, oval or elongate, spire prominent and sharp; whorls convex, each carrying three or more varices, which may be spiny, foliaceous, or tubercular. Aperture ovate; anterior canal more or less long, straight or curved, narrow and tubular, often nearly closed; no posterior canal; outer lip thick, inner lip smooth. Operculum oval, nucleus sub-apical. Eocene to present day. Ex. *M. brandaris*, Recent; *M. frondosus*, *M. tricarinatus* (*asper*), Barton Beds.

Typhis. Similar to *Murex*; small, with hollow spines; anterior canal short and completely closed. Eocene to present day. Ex. *T. pungens*, Barton Beds.

Fusinus (=*Fusus*). Shell without umbilicus, narrow, fusiform, elongate; spire sharp, with many rounded whorls. Ornament of spiral ribs and transverse ribs or folds. Aperture oval; outer lip simple, thin, interior often striated. A long, straight, narrow anterior canal, not closed. Columella smooth, without folds. Operculum oval. Cretaceous to present day. Ex. *F. colus*, Recent; *F. porrectus*, Barton Beds.

Clavilithes (=*Clavella*). Shell thick, usually large, fusiform, nearly smooth (except the earlier whorls, which have transverse and spiral ribs). Whorls often with a posterior carina near the suture. Aperture pyriform, channelled posteriorly, with a long straight anterior canal; outer lip thickened posteriorly. Last whorl contracting rapidly in front. Eocene to present day. Ex. *C. longævus*, Barton Beds; *C. parisiensis*, Middle Eocene.

Mitra. Shell fusiform, thick. Spire elevated, summit acute. Aperture narrow, elongate, with a wide and deep notch in front. Columella with four or five oblique folds, becoming stronger posteriorly; outer lip not reflected, thickened, but not grooved internally; inner lip with callus, thick in front. No operculum. Eocene to present day. Ex. *M. episcopalis*, Recent; *M. fusiformis*, Miocene; *M.* (*Mitreola*) *labratula*, Bracklesham Beds.

Voluta. Shell thick, ovate, with short spire and turbinate protoconch. Last whorl very large; on the posterior part are nodules or spines which are continued anteriorly as transverse (or axial) ribs. Aperture elongate, rather narrow, with an angular channel at the posterior end; broad, and deeply notched at the anterior end; outer lip thick; inner lip with thin callus. Columella with several transverse, only slightly oblique folds, of which the four or five anterior are strong and nearly equal, and the posterior two or three are smaller. Eocene to present day. Ex. *V. musica*, Recent; *V. musicalis*, Eocene.

Volutospina. Shell fusiform, with rather short, conical spire; protoconch small, with sharp apex. Last whorl very large, tapering anteriorly. Whorls step-like owing to the posterior part being flattened or concave; at the angle of the whorls is a row of spines (usually prominent) which are prolonged anteriorly as transverse (or axial) ribs; the latter are usually crossed by spiral ridges. Aperture elongate; at the posterior end one channel at the suture, another at the level of the row of spines; anteriorly the aperture is truncated and slightly notched. Outer lip usually thin; inner lip with thin callus. Columella with four or five very oblique folds, of which the anterior aré stronger than the posterior. Upper Cretaceous to present day. Ex. *V. spinosa*, Eocene; *V. luctatrix*, Barton Beds.

Ancilla (=*Ancillaria*). Shell smooth, oval or oblong; last whorl large; sutures usually covered by callus. Aperture elongate, broadening anteriorly, with a small notch near the suture, and a deep sinuosity at the anterior end which is truncated; columella with callus posteriorly, twisted in front, and with folds anteriorly. Cretaceous to present day. Ex. *A. buccinoides*, Bracklesham, Barton, and Headon Beds; *A. cinnamonea*, Recent.

Turris (=*Pleurotoma*). Shell turreted, fusiform, spire long and sharp, last whorl long. Aperture oval, elongate; outer lip curved, with a deep slit at a short distance from the suture; inner lip smooth. Anterior canal long, straight, narrow. Columella without folds. Operculum horny, ovate, acute, nucleus apical. Upper Cretaceous to present day. Ex. *T. babylonia*, Recent; *T. undata*, Bracklesham Beds. There are numerous sub-genera.

Conus. Shell conical, generally smooth, the last whorl enveloping the greater part of the preceding whorls. Spire short, flattened or conical, with many whorls. Aperture long, narrow, straight, with parallel or sub-parallel borders, ending anteriorly in a truncated canal; outer lip thin, simple, no folds or teeth, notched at the suture. Columella straight, smooth. Operculum horny, much smaller than the aperture. The outer shell is thick, but the inner parts of the whorls become resorbed and very thin. Upper Cretaceous to present day. Ex. *C. marmoreus*, Recent; *C. deperditus*, Bracklesham Beds. There are numerous sub-genera.

Conorbis. Biconical, with elevated spire. Outer lip arched, with a sinus near the suture. Eocene and Oligocene. Ex. *C. dormitor*, Barton Beds.

The Heteropoda are a group of the Prosobranchiata which have become modified for a pelagic mode of life. The foot is laterally compressed so as to form a vertical fin. A shell may be absent, but, when present, it is always thin and light. Only a very few forms have been found fossil.

ORDER II. OPISTHOBRANCHIATA

The visceral nerve-cord (except in a few genera, *e.g. Acteon*) is a simple, untwisted loop. One gill only may be present, but is absent in some forms. The gill is placed behind the heart; there is one auricle only, which is behind the ventricle. An operculum is generally absent. All the Opisthobranchiata are marine; they are divided into two groups:

(1) the *Nudibranchia*, in which there is no mantle, no gill, and no shell in the adult. No examples of this division have been found fossil;

(2) the *Tectibranchia*, which usually possess a mantle, a shell, and a true gill. The following genera are examples of the Tectibranchia.

Acteon. Shell oval, ornamented with spiral pitted striæ or grooves; spire prominent, conical, sharp. Aperture elongate, rounded in front; outer lip sharp; columella with one strong, slightly oblique fold at the anterior end. Cretaceous to present day. Ex. *A. tornatilis*, Coralline Crag to present day.

Avellana. Shell globular, ornamented with spiral striæ or grooves; spire very short. Aperture semi-lunar, curved, entire; outer lip much thickened, reflected externally, dentate internally. Inner lip thickened, with two or three prominent folds. Cretaceous. Ex. *A. incrassata*, Upper Greensand.

Bulla (=*Bullaria*). Shell solid, smooth, sub-globular or ovoid, convolute. Spire concave. Aperture as long as the last whorl, rounded at both ends, widest in front; outer lip sharp. Inner lip with callus. Cretaceous to present day. Ex. *B. ampulla*, Recent; *B. globulus*, London Clay and Bracklesham Beds.

The Pteropods are pelagic Opisthobranchs in which the foot is modified to form two lateral wing-like fins, and the head is not well marked. The shell is conical, urn-like or spirally coiled, but is absent in some forms. Pteropods occur in large numbers near the surface in the open ocean, especially in warm regions; and their shells, which are thin and transparent, form a considerable part of the 'pteropod ooze'—one of the deep-sea deposits found in parts of the Atlantic Ocean. They are regarded as specialised members of the Tectibranch group which have become adapted to a pelagic mode of life.

Living families of Pteropods, represented mainly by recent genera, occur in the Upper Cretaceous and Tertiary formations, but are not known from earlier deposits. In the Palæozoic formations, however, numerous fossils, which have been regarded by various authors as Pteropods, are found. Thus in the Silurian and Devonian rocks large numbers of small conical smooth shells, which closely resemble the living *Styliola*, occur, and may be the remains of Pteropods.

Hyolithes (fig. 133), *Conularia*, *Tentaculites*, *Salterella* and other allied genera are also found in the Palæozoic, beginning as far back as the Cambrian and Ordovician. Their shells, however, are larger and thicker than those of living Pteropods, and in *Conularia* they were, in some cases, attached by the apical end. Some writers have suggested that these Palæozoic genera are allied to primitive Cephalopods, since septa are sometimes present in the shell and possibly also a siphuncle. The discovery by Walcott of the presence of wing-like fins in specimens of *Hyolithes* from the Middle Cambrian supports the view that that genus is really a Pteropod; if that is confirmed it follows that the Pteropods are the oldest group of the Opisthobranchs and cannot have been derived from the later Opisthobranchs which are not known before the Carboniferous period. *Conularia* is considered by Kiderlin to belong to the Scyphozoa (p. 78).

Hyolithes (= *Theca*) (fig. 133). Shell calcareous, straight, rarely curved, pyramidal; its section triangular, elliptical, semi-elliptical or nearly circular; surface smooth or striated; posterior part sometimes crossed by septa. Aperture with an operculum. Cambrian to Permian. Ex. *H. elegans*, Ordovician.

Fig. 133. *Hyolithes* from the Cambrian, showing the operculum. × ⅔.

Conularia. Shell thin, formed of chitin, more or less impregnated with lime; generally straight, pyramidal, with four sides; each angle of the pyramid with a straight groove; each lateral face may have a median longitudinal groove. Apical part of shell sometimes with a few convex septa. Surface smooth, or ornamented with numerous transverse, parallel, angulated ridges, and sometimes with longitudinal ridges. Aperture partly closed by incurved triangular lobes. Ordovician to Lias. Ex. *C. quadrisulcata*, Carboniferous.

Tentaculites. Shell calcareous, thick, solid, in the form of a greatly elongated cone, straight or slightly curved, with circular section; apical part with septa, its end often with a vesicular enlargement. Surface provided with prominent, transverse, parallel rings, and with transverse and longitudinal striæ. Ordovician to Devonian. Ex. *T. anglicus*, Bala Beds.

Other genera, which appear to be allied to the preceding, are *Hyolithellus*, *Salterella*, and *Coleolus* from the Cambrian.

ORDER III. PULMONATA

The mantle-cavity is modified to form a lung and there is no gill. The visceral nerve-cord is an untwisted loop. An operculum is nearly always absent in the adult. The Pulmonata are mainly land and fresh-water forms, and are hermaphrodite. They appear first in the Devonian.

Limnæa. Shell spiral, thin, horny; last whorl very large, rounded; spire sharp. Aperture large, oval, rounded in front. Columella more or less twisted. Peristome sharp, entire. Purbeck Beds to present day. Lives in fresh water. Ex. *L. stagnalis*, Pliocene to present day; *L. fusiformis*, Headon Beds.

Planorbis. Shell discoidal, sinistral, horny, whorls numerous. Aperture oblique; peristome simple, sharp. Jurassic to present day. Fresh water. Ex. *P. corneus*, Red Crag to present day; *P.* (*Planorbina*) *euomphalus*, Headon Beds.

Helix. Shell variable—conical, or globular; smooth except for growth-lines; with or without an umbilicus; aperture oblique, slightly higher than wide. Peristome simple or reflected. Eocene to present day. Lives on land. Ex. *H. pomatia*, Recent. There are numerous sub-genera.

Distribution of the Gasteropoda

Some of the Gasteropoda live on land, others in fresh water, but the majority are marine; they are found in the seas of all parts of the world but are especially abundant in warm regions and in comparatively shallow water. A few forms can exist both on land and in water, *e.g. Ampullaria*, which commonly lives in lakes and rivers, and is also found on land. Some marine genera, such as *Littorina*, *Cerithium*, and *Thais* (*Purpura*), are able to live in fresh as well as in salt water; on the other hand some fresh-water forms are at times found living in the sea, *e.g. Limnæa*, *Theodoxus* (*Neritina*), *Bithynia*, and *Planorbis*; this is especially the case in places where the water is less salt than the main mass of the ocean, as for instance in the Baltic, where we find the genera just mentioned living side by side with *Littorina* and with the marine lamellibranchs *Cardium*, *Tellina*, and *Mya*. The Opisthobranchiata are entirely and the Prosobranchiata mainly marine. Nearly all the Pulmonata are found on land or in fresh water.

In this place a few words may be said with regard to the distribution of the marine Mollusca generally.

These may be divided into two groups belonging to the Plankton and the Benthos respectively.

The *Plankton* includes animals which swim or float either near the surface of the sea or at various distances below it; among the Mollusca the chief forms are the Pteropods, the Heteropods and a few other Gasteropods, as well as many Dibranchiate Cephalopods; the shells in these are either thin and light or altogether wanting. The geographical distribution of the species which live near the surface is determined mainly by the temperature of the water.

The *Benthos* includes animals which are fixed to the sea floor or live crawling on it or swimming just above it. The distribution of the Mollusca in depth depends on the depth of the sea and the accompanying changes in temperature, pressure, light and other physical conditions. The Benthos may be divided into:

(1) The *Littoral zone*, which extends between high and low water marks and is consequently inhabited by animals which can live exposed to air for periods each day. In the European seas this zone is characterised by the abundance of the genera *Littorina*, *Trochus*, *Patella*, *Hydrobia*, *Haliotis*, *Fissurella*, *Solen*, *Mya*, *Donax*, *Cardium*.

(2) The *Continental shelf*. This includes the gradual slope from the low water-mark down to a depth of 100 or some-

times 200 fathoms, and extends to a distance of from 20 to 200 miles from land. It is on this part of the sea floor that most of the terrigenous deposits are laid down. The character of the molluscan fauna is influenced largely by the nature of the sediment on the sea-bottom, some genera (*e.g. Mya*, *Scrobicularia*, *Lutraria*) being found especially on muddy bottoms, others (*e.g. Natica*, *Turritella*, *Cypræa*, *Cardium*) on sandy, and yet others (*e.g. Buccinum*, *Littorina*, *Patella*, *Arca*) on rocky

The upper part of the continental shelf from low water down to about 15 fathoms is known as the *Laminarian zone*. It is characterised by the great abundance of algæ (*Laminaria*, etc.) which afford food for numerous phytophagous molluscs; it is the region into which sunlight penetrates freely, the action of waves is felt, periodic changes of temperature occur, and the salinity is reduced owing to the drainage of fresh water from the land. In the European seas some of the commonest genera are *Trochus*, *Nassaria*, *Rissoa*, *Ostrea*. Nudibranchs are also very numerous.

Below the Laminarian zone the conditions become more uniform and less liable to sudden alteration. The changes in temperature are gradual, and seasonal rather than diurnal; the salinity is more constant, and light diminishes gradually with increasing depth and with it vegetation decreases. At a depth of about 100 fathoms (the lower limit of the continental shelf) the finer terrigenous materials are deposited, forming what is known as the 'mud-line'—a rich feeding ground for animals.

The part below the Laminarian zone down to a depth of about 25 fathoms is known as the *zone of Nullipores or Corallines* on account of the numerous calcareous algæ (Corallinaceæ), and is characterised by the abundance

Turris, *Fusinus*, *Chrysodomus*, *Buccinum*, *Natica*, *Eulima*, *Venus*, *Dosinia*, *Astarte*, *Nucula*, *Arca*, *Lima*, *Pecten*.

Below this is the *zone of Brachiopods and deep-sea Corals*; off Europe *Oculina* is the common coral; Brachiopods and Polyzoa are abundant. Some of the chief molluscs are *Turritella*, *Odostomia*, *Dentalium*, *Tellina*, *Cuspidaria* (= *Neæra*), *Yoldia*.

(3) The *Deep Sea and Abyssal Region* begins at the edge of the continental shelf with the relatively steep 'continental slope' extending down to about 500 fathoms, followed by the more gentle slope to the great deeps of the oceans. In this region light, except in the shallowest parts, is absent, the temperature is very low and nearly uniform at any one spot, currents are not felt, and the pressure of the water becomes very great. The only variation of importance is in the nature of the sediment which consists of fine ooze. The number of animals decreases with the increase of depth. The shells of the molluscs are mostly thin, colourless, transparent and of small size. Scaphopods are numerous; other common forms are *Turris*, *Fusinus*, *Acteon*, *Scaphander*, *Philine*, *Arca*, *Nucula*, *Limopsis*, *Nuculana*, *Lima*, *Pecten*.

Owing to the relative uniformity of the physical conditions the geographical distribution of the deep-sea species, although not unlimited, is greater than the range of the species which live on the continental shelf, especially on its shallower parts.

Of the Mollusca which live in shallow water or at moderate depths some few species have a very wide or almost cosmopolitan distribution, but the majority have a more limited range. In studying the geographical distribution of these Molluscs it is found that a number of areas or provinces can be recognised, each of which is characterised by the abundance of certain genera and species, and by the presence of some species which are either confined to that province or rarely found to extend beyond it; so that the general assemblage of molluscs found in each province possesses characteristic features. Two neighbouring provinces are not, as a rule, separated by a sharp boundary, and but few genera are confined to any one province. In the European and Northern Atlantic region the chief provinces[1] are: the *Arctic*, which includes the polar seas and extends as far south as the north coast of Iceland and the North Cape on the east of the Atlantic and to the shores of Newfoundland on the west; the *Boreal*, extending from the last down to near the southern end of Norway and including Iceland (except the north coast), the Faroe Islands and perhaps the Shetland Islands, and the American coast from the Gulf of St Lawrence to Cape Cod. The occurrence of the Boreal fauna on both sides of the Atlantic is accounted for by the former existence of a shallow coastal region across the North Atlantic along which this province was at that time continuous; the *Celtic*, including the coasts of Southern Sweden, the Baltic, Denmark, Northern France and the British Isles, and the *Lusitanian*, comprising the coasts of the Bay of Biscay, Portugal, the Mediterranean, and North-west Africa, including the Azores, the Canaries and Madeira groups. Altogether some nineteen provinces have been recognised, and these may be grouped into larger regions.

The chief barrier to the geographical extension of species is temperature, and consequently their range is influenced largely by the warm and cold currents in the surface waters of the sea. An example of this is seen in the North Atlantic where, owing to the cold Labrador current, the Arctic province extends much further south on the American coast

[1] For a map of the provinces see Woodward's *Manual of the Mollusca*, or Fischer's *Manual de Conchyliologie*, or A. M. Davies' *Tertiary Faunas*, vol. II, 1934.

than on the European; similarly, owing to the Gulf Stream drift the Boreal province is found further north on the European side than on the American. Another striking instance of the influence of currents is seen off South Africa; the warm water molluscs of the West coast are separated from those of the East coast by the cold water of the Antarctic drift which flows to the coast of Cape Colony, forming a barrier between the West African province and the Indo-Pacific province, thus causing the Cape province to have a special molluscan fauna.

The distribution of shallow-water molluscs is also interrupted by the presence of a wide and deep ocean. Thus the fauna of the Indo-Pacific province which extends from East Africa along the coasts of the Indian Ocean to the Malay Archipelago, Northern Australia and the islands of the Pacific as far as 108° W., is prevented from reaching the American coast by the deep and broad Pacific Ocean.

Some species of molluscs which live in shallow water in cold regions are found to extend to temperate or tropical regions in deeper water where a similar temperature occurs. A few species are independent of both temperature and depth; thus *Venus mesodesma* was found on the shores of New Zealand at 55° F., and was dredged in 1000 fathoms at 37° F. off Tristan da Cunha.

In the Palæozoic and Mesozoic formations gasteropods are generally less abundant than lamellibranchs, but they exceed them at the present day. The earliest forms occur in the Lower Cambrian Beds. Throughout the Palæozoic formations the holostomatous Prosobranchiates are the predominating forms; no gasteropods with a well-developed canal are known to occur until the Trias is reached, but siphonostomatous genera become fairly abundant in the

Oolites, they increase still more in the Cretaceous, and in the Tertiary they are the principal forms. The Heteropoda are represented by a few forms only, the first occurring in the Miocene. The Opisthobranchiata range from the Carboniferous to the present day; they are moderately well represented in the Jurassic and Cretaceous formations, and become more abundant in the Tertiary. Pteropods belonging to living types are found in the Upper Cretaceous and later formations, and earlier forms which may belong to this group occur in the Palæozoic. Marine forms of Pulmonata appear first in the Devonian; non-marine forms (*e.g. Anthracopupa*) are found in the Carboniferous, but are rare until the Purbeck and Wealden periods, and become abundant in the Tertiary deposits.

The most important genera of Gasteropoda found in the different systems are:

Cambrian. *Scenella, Straparollina, Ophileta, Stenotheca.*

Ordovician. *Crytolites, Sinuites, Holopœa, Raphistoma, Cyclonema, Maclurea, Subulites.*

Silurian. '*Pleurotomaria*', *Bellerophon, Poleumita, Holopœa, Cyclonema, Holopella, Platyceras.*

Devonian. '*Pleurotomaria*', *Murchisonia, Bellerophon, Loxonema, Euomphalus, Macrochilina, Capulus.*

Carboniferous. *Metoptoma*, '*Pleurotomaria*', *Murchisonia, Bellerophon, Loxonema, Euomphalus, Naticopsis, Macrochilina, Capulus.*

Permian. *Pleurotomaria, Murchisonia, Loxonema, Macrochilina.*

Trias. *Pleurotomaria, Worthenia, Murchisonia (Cheilotoma), Loxonema, Naticella, Natica, Promathildia.*

Jurassic. *Pleurotomaria, Amberleya, Cirrus, Trochus, Natica, Pseudomelania, Bourguetia, Nerinea, Cerithium, Dicroloma (=Alaria), Malaptera, Purpurina, Purpuroidea.*

Cretaceous. *Pleurotomaria, Architectonica (=Solarium), Turritella. Natica, Viviparus, Cerithium, Epitonium (=Scala), Aporrhais, Dicroloma (=Alaria), Avellana.*

Eocene. *Xenophora, Calyptræa, Natica, Melanatria, Turritella, Cerithium, Tibia (=Rostellaria), Hippochrenes, Rimella, Aporrhais, Cypræa, Cassis, Galeodea, Cymatium, Fusinus, Clavilithes, Sycostoma, Pisania, Ficus, Murex, Typhis, Voluta, Volutospina, Volutilithes, Oliva, Ancilla, Turris (=Pleurotoma), Conus, Conorbis.*

Oligocene. *Nerita, Theodoxus, Natica, Viviparus, Melania, Melanopsis, Cerithium, Potamides, Murex, Fusinus, Ancilla, Turris, Limnæa, Planorbis, Rissoa, Helix, Amphidromus.*

Miocene. *Trochus, Calyptræa, Crepidula, Ampullina, Xenophora, Turritella, Terebralia, Strombus, Ficus, Cypræa, Ocinebra, Euthriofusus, Tudicla, Conomitra, Scaphella, Ancilla. Oliva. Terebra, Conus.*

Pliocene. *Emarginula, Fissurella, Trochus, Epitonium, Turritella, Natica, Littorina, Capulus, Cerithium, Aporrhais, Trivia, Buccinum, Liomesus (=Buccinopsis), Sipho (=Tritonofusus), Chrysodomus, Nassaria, Thais, Trophon, Scaphella, Acteon.*

CLASS III. SCAPHOPODA

The Scaphopoda include only a few genera, which in some respects resemble the lamellibranchs, and in others the gasteropods. The body is elongated in an antero-posterior direction, and is bilaterally symmetrical. The mantle is nearly cylindrical, since its right and left margins are united ventrally; the mantle-cavity is open at both ends. The mantle secretes a nearly straight or slightly curved tubular shell which is also open at both ends, and gradually increases in width from the posterior to the anterior end; the concave side is dorsal. The foot is elongated and cylindrical; it can be protruded through the larger (anterior) aperture of the mantle and shell, and serves as a burrowing organ. The animal is attached to the posterior part of the shell by means of a muscle; an odontophore is present, but the head is rudimentary, and eyes, gills, and heart are absent. The sexes are separate. All the scaphopods are marine, and they usually live buried in sand or mud, with only the small posterior extremity projecting into the water; they range from the shore-line down to a depth of 2500 fathoms; only a few occur in the littoral zone, the majority being found in deeper water. The earliest forms are found in the Ordovician rocks.

Dentalium. Shell conical or sub-cylindrical, tapering posteriorly, slightly curved. Anterior aperture simple, circular; not constricted; posterior aperture smaller, without a fissure. Surface ornamented with longitudinal striæ or ribs. Eocene to present day. Ex. *D. elephantinum*, Recent. Forms closely allied to *Dentalium* occur in many Palæozoic and Mesozoic formations.

CLASS IV. CEPHALOPODA

The Cephalopods are entirely marine and are more highly organised than other molluscs; well-known living forms are the cuttle-fish, the squids, the paper-nautilus and the pearly nautilus, whilst amongst extinct types are belemnites, ammonites and goniatites. Existing forms are always bilaterally symmetrical. The head is well marked and is separated from the body by a constriction; it is especially characterised by the presence of a circle of arm-like or lobe-like processes around the mouth (fig. 134, *e*, *f*); these processes are provided either with sucking-discs or with tentacles, and are used for seizing food, and in locomotion. Behind the head is a muscular tube termed the *funnel* (*d*), which opens in front to the exterior, and behind into the mantle-cavity (*c*); this may be either a perfect tube or may be formed by the apposition of two trough-like lobes. The arms around the mouth have been regarded as part of the foot, and the funnel as representing the remainder of it. The name *Cephalopoda* is due to the view that the fore-foot has grown round the mouth and is divided up into arms or lobes.

On the upper surface of the head there are two large eyes, which, except in *Nautilus*, are almost as highly developed as in vertebrate animals. The mantle is formed by a single fold of the skin, which passes quite round the body; on the upper or dorsal surface the fold is very shallow so that the mantle-cavity exists mainly on the under surface. The feather-like gills (*p*) are placed in the mantle-cavity; in the Dibranchs (cuttle-fishes, etc.) there is one pair, in

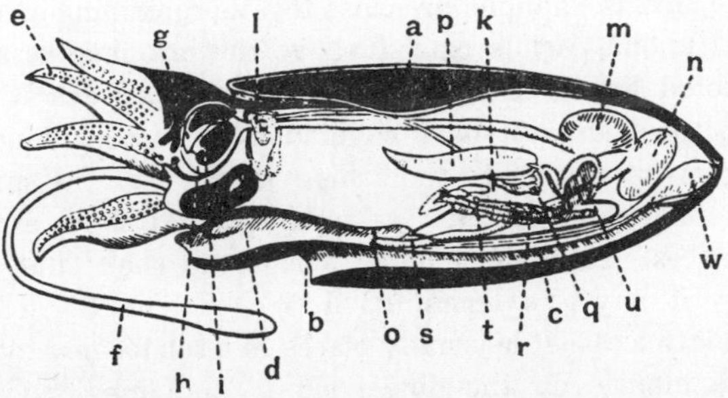

Fig. 134. Diagram of a vertical median antero-posterior section of *Sepia officinalis*. *a*, shell; *b*, mouth of mantle-cavity; *c*, mantle-cavity; *d*, funnel; *e*, arms; *f*, long arm; *g*, the upper beak or jaw; *h*, the lower beak or jaw; *i*, odontophore; *k*, the viscero-pericardial sac; *l*, the nerve-collar; *m*, the crop; *n*, the gizzard; *o*, the anus; *p*, left gill; *q*, ventricle of the heart; *r*, renal glandular mass; *s*, left nephridial aperture; *t*, viscero-pericardial aperture; *u*, branchial heart; *w*, ink-sac. (After Lankester.)

Nautilus there are two. Water flows in at the sides of the mantle-cavity, and can be forced out through the funnel by means of the contraction of the walls of the mantle-cavity. In the Dibranchiate Cephalopods there is a gland, known as the *ink-sac* (*w*), which secretes a black fluid (sepia); the duct from this gland opens with the anus (*o*) into the mantle-cavity; the ink is ejected at times and passes out through the funnel, rendering the water cloudy, and by this means facilitating the escape of the animal from its enemies. Just within the mouth there are two jaws (*g*, *h*) which have the form of a parrot's beak, and are either horny or calcareous. An odontophore (*i*) is also present, but the arrangement of the teeth is less variable than in the gasteropods, and is of little value for systematic purposes.

The heart consists of a median ventricle (*q*), and of lateral auricles, which are either two or four in number, according as there are two or four gills. The nervous system is remarkable in that the ganglia are close together, forming a central mass (*l*); one part is placed above the œsophagus, and is connected by cords with the other part beneath it. This central nervous system is covered by a cartilaginous ring and gives off nerves to the arms, viscera, etc. The sexes of the Cephalopods are always separate, and show external differences. In some genera there is no shell; but when present it may be external (fig. 135) or internal (fig. 134, *a*); in the latter case it is usually placed in a sac formed by folds of the mantle on the dorsal side. The Cephalopoda are divided into three Orders: (1) Nautiloidea, (2) Ammonoidea, (3) Dibranchia. The first and second are sometimes grouped together as the Tetrabranchia.

ORDER I. NAUTILOIDEA

In Palæozoic times the Nautiloid Cephalopods were very abundant, but at the present day the only representative of the group is *Nautilus* (fig. 135). This possesses two pairs of gills, and two pairs of auricles; no ink-sac is present; and the funnel (fig. 135, 9) is not a complete tube, but is formed of two separate parts. Around the mouth are numerous lobe-like processes which are given off from the margin of the head; these represent the arms but do not bear suckers, as is the case in the Dibranchs, but tentacles (8) which can be retracted within sheaths. The hood (fig. 135, 2) is a structure formed by the enlargement of the outer lobe of the foot and serves to close the aperture when the animal withdraws into the shell. The jaws are calcified and are not uncommonly found fossil. The eyes are of simple structure, consisting of a hollow chamber with a pin-hole opening without lens or cornea.

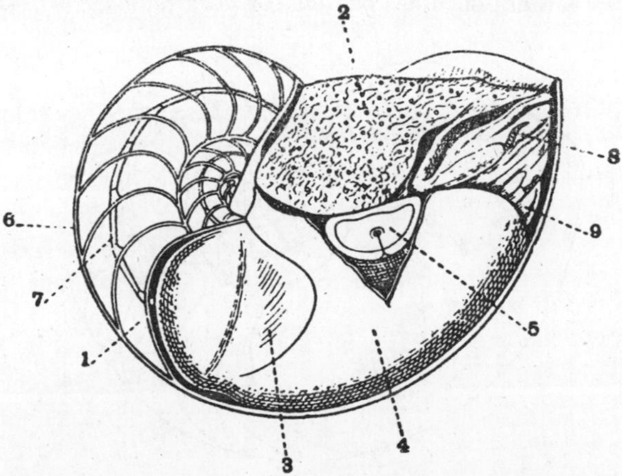

Fig. 135. ***Nautilus pompilius.*** **Half the shell has been removed. 1, last completed chamber; 2, hood part of foot; 3, shell muscle; 4, mantle, cut away to expose the eye (5); 6, outer wall of shell; 7, siphuncle; 8, tentacle-bearing lobes of foot; 9, funnel. (After Graham Kerr.)** $\times \frac{1}{2}$.

A shell is present in all Nautiloids and is always external: it consists of a tube, which tapers to a point at one end, and may be straight, arched, or spiral. In the spiral forms the whorls may be separate, or in contact throughout; commonly they are all in one plane, but in some cases they form a helicoid spiral. In some genera with spiral shells the whorls only just touch, but in others the later whorls partly or entirely overlap the earlier ones. Sometimes, as in

Lituites and *Discitoceras*, the later part of the shell becomes straight and separated from the spiral part.

The interior of the shell, unlike that in most gasteropods, is divided into a number of chambers by means of transverse partitions termed *septa* (fig. 136, *b*); generally the chambers increase in size towards the aperture of the shell. The body of the animal occupies the last or *body-chamber* (*a*), to the

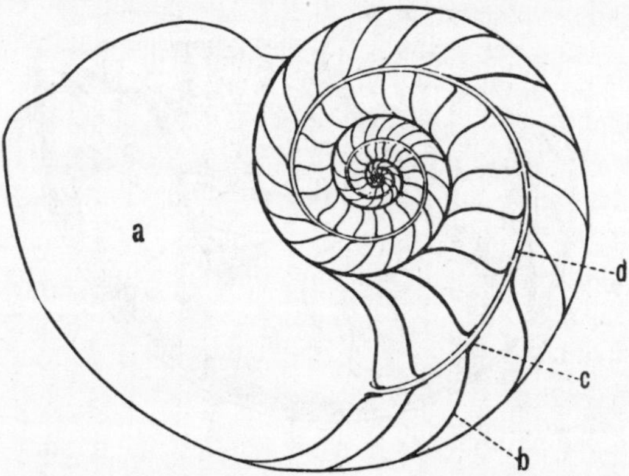

Fig. 136. Section of the shell of *Nautilus pompilius*, Recent. *a*, body-chambers; *b*, septum; *c*, septal neck; *d*, siphuncle. × ⅓.

walls of which it is attached by the mantle and the muscles (fig. 135, 3); in *Nautilus* there are two oval muscular impressions, one on each side, near the last septum and the inner side of the whorl; these impressions are marked by faint concentric lines. The muscles are connected both above and below by a band of fibres called the *annulus*, which likewise leaves a mark on the shell. In *Nautilus* the funnel (fig. 135, 9) is placed at the external margin of the aperture, so that this part of the last chamber is regarded as ventral.

All the chambers, except the body-chamber, are filled with gas, giving buoyancy to the shell. The shell grows by the addition of material at the margin of the aperture; after a certain period the body of the animal moves forward and a new septum is secreted behind it, thus cutting off a new air chamber. This movement occurs after a period of growth, and is not related, as some have supposed, to periods of reproduction, since it is only after the shell is completed that reproduction begins. In *Nautilus* the last air chamber of the completed shell is usually somewhat smaller than the preceding one (fig. 136). All the air chambers are traversed by a slender cord-like prolongation of the posterior end of the body, containing arteries, and known as the *siphuncle* (fig. 135, 7; 136, *d*; 137 *c*). The position of the siphuncle varies in different genera; in *Nautilus* it pierces the septa at or near their centres; in others it may be near either the external or the internal margin of the whorl. In the modern *Nautilus* the siphuncle has only a thin calcareous covering; but in many fossil Nautiloids it is completely invested by a calcareous tube. In some Palæozoic genera (*e.g. Actinoceras*) the interior of the siphuncle is partly filled up with calcareous deposits. The septa are often prolonged in the form of funnels around the siphuncle, so as to insheath it more or less completely; they may be short or may reach from one septum to the next or even further; these funnels are termed *septal necks* (figs. 136, *c*; 137, *d*); in nearly all the Nautiloidea they are directed backwards (retrosiphonate).

The aperture of the shell has, in some cases, a simple margin, being either straight or slightly curved; in others, processes are given off from the external margin or from the sides; in *Nautilus* there is a sinus at the external (ventral) margin where the funnel lies, and the lines of growth on the shell are correspondingly curved. In some fossil Nautiloids

(Phragmoceras) the sinus is at the inner margin of the aperture, which was therefore presumably ventral. In a few forms (e.g. Gomphoceras, fig. 140) the aperture, owing to the inward growth of the margin of the body-chamber, is constricted.

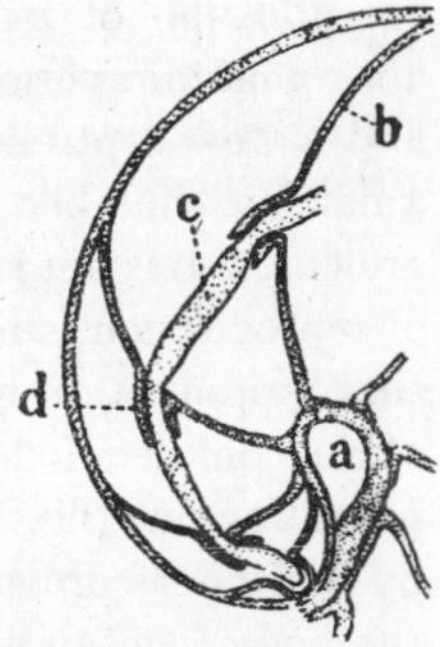

Fig. 137. Median section of the central part of the shell of *Nautilus*. *a*, central perforation; *b*, septum; *c*, siphuncle; *d*, septal neck. Enlarged.

The line where the edge of the septum unites with the outer or tubular part of the shell is known as the *suture-line*; obviously this will only be seen when the shell is removed; but fossil forms frequently occur as casts and in these the sutures are clearly shown. One of the chief characters of the shell in the Nautiloidea is the simple form of the suture-lines: usually they are either straight or only slightly undulating.

The shell which covered the embryo in the Cephalopoda is known as the *protoconch*[1]; in the Nautiloidea it varies in form in different genera and species, and may be saucer-shaped (fig. 137), cup-shaped, conical or ovoid. The siphuncle (*c*) commences in the first chamber as a closed tube.

Orthoceras. Shell straight or occasionally slightly curved, elongate-conical; transverse section usually circular. Septa concave; suture-line straight. Body-chamber large; aperture not contracted or produced into lobes. Siphuncle cylindrical, without internal calcareous deposits; usually central, but sometimes sub-central or ex-central. Ornamentation variable. *Orthoceras*, as defined above, includes numerous species which have been grouped into more restricted genera based mainly on the character of the ornamentation. Tremadoc Beds to Trias.

[1] This corresponds to the protegulum of the Brachiopods and to the prodissoconch of the Lamellibranchs.

Ex. *O. intermedium*, *O. annulatum*, Silurian; *O. undulatum*, Carboniferous Limestone.

Endoceras (fig. 138). Shell straight as in *Orthoceras*. Septal necks extend backwards from one septum to the next. Siphuncle marginal or sub-marginal, cylindrical, very large; the interior with numerous backwardly-directed cones, and an endosiphon (rarely preserved). Ordovician. Ex. *E. proteiforme*.

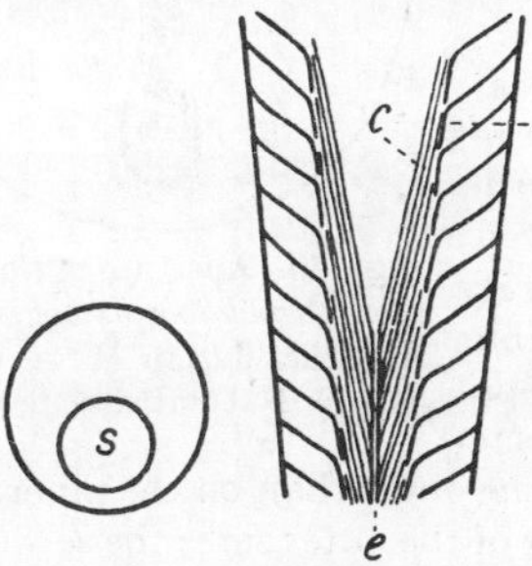

Fig. 138.

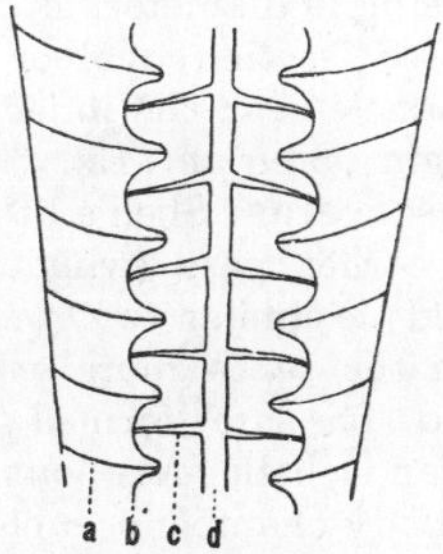

Fig. 139.

Fig. 138. ***Endoceras.*** **Diagrammatic transverse and longitudinal sections.** ***c,*** **cones;** ***e,*** **endosiphon;** ***n,*** **septal neck;** ***s,*** **siphuncle.**

Fig. 139. Diagrammatic section of a portion of the shell of ***Actinoceras.*** ***a,*** **septum;** ***b,*** **siphuncle;** ***c,*** **canals from endosiphon;** ***d,*** **endosiphon.**

Actinoceras (fig. 139). External form similar to the preceding; often very large; section usually elliptical. The siphuncle is large, and inflated between the septa so that each segment is spheroidal, and contains in the interior a large amount of calcareous deposit. In the centre of the siphuncle is a small tube known as the *endosiphon* (*d*), from which radiating tubes (*c*) are given off between the septa and pass to the siphuncle. Ordovician. Ex. *A. bigsbyi*. *Rayonnoceras* is similar to *Actinoceras*, Carboniferous. Ex. *R. giganteum*.

Gomphoceras (fig. 140). Shell ovoid, short, straight or slightly curved; section nearly circular; body-chamber very large, aperture contracted, T-shaped. Septa close together. Siphuncle sub-cylindrical or beaded, sub-central, placed nearer the more convex side of the shell. Surface smooth or with transverse ribs or striæ. *Gomphoceras* has been divided into several 'genera' based mainly on the form of the aperture. Ordovician to Devonian. Ex. *G. ellipticum*, Lower Ludlow.

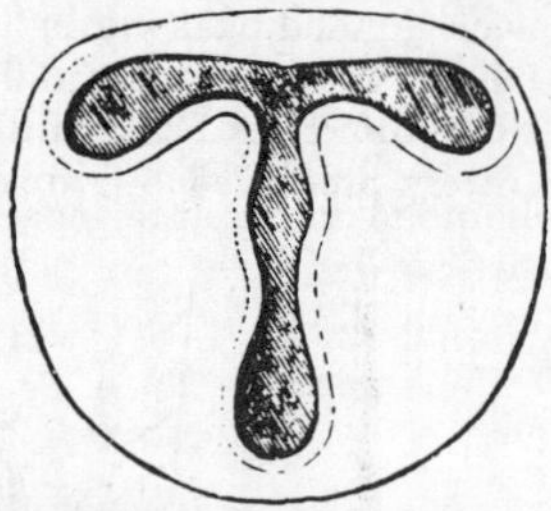

Fig. 140. Aperture of *Gomphoceras* (*Mandaloceras*) *bohemicum* from the Silurian. (From Woodward.) Natural size.

Phragmoceras. Similar to the last, but curved and rapidly increasing in diameter, laterally compressed, section oval or elliptical; siphuncle near the inner (concave) margin. Silurian. Ex. *P. broderipi.*

Ascoceras. Shell a little curved; the earlier part (which is rarely found) is similar to *Orthoceras*, but with the septa more widely separated. The later formed part is saclike and a little more convex on the outer than on the inner side; the body-chamber occupies most of the outer side; the septa join together and then bend round and divide again before reaching the inner side of the shell; the siphuncle of this part is very short. Ordovician, but chiefly Silurian. Ex. *A. bohemicum*, Silurian.

Cyrtoceras. Shell short and stout, curved, with elliptical or sub-trigonal section. Body chamber very short. Siphuncle large, beaded between the septa, usually sub-marginal and near the convex side of the shell; filled with radiating lamellæ; with an endosiphon in the adult. Devonian. Ex. *C. depressum.*

Poterioceras. Shell smooth, fusiform, slightly curved, inflated in the middle, contracted at both ends, but especially at the apical end. Section elliptical in the adult. Siphuncle subcentral or marginal, inflated between the septa. Last chamber large; aperture simple, contracted. Carboniferous. Ex. *P. fusiforme*, Carboniferous Limestone.

Cophinoceras ('*Gyroceras*'). Shell consisting of one or several whorls, coiled in one plane, either just touching or separate; highly ornamented with nodes, ridges, etc. Siphuncle nummuloidal, as in *Cyrtoceras*, usually near the convex margin. Devonian. Ex. *C. ornatum.*

Lituites. The first part of the shell coiled in a plane spiral; usually four whorls either touching or slightly separated. Th later part of the shell separates from the spiral part and is straight and long; it gradually increases in diameter towards the aperture which has two projecting lobes. Siphuncle cylindrical, sub-central. Septa concave. Ordovician. Ex. *L. lituus.*

Nautilus (figs. 135, 136, 137). Shell more or less globose, spiral, whorls few, coiled in one plane, and more or less completely embracing. Umbilicus usually small or absent. Body chamber much larger than the preceding one, aperture simple, with an external sinus. Septa concave, suture-lines more or less undulating. Siphuncle central, septal necks short and directed backwards. Surface of shell smooth or ornamented with striæ or ribs. Trias to present day. Ex. *N. pompilius*, Recent; *N. regalis*, London Clay. The Mesozoic species are now regarded as belonging to separate genera.

Vestinautilus. Shell spiral; whorls thick, partly overlapping, sub-hexagonal in section; external margin broad, rounded or concave, with a ridge or ridges on each side. Umbilicus large. Siphuncle sub-central. Sutures with a backwardly-directed sinus at the external margin. Carboniferous. Ex. *V. cariniferus.*

Discitoceras (=*Discites*). Shell spiral, compressed, discoidal; whorls quadrangular in section, increasing in size gradually, sometimes a little embracing, with the external margin flat or grooved, and the sides flattened. The last part of the shell is separated from the preceding whorl for a short distance. The earlier whorls have longitudinal ribs. Carboniferous Limestone. Ex. *D. leveilleanus.* Other Carboniferous genera with plane spiral shells are *Cœlonautilus*, *Temnocheilus.*

Aturia. Shell discoidal, whorls compressed, completely embracing, with rounded external margin; suture-line zigzag, with a deep angular lobe on each side. Siphuncle near the internal margin; septal necks large and very long, completely covering the siphuncle. Eocene and Miocene. Ex. *A. zic-zac*, Eocene.

Distribution of the Nautiloidea

At the present day the Nautiloidea are represented by only four species of *Nautilus*, which are found in the Indian Ocean and the East Indian Archipelago (from Sumatra to Fiji). *Nautilus* lives in fairly shallow water, either crawling on the sea bottom by means of its tentacles or swimming, but sometimes rising to the surface of the sea. Most of the extinct Nautiloids probably led a similar benthonic mode of life.

This group appears much earlier in the geological series than either the Ammonoidea or the Dibranchia. The early forms are slightly curved cones, subsequently straight and spiral shells appear; but the straight or nearly straight forms predominate in the Palæozoic and Trias.

Volborthella and *Salterella*, found in the Lower Cambrian, are regarded by some authors as primitive Nautiloids, and by others as belonging to the group which includes *Hyolithes* (p. 293). Curved forms, resembling *Cyrtoceras* externally, appear in the Upper Cambrian. In the Ordovician the Nautiloidea increase in importance (*Endoceras*, *Piloceras*, *Lituites*, etc.), and the group attains its maximum development in the Silurian, where the number of species is very great (*Orthoceras*, *Gomphoceras*, *Phragmoceras*, *Ascoceras*, and the helicoid genus *Trochoceras*); it decreases slightly in importance in the Devonian and Carboniferous, and is but poorly represented in the Permian. The chief genera in the Carboniferous are *Orthoceras*, *Rayonnoceras*, *Poterioceras*, *Discitoceras*, *Vestinautilus*, *Cœlonautilus*, *Pleuronautilus*, *Temnocheilus*. The only genus which extends beyond the limit of the Palæozoic period is *Orthoceras*, which is common in the Trias. *Nautilus* occurs first in the Trias, and is abundant in the Jurassic and Cretaceous, when its distribution seems to have been world-wide; in the Tertiary it is relatively rare. *Aturia* appears in the Eocene and Miocene.

ORDER II. AMMONOIDEA

The Ammonoids are quite extinct and include the ammonites, goniatites, etc. The shell is generally coiled into a plane spiral, and as a rule the suture-lines show complicated patterns (fig. 141). The siphuncle is at the margin of the shell—generally at the outer, but in one Upper Devonian group (the Clymeniidæ) at the inner margin; it is usually more slender than in the Nautiloids and does not contain internal calcareous deposits. The septal necks in the ammonites are directed forwards (prosiphonate), except in some of the earliest chambers; in the Clymeniidæ and some goniatites, on the other hand, they point backwards as in the Nautiloids (retrosiphonate); but in the more advanced types of goniatites they are transitional, a small collar-like part projects in front of the septum, but the main part of the septal neck extends backwards.

The form of the suture-lines varies considerably in different genera and is of great importance for systematic purposes. The central part of each septum is flattened or slightly undulose, but the edges become folded or even frilled, often giving rise to very complex suture-lines; by this means greater support is afforded to the outer tubular part of the shell than is the case in the Nautiloids where the sutures are simple. The portions of the suture-line which are convex towards the mouth of the shell are termed *saddles* (fig. 141, *s*), while the intervening concave portions are known as the *lobes* (*l*). In many forms the lobes and saddles exhibit secondary foldings, which may be slight, producing merely a denticulate pattern, or may be deep and provided with other smaller foldings, giving a foliaceous appearance to the suture. The lobes and saddles are nearly always similar on the two sides of the shell: commonly there is first the *external lobe* (fig. 141, *l*) at the external margin, then the *superior* and *inferior* (or *first* and *second*) *lateral lobes* on the sides of the whorl (1 *l*, 2 *l*), and near the inner margin other lobes known as *auxiliary lobes* (*al*) may occur;

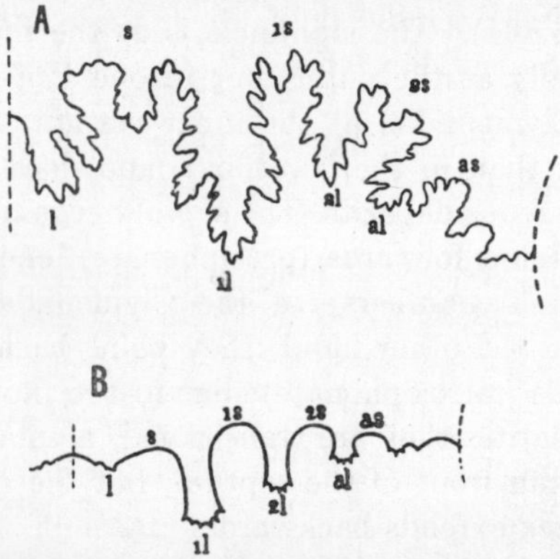

Fig. 141. A. Suture-line of an Ammonite (*Parkinsonia dorsetensis*) from the Inferior Oolite. B. Suture-line of *Ceratites nodosus*, from the Muschelkalk. *l*, one half of the external lobe; 1*l*, 2*l*, superior and inferior lateral lobes; *al*, first auxiliary lobe; *s*, external saddle; 1*s*, 2*s*, superior and inferior lateral saddles; *as*, first auxiliary saddle. In each case the straight line on the left represents the position of the siphuncle at the external margin, and the curved one on the right the line of contact with the next whorl (= umbilical suture).

on the internal margin (opposite to the external lobe) is the *internal lobe*. The saddles are arranged in a similar manner; there are the *external saddle* (*s*), the *lateral saddles* (1 *s*, 2 *s*), and *auxiliary saddles* (*as*). The external lobe is often divided by a median (*siphonal*) saddle (as in fig. 142 B). Sometimes other lobes and saddles (termed *adventitious*) are formed by the subdivision of the external saddle (fig. 153) *or* of the median (*siphonal*) saddle in the external lobe (fig. 142 A, *E*) *or* of the lateral lobe (fig. 142 B, *Ad* 1, 2). With a few exceptions the suture lines of one side of the shell are symmetrical with those of the other side.

There is a relationship between the number of lobes and saddles in the suture-line and the form of the whorl. When the whorls only just touch and are circular or oval in section (fig. 156) usually only the fundamental lobes and saddles are present, but when the whorls overlap (fig. 154)

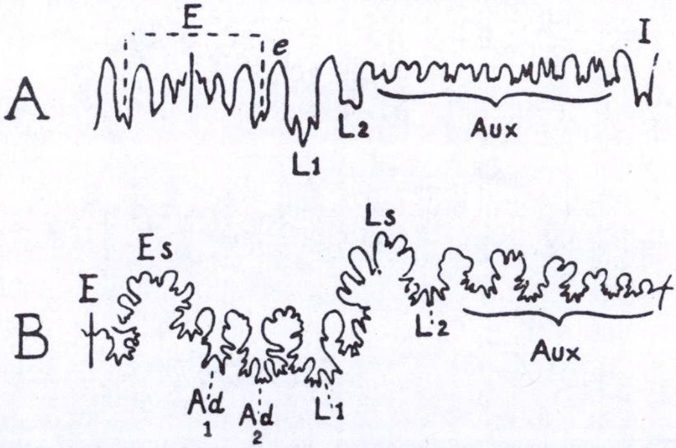

Fig. 142. A. *Pseudosageceras multilobatum*, Trias. (After Nötling.) *E*, external lobe with median saddle divided; *e*, external saddle; *I*, internal lobe; *L* 1, *L* 2, first and second lateral lobes; *Aux*, auxiliary lobes. B. *Coilopoceras springeri*, Chalk. (After Hyatt.) *E*, external lobe; *Es*, external saddle; *Ls*, lateral saddle; *Ad* 1, 2, adventitious lobes; *L* 1, *L* 2, first and second lateral lobes; *Aux*, auxiliary lobes.

there is a tendency to increase the number by the development of auxiliary and sometimes also adventitious lobes and saddles.

The naming of the lobes and saddles according to their position in the adult shell appears to be less satisfactory than a terminology based on the study of the development of the suture-line during ontogeny. From the figures showing the development of the suture-line of an Hoplitid ammonite (fig. 143) it will be seen that the first lateral lobe (*L*) is always a fundamental part of the suture-line (fig. 143, *a*); but the lobe which is formed later at the umbilical suture (fig. 143, *b*, 1) subsequently gives rise by subdivision to a number of lobes (2–8) on the external as well as on the internal side of the umbilical suture. When uncoiling of the shell occurs, as in *Baculites* (fig. 163), the elements of the

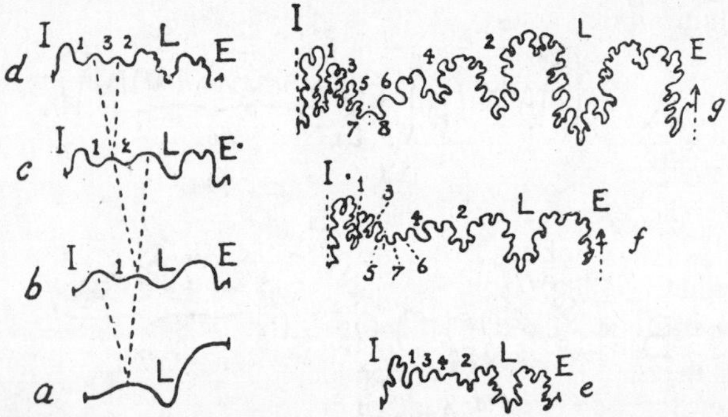

Fig. 143. Development of the suture-line in a Hoplitid Ammonite from the Gault. *a, b*, first and second suture-lines; *c—g*, later sutures; *E*, external lobe; *I*, internal lobe; *L*, lateral lobe; *b*, 1, umbilical lobe; 2—8, lobes resulting from the division of 1. (After Spath.)

suture-line remain the same throughout life, namely, the external lobe, first lateral lobe, second lateral lobe, internal lobe. (Fig. 164, *E*, *L* 1, *L* 2, *I*.)

The form of the successive suture-lines remains almost constant on the adult part of the shell, but on the younger parts the sutures are less complex. The suture-line of the first septum may be straight or only slightly curved, as in the Nautiloids; or it may show a broad external saddle; or a narrow external saddle with a lateral lobe on each side. In the second and later septa the sutures become successively more folded, until the adult form is attained. The development of the suture-line in a typical goniatite is shown in fig. 144.

In the early Ammonoids the suture-lines are comparatively simple (fig. 145); the minor divisions of the lobes and saddles begin to appear in some Carboniferous genera; they increase in importance in the Permian and attain a great development in the Trias where the ammonites commonly possess very complex sutures. The acme of complication is reached in the Upper Triassic genus *Pinacoceras* (fig. 153). The gradual advance in the complexity of the suture-line is shown in a series beginning with *Anthracoceras* from the Carboniferous and ending with *Ussuria* from the Trias (fig. 146).

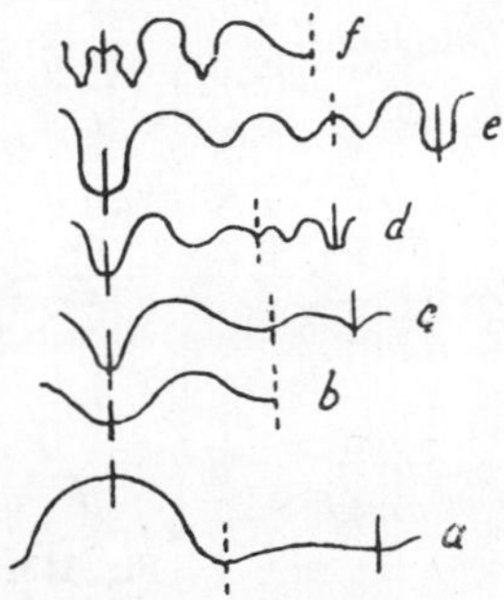

Fig. 144. Development of the suture-line of a Goniatite, *Homoceras diadema*, Carboniferous. *a*, *b*, *c*, first, second and third suture-lines; *d*, *e*, later suture-lines; *f*, adult suture-line. (After Branco.)

The protoconch of the Ammonoids is formed of calcareous material; in a few of the straight or loosely coiled genera it is spherical or ovoid in shape, but in most genera with closely coiled shells it shows a convolute mode of growth and becomes barrel-shaped (fig. 147, *a*). The first septum closes the aperture of the protoconch. The siphuncle (*b*) commences with a bulbous enlargement (the *cæcum*) which projects into the protoconch. The cæcum is attached to the opposite side of the protoconch by strands (the *prosiphon*). In the first few chambers the siphuncle is variable in position and relatively large, but afterwards it gets gradually nearer the external margin of the whorl and becomes relatively smaller.

In the body-chamber of some ammonites and goniatites

and in *Baculites* and *Scaphites*, a pair of calcareous plates, known as the *aptychus* (fig. 148), are occasionally found; in shape they are triangular or nearly semicircular; the

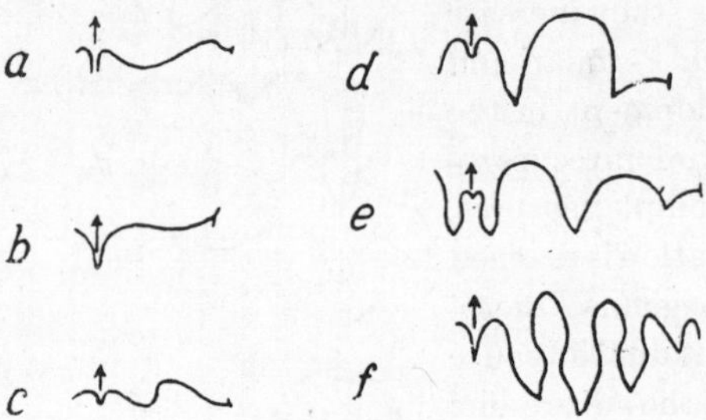

Fig. 145.

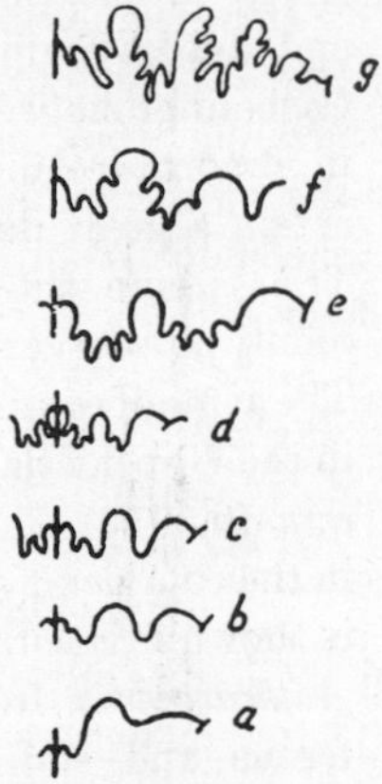

Fig. 146.

Fig. 145. Suture-lines of Goniatitids. The arrow is at the external (siphonal) margin and points towards the aperture of the shell. *a*, *Agoniatites bohemicus*, Devonian; *b*, *Anarcestes plebeius*, Devonian; *c*, *Tornoceras uniangulare*, Devonian; *d*, *Manticoceras intumescens*, Devonian; *e*, *Gāstrioceras listeri*, Coal Measures; *f*, *Prolecanites compressus*, Carboniferous. (After Barrande, Sandberger and Crick.)

Fig. 146. Development of the suture-lines in a series of genera. *a*, *Anthracoceras discus*, Carboniferous; *b*, *c*, *Dimorphoceras gilbertsoni*, Carboniferous; *d*, *Dimorphoceras looneyi*, Carboniferous; *e*, *Thalassoceras gemmellaroi*, Lower Permian; *f*, *Thalassoceras varicosum*, Lower Permian; *g*, *Useuria schamaræ*, Lower Trias. (After Spath.)

margins where the two plates are in contact are straight, the others curved. Since in one ammonite an aptychus was found closing the aperture of the shell, it is probable that it served as an operculum (p. 278) and was attached to a part of the body representing the hood of *Nautilus* (fig.

135, 2). A similar structure, but consisting of chitin, and with the two plates united, is found in the body-chamber of some ammonites.

In a few Ammonoids the shell is either a straight cone (*e.g. Lobobactrites*) or coiled into a helicoid spiral (*e.g. Turrilites*, fig. 165), but in the great majority of the genera all the whorls are in a plane spiral, and in such the form of the shell depends mainly on whether the later whorls grow round

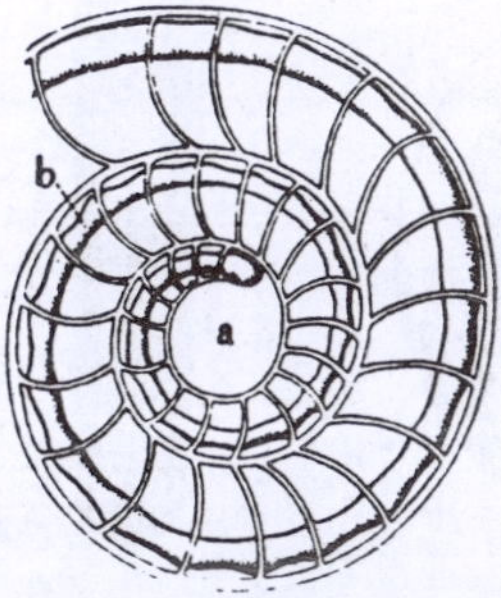

Fig. 147.

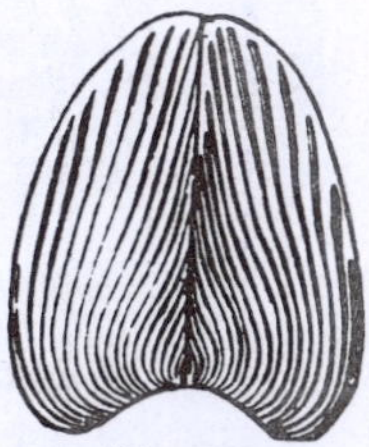

Fig. 148.

Fig. 147. Section, just above the median plane, of the early part of an ammonite—*Promicroceras planicosta*, Lias. *a*, protoconch; *b*, siphuncle. (After Branco.) × 21.

Fig. 148. Aptychus of an ammonite, from the Oxford Clay. (From Woodward.)

the earlier, or are simply in contact with them or slightly separated; in some genera the last whorl partly (fig. 160) or completely (fig. 154) conceals all the previous ones, but in others (fig. 156) the whole of the whorls are visible, and then the umbilicus—which is present on both sides of the shell—is very large. When the diameter of the whorl from side to side (*i.e.* the *thickness*) is greater than the diameter from the internal to the external margin (*i.e.* the *height*) the whorl is said to be *depressed*; the umbilicus is then deep, and if in such cases the later whorls embrace the earlier, then the umbilicus will be both deep and narrow. The whorl is *compressed* when its height is greater than its thickness (fig. 149). In some ammonites the outer whorl begins to uncoil, and this leads on to the *scaphitoid* coiling

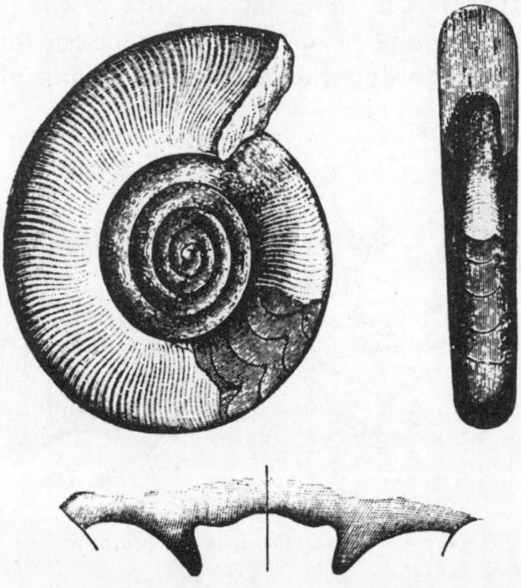

Fig. 149. *Oxyclymenia undulata*, Upper Devonian. The lower figure shows the form of the suture-line; the vertical line indicates the position of the external margin of the whorl; the curved line is the umbilical suture. (After Nicholson.)

in which the body-chamber is quite free (fig. 161), and to *hamitoid* when the shell is bent upon itself in crozier-hook fashion. The extreme of uncoiling is reached in *Baculites* (fig. 163) in which only a small part of the shell is spiral, the remainder being straight.

The surface of the shell may be smooth or ornamented with striæ, ribs, tubercles, or spines; as a rule the ornamentation is much more developed in Mesozoic than in Palæozoic genera. In some ammonites (fig. 158) the external margin of the shell is provided with a ridge or *keel*, and in these forms the ribs of the two sides are not continuous. The keel may be smooth or toothed. In some genera there is either a groove or a flattened margin in place of the keel. The aperture of the shell in the ammonites is frequently produced into lobes at the sides, or into a pointed projection at the external margin (figs. 158, 159). In some of the goniatites there is a sinus at the external margin of the aperture indicating the position of the funnel, but this disappears in the more advanced types, and, with few exceptions, is absent in ammonites.

The character of the ornamentation, the shape of the whorls, the position of the siphuncle, and the form of the sutures, change at different periods in the life of the individual; these changes, which occurred during growth from the protoconch up to the adult, can be traced out by examining the early whorls of the shell. From a study of this development of the individual (ontogeny) attempts have been made to trace out the phylogeny of various types of Ammonoids. As in the case of the Brachiopoda (p. 197) it has been found that some forms, which in the adult state appear to be nearly identical, differ in their development, indicating that they have descended from different ancestors. Similarly, the development of the suture-lines and other features of the shell show that the ammonites have descended from more than one group of goniatites.

Since the Ammonoids are extinct and their soft parts unknown it is impossible to determine whether they possessed one pair of gills as in the Dibranchia or two pairs as in *Nautilus* (Tetrabranchia). As, however, the shell was external and agrees closely in structure with that of the

Nautiloids, and the muscular impressions in the body-chamber are similar to those found in *Nautilus*, we may regard the Ammonoids as closely allied to the Nautiloidea; this view of their relationship receives further support from the resemblance shown by the early Ammonoids (in their relatively simple sutures, backwardly-directed septal necks, the sinus at the external margin of the aperture, etc.) to the Nautiloids. The protoconch of the Ammonoids, however, resembles that of the Belemnites as much as that of *Orthoceras*, but differs from that of *Nautilus*.

A striking feature of the majority of the Ammonites is the bilateral symmetry of the shell. This is characteristic of animals which live with the median plane of the body in a vertical position. Such a position would be likely to follow from the hydrostatic effect of the air chambers; these in life would be uppermost and the body-chamber below. The vertical position and bilateral symmetry are more likely to be retained in swimming molluscs than in those which live crawling on the sea floor. A swimming mode of life would be favoured by the buoyant influence of the air chambers, and those with flattened shells and acute margins would experience less resistance in moving through water than those with convex whorls, broad rounded external margin, and strong ribs. It seems probable that most of the Ammonites were nectonic and lived at no great distance from land, *i.e.* on the continental shelf; they may be abundant even in very shallow water deposits. Some genera, however, such as *Phylloceras*, *Lytoceras* and *Arcestes*, have extremely thin shells, comparable with those of pteropods and other pelagic molluscs. These seem to have been still better adapted for swimming and floating, and may well have led a semi-pelagic existence.

The different modes of coiling seen in such genera as

Crioceras, *Turrilites* and *Baculites*, were probably related to the adoption of a more distinctly benthonic mode of life. This benthonic habit may also account for the partial loss of symmetry in a small number of Ammonites; in these the siphuncle may be shifted from the median plane, and the suture-line of one side of the shell not quite symmetrical with that of the other. In *Nipponites*, from the Upper Chalk of Japan, the early whorls are coiled like *Turrilites* but the later whorls are quite irregular; it has been suggested that this was a fixed form, like *Vermetus* among gasteropods.

The wide geographical distribution of the species of Ammonites accords with a swimming mode of life; but, with an external and chambered shell it is not likely that Ammonites were capable of such rapid motion through water as are the Dibranchs.

A. *Clymeniids.* Siphuncle at the internal margin.

Clymenia. Shell discoidal, with wide shallow umbilicus; whorls numerous, more or less flattened, all visible, but each partly embracing the preceding one; body-chamber generally occupying three-quarters of the last whorl. Aperture with a sinus at the external margin. Suture-lines with a simple rounded lateral lobe, a narrow lobe at the internal margin (below the siphuncle) and a broad saddle at the external margin. Siphuncle on the internal margin; septal necks directed backwards. Surface usually ornamented with transverse growth-lines. Upper Devonian. Ex. *C. lævigata.* Other genera of the Clymeniids, distinguished mainly by the character of the suture-line, are *Cyrtoclymenia*, *Oxyclymenia* (fig. 149), *Cymaclymenia*, *Gonioclymenia*, etc.

B. *Goniatitids.* Siphuncle at the external margin. Septal necks usually directed backwards, but sometimes partly or entirely forward. Suture-line relatively simple. This group comprises forms commonly known as 'goniatites'.

Anarcestes (fig. 145 *b*). Shell with a wide umbilicus, and broad, rounded external margin. Body-chamber long; aperture with a deep external sinus. Sutures very simple; the external lobe funnel-shaped and not divided by a saddle; lateral lobe very flat. Septal necks long, directed backwards. Lower and Middle Devonian. Ex. *A. plebeius.*

Gyroceratites. Shell discoidal; the early whorls not in contact, the later ones contiguous. Siphuncle at the external margin. Sutures very simple, concave on the sides of the whorls, with a funnel-shaped external lobe. Devonian. Ex. *G. gracilis* (= *Mimoceras compressum*).

Agoniatites (fig. 145 *a*). Whorls elevated, increasing in height very rapidly, with the external margin more or less truncated. Umbilicus narrow or of moderate width. Body chamber ½ to ⅔ of the length of the last whorl. Septal necks directed backwards. In the suture-line a narrow external lobe, and a lateral lobe usually rather deep. Lower and Middle Devonian. Ex. *A. expansus.*

Manticoceras (= *Gephuroceras*) (fig. 145 *d*). Shell with or without umbilicus; external margin rounded. Body chamber short; with a sinus at the external margin of the aperture. Septal necks short, directed forwards. In the suture-line a deep lateral lobe, a wide saddle, and a broad external lobe divided by a median saddle. Upper Devonian. Ex. *M. intumescens.*

Tornoceras (fig. 145 *c*). Shell smooth, with rounded external margin; umbilicus closed. Suture-line with the external and lateral saddles rounded and undivided; the siphonal lobe small and funnel-like; lateral lobes angular or rounded. Upper Devonian. Ex. *T. simplex.*

Goniatites (restricted) (figs. 150, 151). Shell smooth or striated, whorls generally wide and embracing, with rounded external margin; umbilicus small or closed. Septal necks short, directed backwards but usually also with a small part projecting forwards. External lobe divided by a small saddle; external saddle narrow; lateral lobe angular and deep; lateral saddle broad, rounded, and undivided. Carboniferous Limestone. Ex. *G. sphæricus.*

Gastrioceras (fig. 145 *e*). Shell with tubercles at the margin of the umbilicus. External margin broad, rounded. External lobe broad and deep, divided by a saddle; first lateral lobe deep, tongue-shaped, angular; second lateral lobe small, angular; saddles rounded. Coal Measures. Ex. *G. listeri.*

Prolecanites (fig. 145 *f*). Shell smooth or striated, flattened, with a large umbilicus. Lobes and saddles numerous. External lobe not divided; two or three lateral lobes, sharp; lateral saddles narrow and rounded. Carboniferous Limestone. Ex. *P. mixolobus, P.* (*Merocanites*) *compressus.*

Fig. 150.

Fig. 151.

Fig. 150. *Goniatites sphæricus,* from the Carboniferous Limestone. The shell has been dissolved exposing the sutures. Side view.

Fig. 151. Front view showing the siphuncle at the external (upper) margin. (From Woodward.) × $\frac{2}{3}$.

In all the other genera (ammonites and uncoiled ammonitids) the siphuncle is at the external margin, the septal necks are directed forward in the adult, and the suture-lines are more complex than in the goniatitids. These include a very large number of genera of which only a few representative forms can be given here. For convenience they may be divided into (C) Triassic Ammonites, (D) Jurassic and Cretaceous Ammonites, (E) Uncoiled Ammonitids.

C. Triassic Ammonites.

Ceratites (figs. 141 B, 152). Shell discoidal; on the sides are ribs which often bear tubercles near the umbilical and external margins; external margin broad, convex or flattened; umbilicus moderately large; body-chamber short. Saddles rounded, lobes denticulate; external lobe broad and short. Trias (Muschelkalk). Ex. *C. nodosus.*

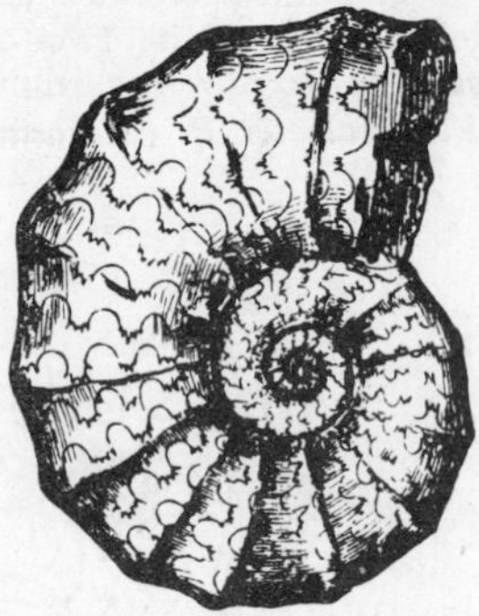

Fig. 152. *Ceratites nodosus*, from the Muschelkalk. The shell has been removed, exposing the sutures. × ⅔.

Trachyceras. Shell flattened, highly ornamented with ribs which bear tubercles or spines arranged in spiral rows; at the external margin is a groove; umbilicus generally small. Body-chamber short. Sutures simple, lobes and saddles toothed. Trias. Ex. *T. aon.*

Arcestes. Shell smooth or striated, nearly globular, with thick whorls; umbilicus small or closed; body-chamber very long. Lobes and saddles numerous and foliaceous, arranged in a straight row and gradually decreasing in size from the external to the internal margin; there are two lateral lobes and many auxiliary lobes; saddles with narrow stems and fine branches. Trias. Ex. *A. intuslabiatus.*

Pinacoceras. Shell large with small umbilicus and very acute external margin. Suture-line (fig. 153) with numerous adventitious and auxiliary lobes and saddles. Upper Trias. Ex. *P. metternichi.*

D. Jurassic and Cretaceous Ammonites.

Phylloceras (figs. 154, 155). Shell smooth or with fine striæ or gentle folds, never with tubercles; external margin rounded; umbilicus very small or closed. Saddles and lobes numerous; saddles divided, the extremities being rounded Jurassic to Cretaceous. Ex. *P. heterophyllum*, Upper Lias.

Lytoceras (fig. 156). Shell smooth or ornamented with transverse striæ, and often with laminar projections (varices)

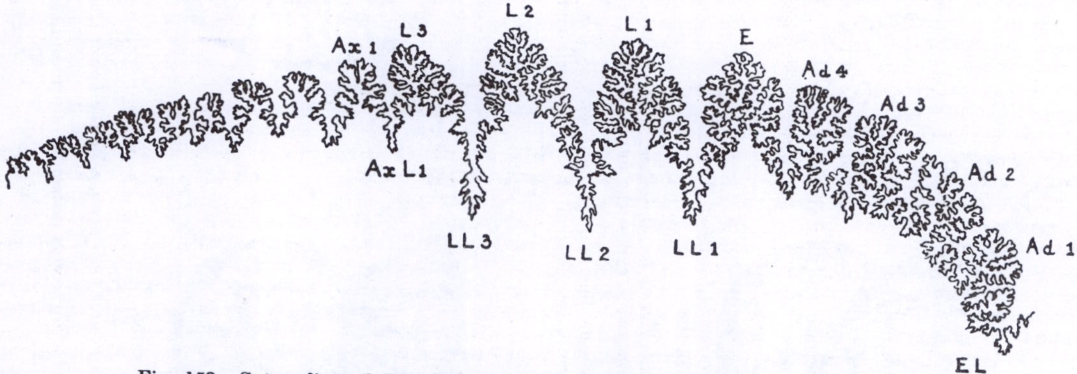

Fig. 153. Suture-line of *Pinacoceras metternichi*, Upper Trias. *EL*, external lobe; *LL* 1, 2, 3, lateral lobes; *Ax L* 1, first auxiliary lobe; *Ad* 1–4, adventitious saddles; *E*, external saddle; *L* 1–3, lateral saddles; *Ax* 1, first auxiliary saddle. (After Diener.)

Fig. 154.

Fig. 155.

Fig. 154. *Phylloceras heterophyllum*, from the Lias. A part of the shell has been removed to expose the sutures. $\times \frac{1}{4}$.

Fig. 155. Suture-line of *Phylloceras heterophyllum*, from the Lias. The arrow indicates the position of the siphuncle and points towards the aperture of the shell. (From Woodward.) Natural size.

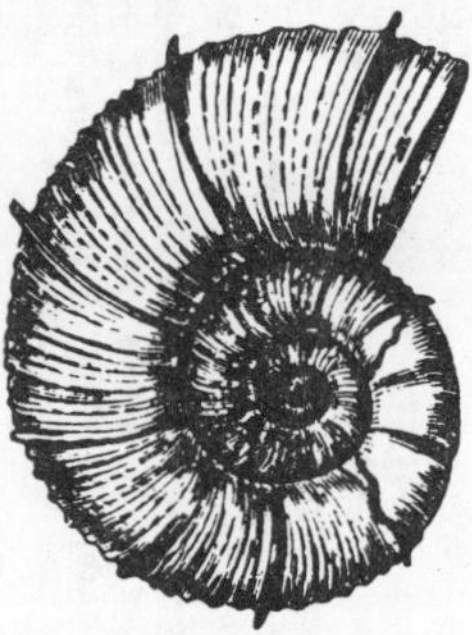

Fig. 156.

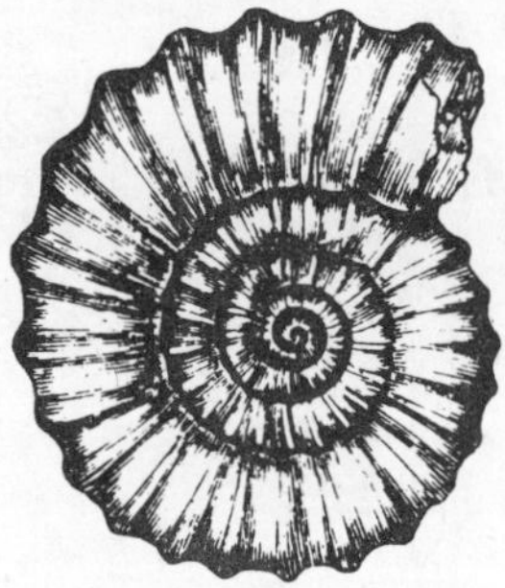

Fig. 157.

Fig. 156. *Lytoceras fimbriatum*, from the Lower Lias. $\times \frac{1}{4}$.

Fig. 157. *Androgynoceras maculatum*, from the Lower Lias. $\times \frac{1}{2}$.

placed at intervals. Whorls rounded, and only slightly or not at all embracing; aperture usually simple. Suture-line deeply and finely divided, consisting of an external lobe, two lateral lobes, and a narrow internal lobe; of an external saddle and two lateral saddles. Lateral lobes and saddles nearly symmetrically divided. Lias to Cretaceous. Ex. *L. fimbriatum*, Lower Lias.

Psiloceras. Shell discoidal, umbilicus large; whorls increasing in size very slowly, external border rounded; surface smooth or striated, occasionally with ribs. Sutures not much divided. Lower Lias. Ex. *P. planorbis*.

Schlotheimia. Shell flat, discoidal, umbilicus usually large. Ribs strong, curved, often bifurcated in the adult, bending forward at the external margin, where they meet at an angle, but are often interrupted by a slight furrow or smooth band at the margin. Sutures deeply divided; superior lateral lobe generally deeper than the external lobe. Lower Lias. Ex. *S. angulata*.

Arietites. Shell discoidal, umbilicus large; whorls numerous, only slightly embracing, with the external border flattened and provided with a keel having a groove on each side of it. Surface with strong simple ribs, which are straight or bent near the margin. Body-chamber occupying from one to one and a quarter whorls. Suture-line not much divided, with two lateral lobes and one auxiliary lobe. Lower Lias. Ex. *A. turneri*.

Oxynoticeras. Shell much flattened; umbilicus small; external margin sharp or keeled; surface smooth or striated. Suture-line not deeply divided; external saddle large, divided; auxiliary lobes present. Lower Lias. Ex. *O. oxynotum*.

Androgynoceras (fig. 157). Shell discoidal, umbilicus large; whorls rounded, ornamented with simple ribs which are continuous over the external margin. Lobes moderately divided; superior lateral lobe larger than the others. Lower Lias. Ex. *A. maculatum*.

Amaltheus. Shell flattened; with a keel, which is toothed or rope-like; umbilicus generally small; surface smooth, or with striæ, or simple or spiny ribs. Aperture with a long process at the external margin. Lobes and saddles deep and much divided, with several auxiliary lobes. Middle Lias. Ex. *A. margaritatus*.

Harpoceras (fig. 158). Shell flattened, with a prominent even keel; umbilicus small or of moderate size. Sides of shell with sickle-shaped undivided ribs. Aperture with projections. Suture-line moderately strongly divided; superior lateral lobe deep. Upper Lias. Ex. *H. serpentinum.*

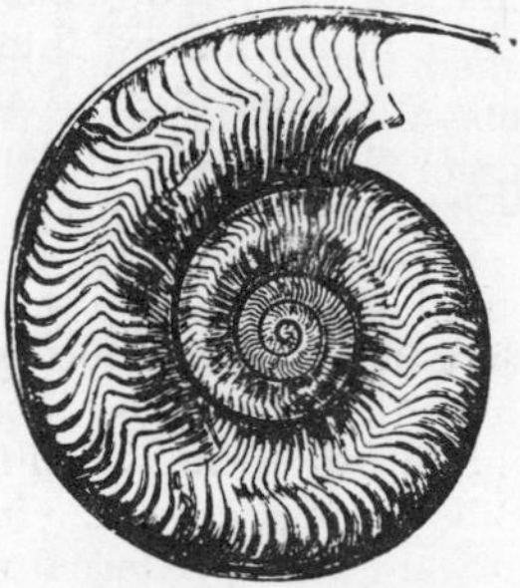

Fig. 158. *Harpoceras serpentinum*, from the Upper Lias. × $\frac{1}{6}$.

Hildoceras. Similar to the last, but with wide umbilicus. Whorls low, subquadrate in section, with broad external margin and usually a deep furrow on each side of the keel and on the sides of the whorl. Ribs distinctly sickle-shaped in most cases. Upper Lias. Ex. *H. bifrons.*

Dactylioceras. Whorls numerous, rounded, only a little embracing; umbilicus large. Ribs numerous, at first straight, afterwards bifurcating, continued over the external margin; without tubercles. Body-chamber long. Suture-line moderately divided; external lobe larger than the superior lateral. Upper Lias. Ex. *D. commune.*

Ludwigia. Shell discoidal, with fairly wide to comparatively narrow umbilicus. External margin flattened or rounded, with a median keel. Ribs very angular, sub-dividing, and strongly reclined on the outer half of the side of the whorl; starting singly from elongated umbilical nodes. Suture-line comparatively simple, with bifid external saddle. Inferior Oolite. Ex. *L. murchisonæ.*

Sonninia. Shell discoidal, with fairly wide to comparatively narrow umbilicus. Whorls high, with oval section and hollow keel; spines at the middle of the side, with or without ribs; the ribs often bifurcating on the outer half of the side of the whorl. Suture-line highly frilled. Inferior Oolite. Ex. *S. propinquans.*

Oppelia. Shell discoidal, much compressed, with small umbilicus; external margin acute, but sometimes rounded on the body-chamber. Ribs sickle-shaped, generally feeble and often confined to crescents on the outer part of the side of the whorl. Suture-line fairly complex, with prominent lateral saddle, and generally numerous lobes and saddles. Inferior Oolite. Ex. *O. subradiata.*

Stephanoceras (fig. 159). Whorls thick, external margin rounded. Umbilicus large. Surface with straight ribs, which bifurcate on the sides of the whorls and often bear tubercles where they bifurcate. Body-chamber long. Aperture often with lobe-like projections. Suture-line deeply divided; external lobe large; inferior lateral lobe and auxiliary lobes small. Inferior Oolite. Ex. *S. humphriesianum.*

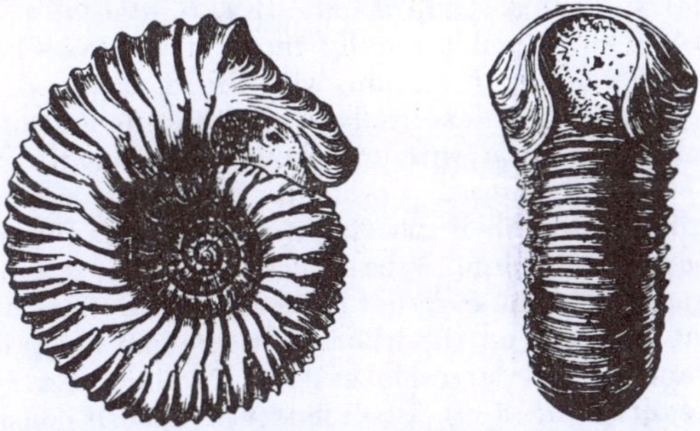

Fig. 159. *Stephanoceras* (*Normannites*) *braikenridgei*, from the Inferior Oolite, showing lappets of aperture. × $\frac{2}{3}$.

Parkinsonia (fig. 141 A). Shell discoidal, umbilicus large; ribs nearly straight, sharp, bifurcating near the external border, but interrupted at the external margin by a groove. Aperture with processes from the sides. Suture-line much divided; external and superior lateral lobes deep; saddles broad. Inferior Oolite. Ex. *P. parkinsoni.*

Macrocephalites. Shell inflated, whorls largely embracing, external margin rounded, umbilicus small. Ribs numerous, dividing near the umbilicus, continued over the external margin, and without tubercles. Suture-line deeply divided. Cornbrash and Kellaways Clay. Ex. *M. macrocephalus.*

Kosmoceras. Shell flattened; umbilicus moderately large. Ribs numerous, bifurcating at the middle of the sides of the shell, where there is generally a row of tubercles, and ending in a tubercle or spine at the external margin; sometimes with tubercles at the edge of the umbilicus. Aperture with long lateral projections (apophyses). Suture-line much divided; the external lobe much shorter than the superior lateral lobe. Oxford Clay. Ex. *K. rotundum*.

Peltoceras. Umbilicus large; whorls generally quadrate, with broad external margin, the inner whorls with numerous continuous ribs, most of which divide near the external margin, across which they extend; the ribs on the later whorls with two rows of tubercles on the sides, one near the outer, the other near the inner margin. Suture-line moderately divided, with large external saddle. Oxford Clay. Ex. *P. athleta*.

Aspidoceras. Shell generally more or less loosely coiled. Whorls sub-quadrate in section, with broad flattened external margin. Spines at the external part of the side and often also at the umbilical margin; with or without connecting ribs. Suture-line comparatively simple. Corallian. Ex. *A. perarmatum*.

Perisphinctes. Shell discoidal, external margin rounded; umbilicus generally large. Ribs straight, continuous, bifurcating once or more near the external border. Constrictions are often present at intervals on the whorls. Suture-line much divided; external and superior lateral lobes large. Corallian. Ex. *P. biplex*.

Cardioceras (fig. 160). Shell flattened; whorls considerably embracing, with strong curved ribs which bifurcate, and bending forward near the external edge join the notched keel; short ribs often intercalated on the external part. Lobes and saddles moderately divided; two short lateral lobes, and two or three auxiliary lobes; internal lobe with a single point. Top part of Oxford Clay (of Oxford district) and Corallian. Ex. *C. cordatum*.

Fig. 160. *Cardioceras cordatum*, from the Oxford Clay. $\times \frac{5}{8}$.

Hoplites. Shell flattened or inflated; umbilicus usually small. Ribs curved, bifurcating, and generally bearing a row of tubercles near the external margin and another near the umbilicus or at the middle of the sides; external margin flattened or deeply grooved. Suture-line finely divided. Gault. Ex. *H. dentatus*.

Acanthoceras. Whorls thick; umbilicus large; ribs simple or bifurcated, with rows of tubercles at the sides and margin;

external margin broad with a median row of tubercles. Saddles broad. Cenomanian. Ex. *A. rhotomagense*, Lower Chalk.

Schlœnbachia. Shell with usually small umbilicus; external margin with a median smooth keel; surface with strong ribs, which are slightly curved forwards and often bear tubercles. External and superior lateral saddles broad; one auxiliary lobe. Cenomanian. Ex. *S. varians*, Lower Chalk.

E. Uncoiled Ammonitids.

The shell is more or less uncoiled, or coiled into a helicoid spiral. These are believed to have been derived from Lytoceratid ammonites.

Macroscaphites (fig. 161). Similar to *Lytoceras*. Discoidal; the last whorl produced and then bent back in the form of a hook. Upper Neocomian. Ex. *M. ivani*.

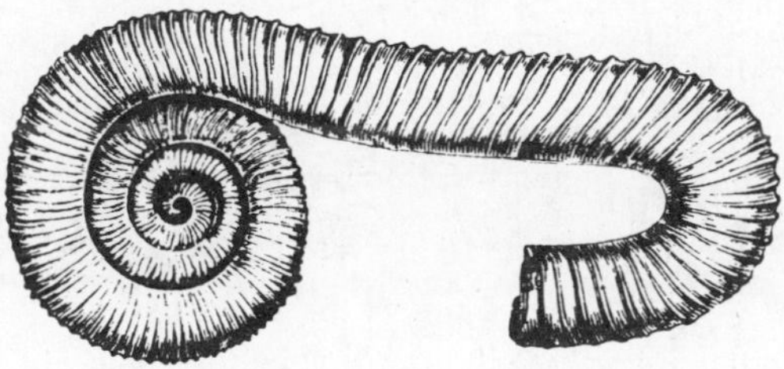

Fig. 161. *Macroscaphites ivani*, from the Lower Cretaceous. × ½.

Crioceras (fig. 162). Shell coiled in a plane spiral; the whorls not in contact. Surface ornamented with ribs which in some cases bifurcate and often bear tubercles and spines. Suture-line with four lobes. Neocomian. Ex. *C. duvali*.

Ancyloceras. Like *Crioceras*, but tuberculate, the last whorl produced in a straight line and then bent back in the form of a hook. Aptian. Ex. *A. matheronianum*.

Hamites. Shell bent upon itself three times, the parts not in contact; body-chamber long. Suture-line similar to *Lyto*

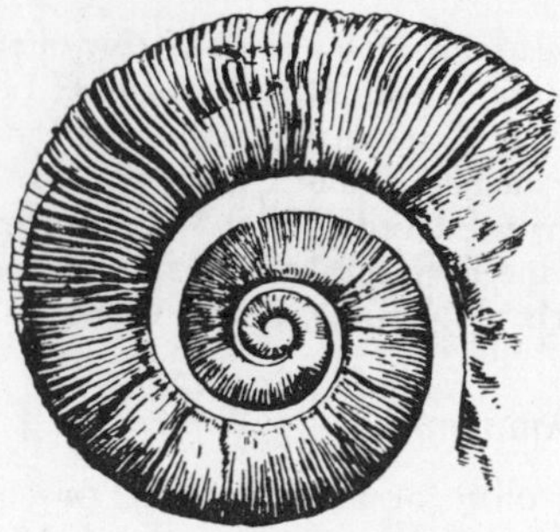

Fig. 162. *Crioceras duvali*, Neocomian. (After Sarasin and Schondelmayer.) $\times \frac{1}{2}$.

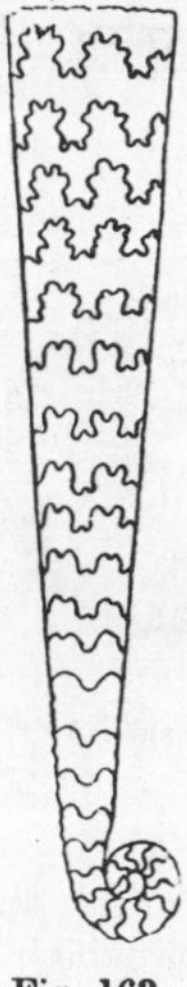

Fig. 163.

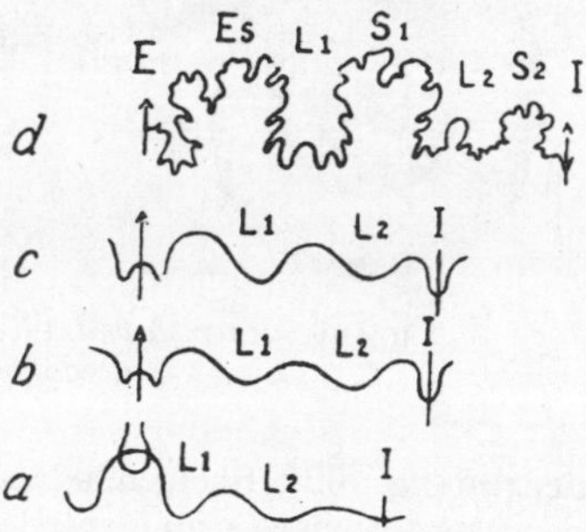

Fig. 164.

Fig. 163. Part of *Baculites chicoensis*, Chalk. (After Perrin Smith.) $\times 5$.

Fig. 164. *Baculites chicoensis*, Chalk. *a*, first suture-line with siphonal cæcum; *b*, second; *c*, sixth suture-line; *d*, adult suture-line of *Baculites capensis*; *E*, external lobe; *L* 1, *L* 2, first and second lateral lobes; *I*, internal lobe; *Es*, external saddle; *S* 1, first lateral saddle; *S* 2, internal (dorsal) saddle. (*a–c*, after Perrin Smith; *d*, after Spath.)

ceras, with the lateral lobes deeply divided. Surface smooth, or ornamented with ribs. Gault. Ex. *H. maximus*.

Baculites (figs. 163, 164). Shell straight (except a small spiral part at the apex, which is the first-formed part of the shell), elliptical in section. Suture-line with the lobes symmetrically divided. Upper Cretaceous. Ex. *B. vertebralis* (=*faujasi*), Chalk.

Scaphites. Shell coiled in a plane spiral; the whorls in contact and embracing, except the last, which is free from the spiral and then recurved in the form of a hook. Surface ornamented with bifurcated ribs which often bear tubercles. Suture-line generally much divided. Gault to Middle Chalk. Ex. *S. æqualis*, Lower Chalk.

Turrilites (fig. 165). Shell helicoid-spiral, turreted, usually sinistral, all the whorls in contact. Surface ornamented with transverse ribs or tubercles. Gault to Chalk. Ex. *T. costatus*, Chalk.

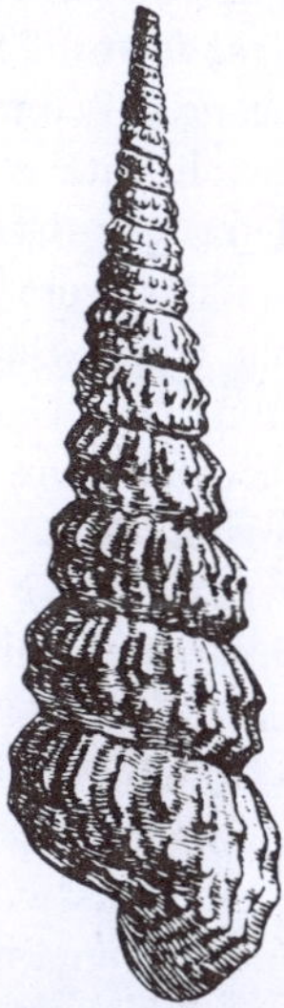

Fig. 165. *Turrilites costatus*, Lower Chalk. × ½.

Distribution of the Ammonoidea

The Ammonoidea have a shorter geological range than the Nautiloidea, the earliest representatives being found in the Lower Devonian and the latest in the Chalk. The most primitive of the Ammonoids, and apparently the ancestral form of the goniatites, is *Agoniatites* from the Lower Devonian; it differs but little from some Silurian Nautiloids (*Barrandeoceras*) except in the greater sinuosity of the superior lateral lobe, in the external position of the siphuncle, and in the more closely coiled early whorls; but similar closely coiled whorls are seen in the Lower Ordovician Nautiloid *Trocholites*. *Lobobactrites* ('*Bactrites*'), found in the Middle Devonian, has often been regarded as the ancestral form of the Ammonoids, but is now believed to be a secondarily uncoiled goniatite; it possesses a straight tapering shell, with a simple superior lateral lobe (like fig. 145, *a*).

Throughout the Devonian and Carboniferous periods the goniatites were dominant, with the Clymeniids abundant in the Upper Devonian. In the history of the various groups of goniatites there is a general tendency for the suture-line to become more folded, for the septal necks to change from retrosiphonate to prosiphonate, and for the sinus in the external margin of the aperture to decrease in size and ultimately to disappear. The various goniatite stocks that persisted into the Permian gradually took on ammonite characters, until in the Upper Trias the maximum of development was reached. Although most of the ammonites in the Trias evolved complicated suture-lines, some, like *Ceratites*, remained relatively simple. The number of genera in the Trias is very large, but not one seems to have survived into the Jurassic period, and only one family (the Phylloceratidæ) passed from the Trias into the Jurassic, so that apparently all the later ammonites must have been derived either directly or indirectly from that family. Two families, the Phylloceratidæ and the closely allied Lytoceratidæ, are remarkable in that they persist with but little change throughout the whole of the Jurassic and Cretaceous periods. Numerous other groups of ammonites were evolved in the Jurassic and Cretaceous, but the possibilities of variation were limited and the same types of whorl-section, ornamentation, and suture-line were often reproduced in the various families and at different horizons. In the Trias and Jurassic there are some genera in which the shell has become secondarily straight (*Rhabdoceras*, *Acuariceras*), partly uncoiled (*Choristoceras*), or coiled into a helicoid spiral (*Cochlo-*

ceras, *Spiroceras*); this feature becomes much more marked in the Cretaceous where we get such genera as *Hamites*, *Macroscaphites*, *Baculites*, *Crioceras* and *Scaphites* (figs. 161–163)—forms which are believed to have been derived from various Lytoceratid ammonites. In the Cretaceous there is a tendency in some stocks for the suture-line to become simplified, and occasionally it returns to the type seen in *Ceratites*. On the other hand in *Indoceras*, the latest of all ammonite genera, there is no sign of simplification; its suture-line with 75 elements (lobes and saddles) recalls the acme of specialisation among the Triassic ammonites.

The distribution of the more important genera of Ammonoids is given in the following table.

Lower Devonian. *Agoniatites, Anarcestes.*

Middle Devonian. *Agoniatites, Tornoceras, Anarcestes, Gyroceratites, Lobobactrites.*

Upper Devonian. *Clymenia* and allied genera. *Tornoceras, Cheiloceras, Sporadoceras, Manticoceras* (*Gephuroceras*), *Beloceras.*

Lower Carboniferous. *Aganides, Münsteroceras, Pericyclus, Protocanites.*

Middle Carboniferous. *Goniatites, Homoceras, Nomismoceras, Dimorphoceras, Prolecanites, Pronorites.*

Upper Carboniferous. *Gastrioceras, Schistoceras.*

Permian. *Agathiceras, Cyclolobus, Medlicottia, Thalassoceras, Popanoceras, Xenodiscus, Paralecanites.* Found chiefly in India, Russia, Sicily and Texas.

Lower Trias. *Ophiceras, Meekoceras, Koninckites, Pseudosageceras, Tirolites, Dinarites.*

Middle Trias. *Ceratites, Gymnites, Ptychites, Proarcestes, Monophyllites, Balatonites.*

Upper Trias. *Trachyceras, Celtites, Tropites, Cladiscites, Halorites, Arcestes, Joannites, Pinacoceras, Discophyllites, Megaphyllites, Lobites, Choristoceras, Cochloceras, Rhabdoceras.*

Lower Lias. *Psiloceras, Schlotheimia, Coroniceras, Arietites, Asteroceras, Oxynoticeras, Xipheroceras, Deroceras, Echioceras, Polymorphites, Acanthopleuroceras. Liparoceras, Androgynoceras*

Middle Lias. *Amaltheus, Pleuroceras.*

Upper Lias. *Harpoceras, Hildoceras, Dactylioceras, Grammoceras, Dumortieria, Hammatoceras.*

Lower Oolites. *Leioceras, Ludwigia, Ludwigella, Sonninia, Oppelia, Stephanoceras, Sphæroceras, Parkinsonia, Procerites.*

Middle Oolites. *Macrocephalites, Cadoceras, Quenstedtoceras, Cardioceras, Hecticoceras, Reineckeia, Kosmoceras, Peltoceras, Aspidoceras, Perisphinctes.*

Upper Oolites. *Pictonia, Rasenia, Aulacostephanus, Amœboceras, Pavlovia, Gigantites.* The following are not found in England: *Virgatites, Virgatosphinctes, Streblites, Haploceras, Simoceras, Berriasella.*

Lower Cretaceous. *Deshayesites, Parahoplites, Cheloniceras, Hoplites, Douvilleiceras, Desmoceras, Puzosia, Mortoniceras, Hamites, Anisoceras.* The following are chiefly foreign: *Spiticeras, Polyptychites, Holcostephanus, Holcodiscus, Acanthodiscus, Lyticoceras, Barremites, Tropæum, Macroscaphites, Crioceras, Heteroceras, Ancyloceras.*

Upper Cretaceous. *Schlœnbachia, Acanthoceras, Pachydiscus, Parapachydiscus, Parapuzosia, Prionotropis, Turrilites, Scaphites, Cyrtochilus, Baculites.* The foreign genera are far more numerous and include many "Pseudoceratites", *i.e.* Ammonites in which the suture-lines are reduced to the simplicity of Triassic *Ceratites*, but these had already started in the Lower Cretaceous.

ORDER III. DIBRANCHIA

The Dibranchia are represented at the present day by the cuttle-fishes, the squids, the calamaries, octopuses, paper-nautilus, etc.; they are of much less importance geologically than the Nautiloids and Ammonoids, the only really common fossil forms being *Belemnites* and its allies. Some of the modern cuttle-fishes attain a length of forty feet or more.

The Dibranchia (fig. 134) have a sac-like or elongated body, and possess one pair of gills only, and one pair of auricles. The number of arms is limited to eight or ten; and on the inner side—that facing the mouth—they are provided with rows of sucking-discs, which sometimes possess horny hooks. The jaws are not calcified, and are consequently seldom preserved in fossil specimens. An ink-sac is always present, and is sometimes found fossil. The funnel is in the form of a complete tube. The eyes are highly developed.

A shell is absent in some forms; when present it is (except in *Argonauta*) internal, being covered by folds of the mantle, and may be either horny or calcareous. In some cases (*Sepia*) it has the form of an oval flattened body, known as the *cuttle-bone*, which is composed mainly of laminated calcareous material with spaces between the laminæ. In the squids the shell is lamellar in form and consists of horny material; it is termed the *pen* or *gladius*. The shell in the cuttle-fishes and squids is placed on the upper or dorsal side of the body in a sac formed by the mantle. In *Spirula* the shell resembles that of a Tetrabranch, but is internal, being almost entirely covered by the mantle; it is situated at the posterior end of the body, and consists of a tube coiled in a plane spiral and divided into chambers by septa, which are traversed by a siphuncle placed near the inner margin; the whorls are not in contact, and a calcareous protoconch is present. The shell in the paper-nautilus (*Argonauta*) is of quite a different nature to that found in other Dibranchs; it is external and spiral, but not chambered, and is without muscular attachments; it is secreted by the terminal portions of the two anterior arms, and is found only in the female, serving for the reception of the eggs.

The Dibranchia are divided into two sub-orders: (1) the Decapoda, (2) the Octopoda.

SUB-ORDER 1. *DECAPODA*

There are ten arms, eight of equal length and two longer than the others. The latter can be more or less completely retracted within pits; their free ends are swollen and suckers are usually borne on those ends only. The suckers are stalked and are provided with a horny ring. An internal shell is always present.

Belemnites (figs. 166–170). The shell consists of three parts—the guard or rostrum (fig. 166, *a*), the phragmocone (*b*), and the pro-ostracum (fig. 167, *d*).

The *guard* is solid and is much more commonly preserved than the other parts; it varies considerably in shape and size, being cylindrical, club-shaped, fusiform, conical, etc. The original form seems to have been a short cone, such as is seen in *Prototeuthis acutus* from the Lower Lias. The end which was directed away from the mouth is always pointed and at the other end there is a conical cavity or alveolus. The guard varies in length from one to fifteen inches. When sliced transversely or longitudinally it is seen to be formed of a number of layers (growth-layers) arranged concentrically around an axial line, which is not quite central but is placed nearer the under surface; it is around this line that the first layers were secreted; the layers become somewhat thicker towards the pointed end and thinner towards the broad end of the guard. Each layer is formed of minute prisms of calcite, which are placed perpendicular to the axial line, thus producing a radiating fibrous appearance in cross-sections. The surface of the guard is sometimes smooth, or it may be granular, or furnished with ramifying vascular impressions; in some species there is a longitudinal groove on the under surface, and there may also be grooves on the sides or on the upper surface (dorsum).

The *phragmocone* (figs. 166, *b*; 167, *a*; 170) is a hollow cone, part of which fits into the alveolus at the broad end of the guard; it is divided into chambers by septa which are concave in front; a slender and beaded siphuncle (fig. 166, *c*) traverses the chambers at the under margin; at the pointed end of the phragmocone is a globular or ovoid protoconch formed of calcareous material (figs. 167, 170). The phragmocone is homologous with the entire shell of a Nautiloid or an Ammonoid; in

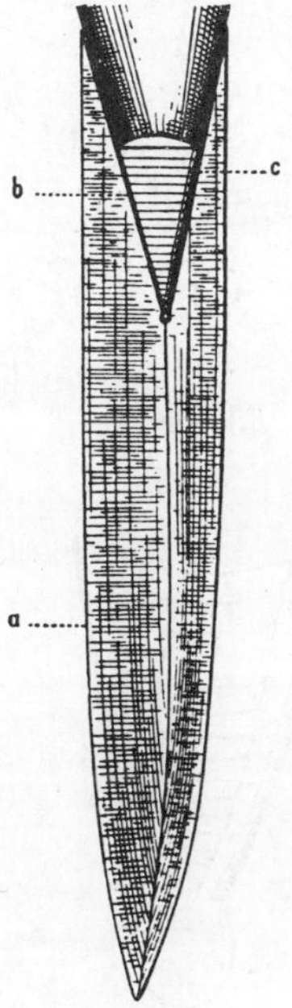

Fig. 166.

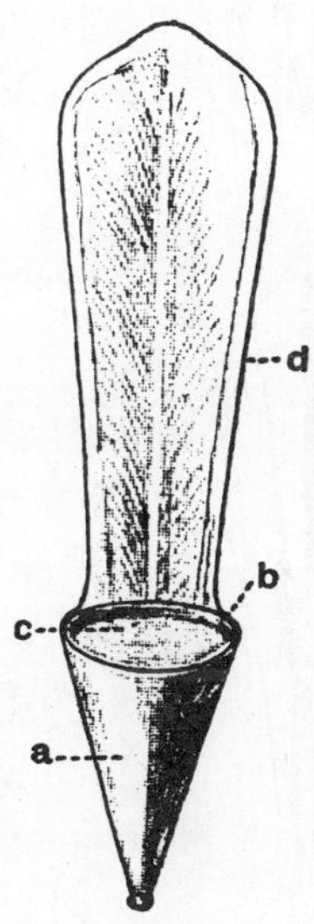

Fig. 167.

Fig. 166. Longitudinal section of *Belemnites*, from the Oxford Clay. *a*, guard; *b*, phragmocone with protoconch at the apex; *c*, siphuncle. $\times \frac{1}{2}$.
Fig. 167. Phragmocone and pro-ostracum of *Belemnites*, from the Lias. Restoration by G. C. Crick. *a*, phragmocone with protoconch at the apex; *b*, front border of phragmocone; *c*. last septum of phragmocone; *d*, pro-ostracum. $\times \frac{2}{3}$.

its conical form and straight suture-lines it resembles *Orthoceras*, but the siphuncle is more slender and nearer the margin than in most forms of *Orthoceras*. The wall of the phragmocone (sometimes termed the *conotheca*) is very thin, and in well-

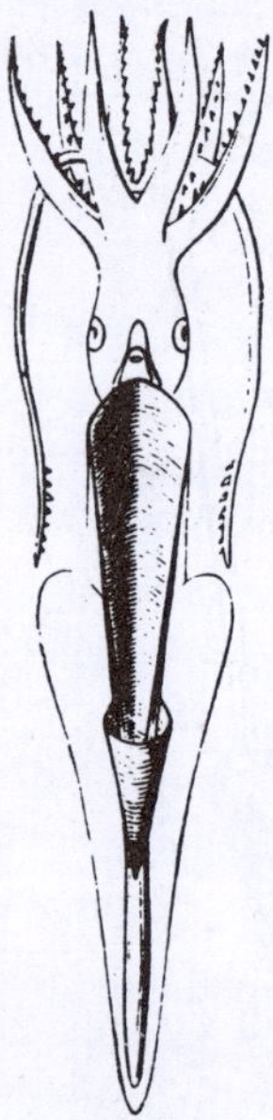

Fig. 168.

Fig. 169.

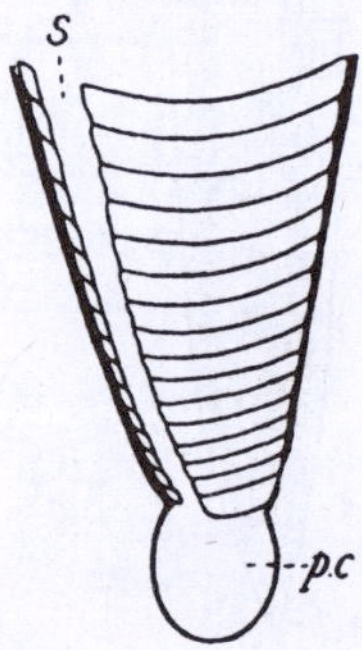

Fig. 170.

Fig. 168. D'Orbigny's restoration of a Belemnite (under surface), showing the probable positions of the guard, the phragmocone, and the pro-ostracum.

Fig. 169. *Belemnites.* Lias, Lyme Regis. Original in the Sedgwick Museum, Cambridge. Showing hooks indicating the presence of eight arms (a–h). $\times \frac{3}{4}$.

Fig. 170. Longitudinal section of part of the phragmocone of a Belemnite (*Cuspiteuthis*) from the Lias. *pc*, protoconch; *s*, siphuncle. Enlarged. (After Christensen.)

preserved specimens the upper part is produced in front into a large laminar expansion (fig. 167, *d*); this prolongation is known as the *pro-ostracum*, and correspon to the 'pen' of the squids, and it may represent the dorsal part of the body-chamber of *Orthoceras*. The head of the Belemnite was immediately in front of the pro-ostracum. The arms were provided with horny hooks, which are sometimes preserved fossil (fig. 169); there was a double row of hooks on each arm, but only eight double rows have yet been found in any specimen; two other arms, with or without hooks, may have been present. The ink-sac, near the base of the pro-ostracum, and the mandibles have also been found in some specimens. The probable positions of the guard, phragmocone and pro-ostracum in the body of the Belemnite are shown in fig. 168, from which it is seen that the guard formed a relatively small part of the entire length of the animal. The guard was at the posterior end of the body and it probably served to counteract the buoyant tendency of the phragmocone, so that the animal could maintain a horizontal position best suited for swimming.

Belemnites range from the Lower Lias to the Upper Cretaceous and include a very large number of species. These have been divided into groups, now regarded as genera, based mainly on the form of the guard, the number and position of the grooves, and the apical angle of the phragmocone. Some of the more important of these are: *Acroteuthis*, Neocomian, ex. *A. subquadrata*; *Actinocamax* and *Belemnitella* (see below); *Belemnopsis*, Inferior Oolite to Neocomian, ex. *B. canaliculatus*; *Cylindroteuthis*, Upper Jurassic, ex. *C. puzosianus*; *Dactyloteuthis*, Upper Lias only, ex. *D. irregularis*; *Hibolites*, Inferior Oolite to Aptian, ex. *H. hastatus*; *Belemnites*, Lias, ex. *B. paxillosus*; *Megateuthis*, Upper Lias and Inferior Oolite, ex. *M. gigantea*; *Prototeuthis*, Lower and Middle Lias, ex. *P. acutus*; *Neohibolites*, Neocomian to Cenomanian, ex. *N. listeri*, *N. semicanaliculatus*; *Oxyteuthis*, Neocomian to Aptian, ex. *O. brunsviciensis*; *Pachyteuthis*, Upper Oolites, ex. *P. excentricus*; *Hastites*, top of Lower Lias to Inferior Oolite, ex. *H. clavatus*.

Belemnitella. Guard cylindrical, with a slit at the under side of the alveolus. Distinct vascular impressions on the under surface of the guard. Upper Chalk. Ex. *B. mucronata*.

Actinocamax (= *Atractilites*). Alveolus of guard either conical or broadly funnel-shaped, and either in contact with the protoconch only or surrounding only the apical part of the phragmocone. Front part of guard often fragile and foliaceous owing to imperfect calcification. Chalk. Ex. *A. verus*, Lower Chalk

Forerunners of the Belemnites are found in the Trias, where they are represented by *Atractites*, *Aulacoceras*, *Dictyoconites* and other genera. In these the phragmocone is long and slender, the septa are concave and far apart, and the siphuncle is at the margin. It resembles closely the shell of the more elongate forms of *Orthoceras*, in some of which the siphuncle is near the margin. In *Atractites* (fig. 171 A), which ranges from the Lower Trias (perhaps also Permian) to the Upper Lias, there is a small guard with a very large phragmocone. In some species of this genus the septa are less widely separated, and the phragmocone approaches more nearly that of the Belemnites. In *Aulacoceras* (fig. 171 B) the guard is long, with a very deep alveolus, so that the main part of it forms a sheath over the very long phragmocone. A pro-ostracum was probably present in these genera, but its presence has not yet been proved.

Another allied group is represented by *Phragmoteuthis* in the Trias, and by *Belemnoteuthis* in the Oxford Clay. In *Phragmoteuthis* (fig. 171 C) the phragmocone is short and blunt, and the guard seems to be represented by a thin brown covering only; the pro-ostracum is twice the length of the phragmocone. *Belemnoteuthis* is similar, but with a definite guard formed of a fibrous layer which becomes thicker towards the point. The phragmocone is more elongated and the pro-ostracum relatively shorter than in *Phragmoteuthis*.

In the Tertiary the Belemnite group is represented by *Belemnosella*, *Belemnosis*, *Beloptera*, *Belosepia* and *Vasseuria* in the Eocene; *Spirulirostra* in the Oligocene and Miocene; *Spirulirostridium* in the Oligocene, and *Spirulirostrina* in the Miocene.

In these genera there is a tendency for the guard to become relatively smaller than in most of the Belemnites, and for the posterior part of the phragmocone to be curved or spiral. In *Belemnosis* the phragmocone is slightly curved. In *Spirulirostra* (fig. 171 D) the curvature is more marked, and in this

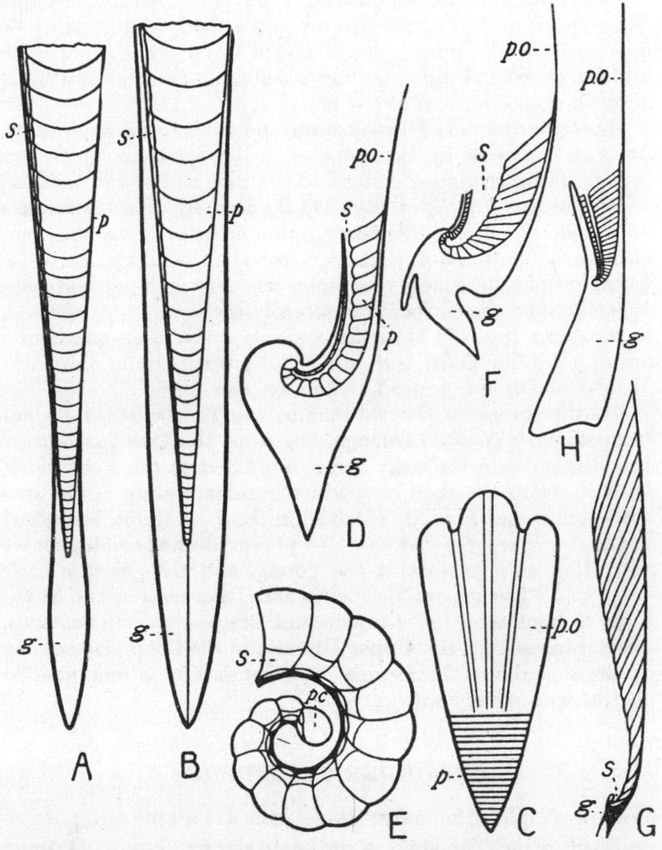

Fig. 171. A, diagrammatic longitudinal section of *Atractites*, Trias. B, *Aulacoceras*, Trias. C, *Phragmoteuthis bisinuata*, Trias (based on Suess). $\times \frac{1}{2}$. D, *Spirulirostra*, Miocene. E, *Spirula*, Recent. F, *Belosepia*, Eocene. G, diagrammatic section of *Sepia* (based on Appellöf). H, *Beloptera* (based on Naef). *g*, guard; *p*, phragmocone; *pc*, protoconch; *po*, proostracum; *s*, siphuncle.

respect it approaches *Spirula* (fig. 171 E) in which the phragmocone consists of a spiral of two and a half whorls, with the siphuncle at the inner margin and a large ovoid protoconch, but the guard and pro-ostracum are absent. The shell is situated near the posterior end of the body.

Belosepia (fig. 171 F) makes an approach to *Sepia* owing to the rapid increase in the diameter of the siphuncle which thus becomes funnel-shaped instead of tubular, and to the obliquity of the septa. In *Sepia* (fig. 171 G) the siphuncle is reduced to a hollow and the guard to a spine or mucro; the shell consists almost entirely of what appears to be the upper side of the phragmocone in which the septa are very oblique and close together; the pro-ostracum is scarcely distinct.

Beloptera (fig. 171 H) retains a relatively large guard of a stumpy log-like form with wing-like projections at the sides. It does not appear to lead on to any later form.

Another group of the Dibranchs, the Teuthoidea, to which the calamary (*Loligo*) belongs, begins in the Lias (perhaps in the Rhætic) and the early forms are allied to the Belemnites. In this group the shell or gladius consists mainly of the pro-ostracum, which is partly calcified in the fossil forms, but horny in the living representatives. The phragmocone is rudimentary and often only present in the young, and the guard is little developed. The genera *Belopeltis* and *Beloteuthis* occur in the Lias; *Geoteuthis* in the Jurassic and Cretaceous. *Plesioteuthis*, which ranges from the Upper Jurassic to the Chalk, is common in the Solenhofen Limestone, and consists of a long slender gladius with a very small guard.

SUB-ORDER 2. *OCTOPODA*

There are eight arms only; the suckers are sessile and possess no horny ring. The shell is rudimentary or absent. *Octopus* and *Argonauta* are well-known examples of this group. The Octopoda, as might be expected from the general absence of a shell, are very poorly represented in the fossil state; the earliest known form is *Palæoctopus* from the Chalk of Lebanon. *Argonauta* has been found in the Pliocene Beds.

Distribution of the Dibranchia

The Dibranchia are more numerous and more varied in existing seas than they were at any former period. Some forms are pelagic, others abyssal, but the larger number are found in littoral regions and are distributed in provinces similar to those of other molluscs (p. 298); typical littoral genera are *Octopus*, *Sepia* and *Loligo*.

The sub-order Decapoda is represented in the Trias by *Atractites*, *Aulacoceras*, *Dictyoconites* and *Phragmoteuthis*. In the Jurassic and Cretaceous Belemnites are the chief forms, and are especially abundant in argillaceous beds. Other Jurassic genera are *Geoteuthis* and *Beloteuthis* in the Lias; *Belemnoteuthis* and *Plesioteuthis* in the Upper Jurassic. *Belemnitella* and *Actinocamax* are limited to the Chalk. In the Tertiary Dibranchs are relatively rare, but several genera are found: *Belosepia*, *Sepia*, *Belemnosella*, *Belemnosis*, *Beloptera*, *Spirulirostra*, and *Spirulirostrina*. The living form *Spirula* is not definitely known to occur fossil. The sub-order Octopoda is represented by *Palæoctopus* (Chalk) and *Argonauta* (Pliocene).

PHYLUM ARTHROPODA

Classes	*Sub-Classes*	*Orders*
1. Crustacea	1. Trilobita	
	2. Branchiopoda	
	3. Ostracoda	
	4. Copepoda	
	5. Cirripedia	
	6. Malacostraca ...	1. Leptostraca
		2. Syncarida
		3. Peracarida
		4. Eucarida
		5. Hoplocarida
2. Onychophora (not fossil)		
3. Myriapoda		
4. Insecta		
5. Arachnida	1. Merostomata ...	1. Xiphosura
		2. Eurypterida
	2. Euarachnida... ...	1. Scorpionida
		2. Pedipalpi
		3. Araneida
		4. Pseudoscorpionida
		5. Phalangida
		6. Acarina
		7. Anthracomarti

The Arthropods have a bilaterally symmetrical body, formed of a series of segments (or somites), but the segments are not all alike, and some of those in front are fused together. Some, or all of the segments, bear a pair of jointed appendages or limbs, those near the mouth being modified to serve as jaws. A chitinous exoskeleton is always present, and is often strengthened by the deposition of carbonate or phosphate of lime; between the segments the integument remains soft and flexible, so that movement of the parts of the body is rendered possible. A heart is found in most forms; it is placed dorsally, and is provided with paired slits, termed ostia. The body-cavity contains blood. In some forms respiration takes place by means of the general surface of the body; others are provided with special organs—gills (or branchiæ), tracheæ, or lung-books. The gills are generally thin projections of the skin borne by some of the appendages; the tracheæ are long, branching tubes, filled with air, which penetrate all parts of the body and open to the exterior; the lung-books are chambers containing leaf-like folds of the skin. The nervous system consists of a supra-œsophageal ganglion or brain, connected by a ring round the œsophagus with two ventral longitudinal cords, usually provided with ganglia, and placed beneath the intestine. The sexes are separate in the majority of forms.

The Arthropoda are divided into the following Classes: (1) Crustacea, (2) Onychophora, (3) Myriapoda, (4) Insecta, (5) Arachnida. The Onychophora include one genus only—*Peripatus*, which has not been found fossil.

CLASS I. CRUSTACEA

The Crustacea are mainly aquatic animals, and are abundant as fossils. They breathe by means of gills, or, when the exoskeleton is thin, through the general surface of the body and by some of the limbs. The chitinous exoskeleton is frequently hardened by a calcareous deposit—hence the name Crustacea. Segmentation is usually well marked, but in the Ostracoda is shown by the appendages only. The exoskeleton of a segment or somite consists of a dorsal part, the *tergum*, and a ventral part, the *sternum*. In the higher Crustacea, three regions may be distinguished in the body: the head, the thorax, and the abdomen; but in the lower forms the trunk is often not clearly differentiated into the thorax and abdomen. The segments of the head are fused together, and include an anterior embryonic region to which the eyes belong, and five segments which bear appendages. Externally the segments (except in Trilobites) are indicated only by the appendages. The number of segments in the trunk (thorax and abdomen) is variable in the lower Crustacea, but is constant in the Malacostraca. In many forms some or all of the segments of the thorax fuse with those of the head, forming a *cephalothorax.* In many Crustacea there is a dorsal shield or *carapace* which covers part, or sometimes the whole, of the body, and originates as an outgrowth from the posterior margin of the dorsal covering of the head. The head bears five pairs of appendages, viz.: the antennules, the antennæ, the mandibles, and two pairs of maxillæ; the first two pairs are in front of the mouth, the last three behind it. The thorax is also provided with appendages, and often the abdomen too. The mandibles and maxillæ, and frequently some of the anterior thoracic appendages, serve as jaws. The Crustacean appendage is typically biramous, consisting of a basal part (the *protopodite*) bearing two branches—the inner called the *endopodite*, and the outer termed the *exopodite* (fig. 172, *en, ex*). The protopodite usually consists of two segments—a proximal or *coxopodite* (2) and a distal or *basipodite* (3). In some cases the exopodite disappears and the limb becomes *uniramous*.

Another form of appendage, which is common in the lower Crustacea, is the *phyllopod* type (fig. 173). In this the limb is usually broader and flatter than in the biramous type and the cuticle is thin. It consists of an axial part (fig. 173, 1–5) bearing a row of lobes or *endites* on the inner side (2′–5′, 6) and other lobes or *exites* on the outer side (*br*, *flb*). The basal endite commonly serves as a jaw or *gnathobase* (*gn*).

The mouth is on the under surface of the head, and the anus is on the last segment (the telson) of the body. Eyes are generally present, commonly a pair of compound eyes, and sometimes a median simple eye; in many Crustacea the former are placed on movable stalks. The sexes are separate except in most of the cirripedes and in some parasitic isopods. In the Malacostraca the genital apertures are on

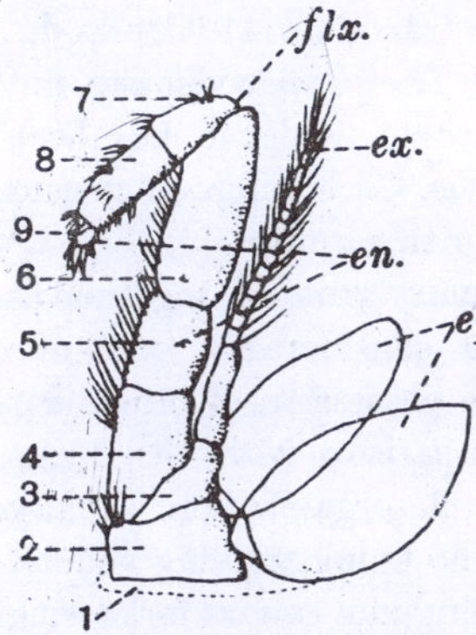

Fig. 172.

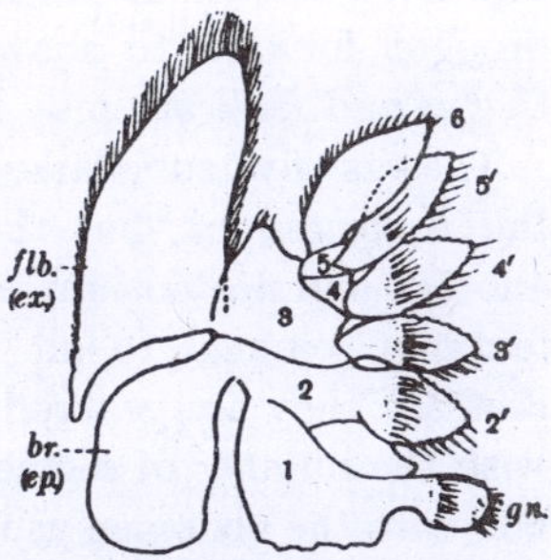

Fig. 173.

Fig. 172. The second thoracic limb of *Anaspides*. *en*, endopodite; *ep*, epipodite; *ex*, exopodite; 1–9, segments of endopodite; 2, coxa or coxopodite; 3, basis or basipodite. (After Calman.)

Fig. 173. Tenth thoracic limb of *Apus*. *br* (*ep*), branchia (epipodite); *flb* (*ex*), flabellum (exopodite); *gn*, gnathobase; 1–5, segments of the limb; 2′–5′, 6, endites.

the sixth thoracic segment in the male, and on the eighth in the female; in the lower Crustacea (Entomostraca) the position of the apertures is variable.

In some Crustacea development is direct, that is to say, the young individual has the same form as the adult; but generally this is not the case, the young undergoing metamorphosis before reaching the adult stage. The two chief larval forms are known as the *nauplius* and the *zoœa*. In the nauplius the body is unsegmented, and possesses three pairs of appendages representing the two pairs of antennæ and the mandibles. In the zoæa stage some of the thoracic appendages are present also, and the abdomen is segmented but possesses no appendages.

The Crustacea are divided into six sub-classes: (1) Trilobita, (2) Branchiopoda, (3) Ostracoda, (4) Copepoda, (5) Cirripedia, (6) Malacostraca. The Copepods are not definitely known as fossils, but some evidence has been brought forward to show that the Carboniferous genera *Cyclus* and *Halicyne* may belong to this group.

The first five sub-classes are usually grouped together as the *Entomostraca*, but they differ considerably from one another and are not united by the possession of important features common to all. In comparison with the Malacostraca they are generally of simple organisation, usually with the number of segments in the trunk varying widely, and with the abdomen usually ending in a caudal fork; with the exception of the Trilobita they are generally of small size, and without a clear differentiation of the trunk into thorax and abdomen. A median unpaired eye is usually present.

SUB-CLASS I. *TRILOBITA*

The Trilobites derive their name from the fact that the body is divided into three parts, by means of two furrows, which extend from the anterior to the posterior extremities; this trilobation is usually conspicuous, but in a few genera (*e.g. Homalonotus*, *Illænus*) it is indistinct or almost obsolete. The body is oval in outline, and flattened from above downwards; it consists of the *head* (fig. 174 A), the *thorax* (B), and the *pygidium* (C). The segments of the head and of the pygidium are fused together, but those of the thorax remain free. Traces of the alimentary canal are sometimes found in the middle or axial part of the Trilobite.

The dorsal surface of the body is protected by a strong, calcareous exoskeleton. The part which covers the head is known as the *head-shield* or *cephalic shield*, and is usually

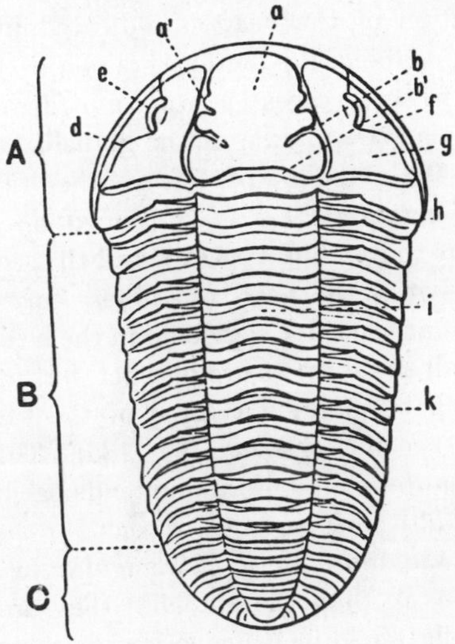

Fig. 174. *Calymene tuberculata*, from the Wenlock Limestone. Dorsal surface. A, head; B, thorax; C, pygidium. *a*, glabella; *a'*, axial furrow; *b*, one of the glabella furrows; *b'*, neck-furrow, behind which is the neck-ring; *d*, facial suture; *e*, eye; *f*, free cheek; *g*, fixed cheek; *h*, genal angle; *i*, axis of thorax; *k*, pleura. Natural size.

semicircular or triangular in shape; in it may be distinguished a median and two lateral portions; the former is the more convex and is termed the *glabella* (*a*), the latter are the *cheeks*. The glabella is marked off from the cheeks by means of a furrow on each side, known as the *axial furrow* (*a'*). The form and relative size of the glabella vary in different genera; in some it extends quite to the anterior margin of the head-shield, in others only a part of the way (fig. 184); sometimes it is wider behind than in front; in other cases it is wider anteriorly, or it may be of uniform width throughout; its convexity also varies considerably—it may be nearly flat, but is sometimes pear-shaped or spheroidal. The segmentation of the head is indicated by transverse furrows on the glabella (*b*)—often three on each side, occasionally four; in some cases the opposite furrows from the two sides meet at the middle of the glabella. On the posterior part of the glabella there is another furrow, which extends quite across it and is continued on the cheeks; this is known as the *neck-furrow* (*b'*), and the segment of the glabella behind it is the *neck-ring*. These furrows indicate the existence of at least five segments in the head. In primitive trilobites all the furrows are distinct, but in later forms there is often a tendency for some of the furrows to be reduced or to become obsolete (fig. 185). This reduction starts with the anterior furrow and extends backwards until, in a few cases, all the furrows disappear.

The cheeks are more or less triangular in shape, and usually less convex than the glabella; they are frequently bordered by a flattened or concave margin which in *Trinucleus* and *Harpes* is very broad. The posterior angles of the cheeks, known as the *genal angles* (*h*), may be rounded (*e.g. Calymene*), but are often pointed or produced into spines, the *genal spines* (*e.g. Paradoxides*, fig. 183). Each cheek is usually divided into two portions by a suture (the *facial suture*, *d*); the inner part—that between the facial suture and the glabella—is termed the *fixed cheek* (*g*); the outer part, known as the *free cheek* (*f*), is slightly movable on the fixed cheek. The course of the facial suture varies in different forms: it may commence on the posterior border inside the genal angle (the *opisthoparian* type, fig. 186), or at or near the genal angle (*h*), or on the lateral border in front of the genal angle (the *proparian* type, fig. 187); it passes inwards to the eye and then bends forwards, and may be continuous with the suture of the other cheek in front of the glabella, or it may cut the anterior margin of the head-shield, in which case it is sometimes united with the suture of the other side on the inferior surface of the head (fig. 176, *d*). When the sutures are continuous in front of the glabella it is evident that the cheeks will also be continuous. Since the position of the facial suture varies in different genera the relative sizes of the fixed and free cheeks will obviously vary too; thus in *Illænus* the free cheek is very narrow, in *Phillipsia* (fig. 190 B, C) very broad. The facial suture was probably of use in ecdysis. Owing to the fusion of the fixed and free cheeks the facial suture is sometimes absent, *e.g.* some species of *Acidaspis* this is probably also the case in *Agnostus*, *Microdiscus*, *Olenellus* and a few other genera. When the facial sutures cut the posterior border of the head-shield the genal spines belong to the free cheek; but when they cut the lateral border the genal spines are continuous with the fixed cheek. The term *cranidium* is used for the part of the cephalic shield enclosed by the facial sutures, that is the glabella and fixed cheeks.

The compound eyes (fig. 174, *e*) are on the upper surface of the head, one on each free cheek in the angle made by the facial suture; they are more or less conical with the summit truncated or rounded, and with the visual surface on the external part. The eyes usually consist of a large number of lenses—in *Remopleurides* the number is stated to be 15,000. Usually the lenses are biconvex or globular and adjacent to one another, but in *Phacops* and its allies the eyes are more highly developed, the lenses being separated by portions of the cephalic shield so that each appears to rest in a separate socket. The eye is entirely on the free cheek, but rests on a buttress or lobe on the adjacent part of the fixed cheek (the *palpebral* lobe). In a few Trilobites the eyes appear to be of a simpler type; for example, in *Harpes* each eye usually consists of two or three lenses only, and in some species of *Trinucleus* of a single lens; but it is probable that in such cases the eye is merely a degenerate form of compound eye. In a few Trilobites (*Agnostus, Microdiscus, Ampyx, Conocoryphe*, some species of *Acidaspis, Phacops*, etc.) eyes are absent; in such cases it is probable that the visual organs have been lost through disuse, just as is the case with some Crustacea at the present day which live at great depths in the sea or in other places where no light can penetrate. Thus it is found that in some of the later forms of *Phacops* the eyes are reduced or have disappeared entirely. When eyes are absent the facial sutures also are usually wanting. In *Cyclopyge* (fig. 175) the eyes are unusually large, occupying the greater part of the free cheeks, and sometimes extending on to the ventral surface; it is probable that this Trilobite was a pelagic animal which swam near the surface of the sea at night, but sank to considerable depths, where there was but little light, during the daytime. In many of the Cambrian Trilobites the eye itself is not found, but since the palpebral lobe is present it is reasonable to infer that it supported the visual organ, and that the absence of the latter is due to imperfect preservation; this view is supported by the recent discovery of the surface of the eye in a specimen of *Olenellus* from the Lower

Fig. 175. *Cyclopyge* [*Æglina*] *binodosa*, Arenig Beds. Natural size.

Cambrian. In *Olenellus* and its allies (fig. 182) the eye is crescentic in form and comes off from the side of the glabella; but when the eye is separated from the glabella there is, in many Cambrian and a few later Trilobites, a thread-like ridge, called the *eye-line*, which extends from the eye to the glabella (fig. 189, *o*—*n*).

In some Trilobites a small tubercle-like projection is found on the middle line of the front part of the glabella; this is probably a visual organ and seems to possess a structure similar to that seen in the median unpaired eye of the Branchiopods and Ostracods.

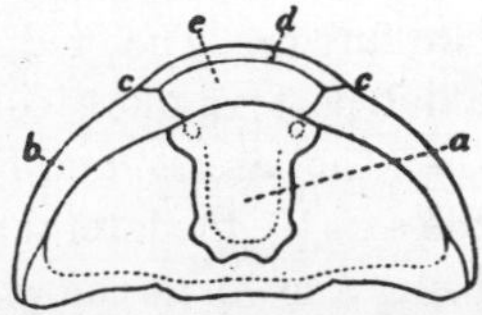

Fig. 176. *Calymene tuberculata*, Silurian. Ventral surface of head. *a*, hypostome; *b*, marginal rim; *c*, facial suture; *d*, transverse suture; *e*, rostral plate. Natural size. (After Barrande.)

The head-shield is continued on the under surface of the head as a reflexed border or marginal rim (fig. 176, *b*); sometimes the facial sutures (*c*) are continued across this border, and they may be joined by a transverse suture (*d*). Attached to the border in the median line is a plate (*a*), usually oval or shield-shaped, situated in front of and below the mouth and known as the *hypostome* or *labrum* (fig. 177). Just behind the mouth is the small lower lip-plate or *metastoma* (fig. 179 A, *m*), which, up to the present time, has been found in *Triarthrus* only.

Fig. 177. Hypostome of *Asaphus tyrannus*, from the Llandeilo Beds. × ½.

In many Trilobites a small oval or elliptical area, sometimes slightly raised like a tubercle, in other cases depressed, is found on each side of the hypostome just behind the middle of its outer surface (fig. 177); these *maculæ* are sometimes entirely smooth, but in other cases a part, or the whole of the surface, shows a structure similar to that of the compound eyes on the dorsal surface of the head, and such may have been visual organs. Maculæ are not known to occur in any other Crustacea.

The thorax (fig. 174 B) consists of a series of segments, which vary in number from two to forty-two, and are movable upon one another, in some cases sufficiently to enable the animal to protect itself by rolling up like a woodlouse. Each segment is divided into a median and two lateral parts by means of two furrows. The median or axial part is more convex than the lateral, and forms the axis (*i*), the lateral parts being known as the *pleuræ* (*k*). The anterior part (fig. 178, *c*) of the axis of each segment is not visible when the animal is unrolled, since it bends down and is overlapped by the preceding segment, for which it forms an articular surface. The pleuræ in some genera possess a longitudinal ridge, in others a groove (*h*), or both ridge and groove may occur; a few forms have plane pleuræ. Each pleura, at some distance from the axis, is curved downwards and usually also backwards; the point where this curvature occurs is known as the fulcrum (*e*); sometimes the outer part of each pleura overlaps the anterior part of the succeeding one, and then the front part of the pleura beyond the fulcrum may be smooth and flattened so as to form an articulating surface of *facet* (*f*). The terminations of the pleuræ are in some cases rounded (fig. 178), in others pointed or produced into spines (fig. 183).

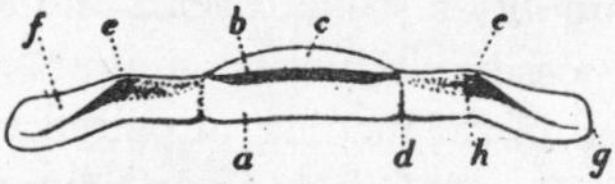

Fig. 178. Dorsal surface of a thoracic segment of *Asaphus expansus*. *a*, ring of axis; *b*, groove; *c*, articular portion; *d*, furrow between axis and pleura; *d*—*g*, pleura; *e*, fulcrum; *f*, facet; *h*, groove on pleura.

The pygidium (fig. 174 C) is commonly triangular or semicircular in shape, and is formed of a variable number segments, which do not differ in any essential respect from those of the thorax but are fused together and immovable; on the dorsal surface the segmentation is shown by grooves only. The pygidium, like the thorax, is divided into a median part or axis, and lateral or pleural portions. The conspicuous grooves on the lateral portions represent the grooves on the pleuræ, and not the divisions between the segments. The axis may reach quite to the posterior extremity or only part of the way, and it tapers more rapidly than the axis of the thorax; in *Bronteus* it is very short. The margin of the pygidium may be even or entire, or may be provided with a posterior spine or with lateral spines. This margin is bent under so as to form a border on the ventral surface similar to that on the ventral surface of the head. In a few primitive Trilobites all the segments behind the head are free so that there is no differentiation into thorax and pygidium.

For a long time the appendages of the Trilobites were unknown. In the great majority of specimens, when the under surface is exposed, the only parts which are found to be preserved are the hypostome and the reflexed borders of the dorsal exoskeleton. But in rolled-up specimens of *Calymene* and *Cheirurus*, Walcott showed, by means of thin sections, that jointed appendages are present on the head, thorax and pygidium, and that the ventral surface of the body is formed of a thin, uncalcified cuticle strengthened by transverse arches.

Subsequently specimens in which the body is not rolled up, showing clearly the ventral surface with the appendages, were obtained from the Utica Slate (Ordovician) near Rome (New York) and from the Middle Cambrian deposits of British Columbia. The most important of these belong to the genus *Triarthrus* from the Utica Slate (fig. 179). Each segment of the body, excluding the last (or anal), is found to bear one pair of appendages, which, with the exception of the first, are biramous. On the head there are five pairs

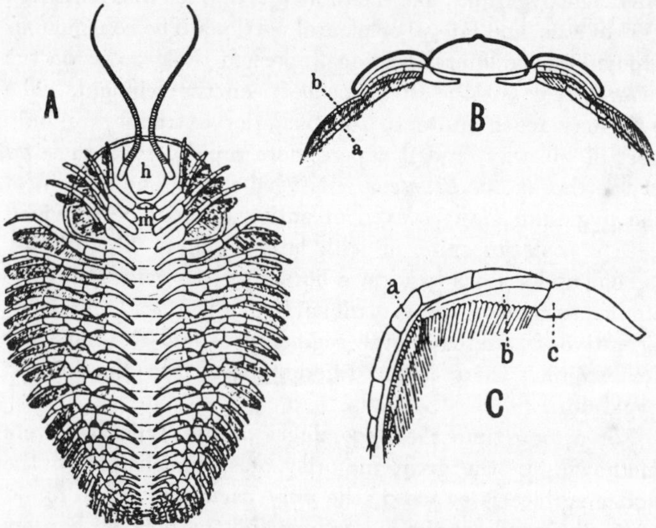

Fig. 179. *Triarthrus becki*, from the Utica Slate (Ordovician) near Rome, New York. (After Beecher.)

A. View of the ventral surface showing appendages, etc. *h*, hypostome; *m*, metastoma. $\times \frac{2}{3}$.

B. Diagrammatic section through the secona thoracic segment. *a*, endopodite; *b*, exopodite.

C. Dorsal view of second thoracic leg. *a*, endopodite; *b*, exopodite; *c*, protopodite with gnathobase. Enlarged.

of appendages. The *first* are the long antennæ which are attached on each side of the hypostome (*h*) and consist of a large basal joint bearing a flagellum formed of numerous short conical joints; these appear to be the only appendages in front of the mouth, and may represent the antennules of other Crustacea. The remaining four pairs of appendages of the head are biramous and all appear to have nearly the same form but increase in size backwards; the *second* pair may represent the antennæ, the *third* the mandibles, and the *fourth* and *fifth* pairs the maxillæ of other Crustacea. Each maxilla consists of a large basal joint (the *protopodite*) which bears a stout *endopodite* and a slender *exopodite*; the latter carries a row of hairs or *setæ*; the inner edge of the protopodite is toothed and served as a jaw (gnathobase); whilst the endopodite and exopodite assisted in locomotion.

The appendages of the thorax are long, but gradually decrease in size backwards, and consist of a protopodite (fig. 179 C, *c*) bearing the endopodite (*a*) and the exopo dite (*b*) which are of nearly equal length. The endopodite is formed of six joints, and probably served as a swimming organ. The exopodite consists of a long basal joint followed by a part consisting of numerous short joints; it bears setæ along its posterior edge and was probably adapted for crawling. The inner prolongations of the protopodites served as gnathobases. The limbs in each pair are widely separated, and in each segment the ventral cuticle between their bases is strengthened by a median longitudinal ridge and one or two oblique ridges on each side. On the posterior part of the thorax some of the joints of the endopodites become flattened.

The appendages of the pygidium are similar to those on the posterior part of the thorax, but are more distinctly leaf-like owing to the flattening and expansion of the first segments of the endopodite which bear setæ; the exopodite is slender. The anal opening is on the last segment (or telson) near the end of the pygidium.

In specimens of *Neolenus* from the Middle Cambrian of British Columbia, Walcott has discovered the caudal fork;

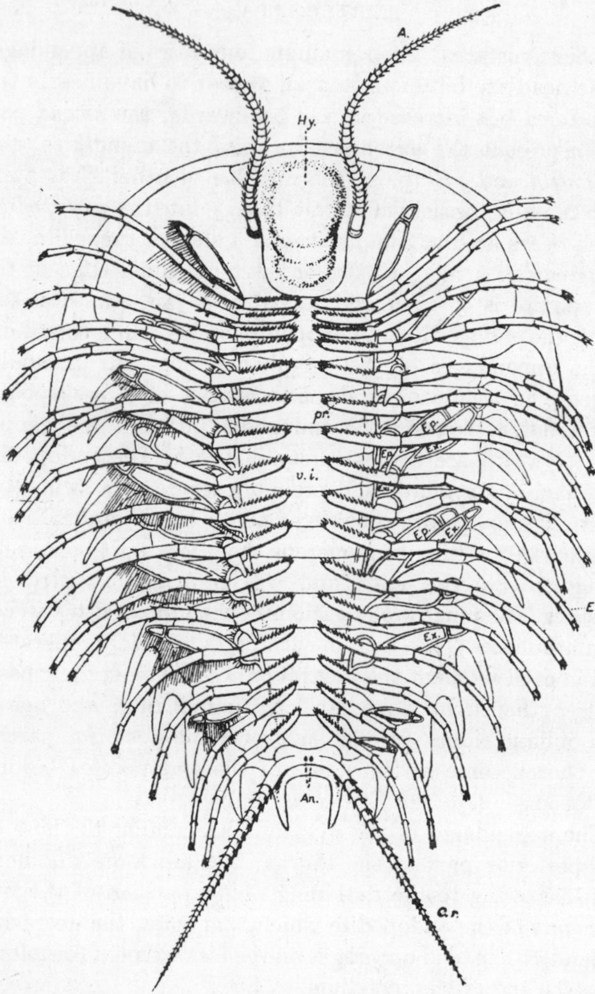

Fig. 180. *Neolenus serratus*, Middle Cambrian. Restoration of ventral surface. *A*, antennules; *An*, anus; *C.r.* caudal rami; *En*, endopodite; *Ep*, epipodite; *Ex*, exopodite; *Hy*, hypostome; *pr*, protopodite; *v.i.* ventral integument. The setæ have been omitted from the appendages on the right-hand side of the figure. (After Walcott.) $\times \frac{3}{4}$.

it consists of a pair of jointed filaments coming off from the end of the pygidium (fig. 180, *C.r.*). In addition to the genera mentioned some of the appendages have been found also in *Olenellus*, *Trinucleus* and Devonian species of *Phacops*.

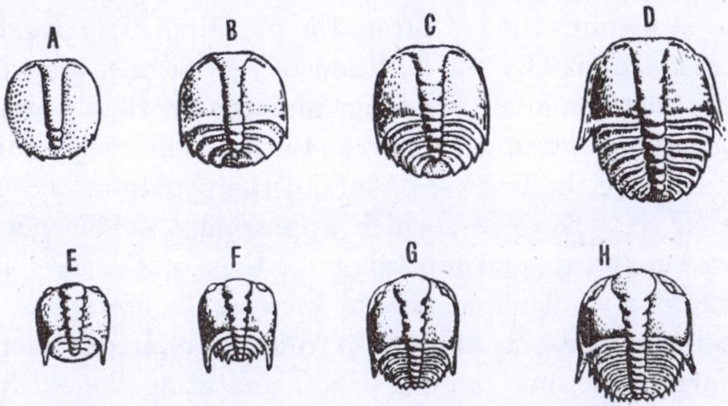

Fig. 181. Development of Trilobites. (After Barrande.)

A—D. *Sao hirsuta*, Cambrian, Bohemia. A, protaspid stage, ×12. B, meraspid stage, with three segments in the pygidium, ×12. C, with more distinct glabella furrows and four segments in the pygidium, ×12. D, with four thoracic segments and pygidial segments, ×10.

E—H. *Phacops* (*Dalmanitina*) *socialis*, Ordovician, Bohemia. × about 8. E, earliest meraspid stage, with eyes at the margin, and three pygidial segments. F, later stage, with more distinct furrows on the glabella, and four pygidial segments. G, with eyes moved inward, and narrow free cheeks; with both thoracic and pygidial segments. H, free cheeks relatively larger and five thoracic and three pygidial segments.

In some fine-grained deposits, especially in the Lower Palæozoic rocks of Bohemia, the larval forms of Trilobites are found well preserved, and by obtaining specimens of different ages it is possible to trace out the changes which occurred in the development of the individual. In the earliest or *protaspid* stage (fig. 181 A), the body is very convex or nearly globular with a discoid or ovate outline, and consists of a large cephalic region and a small pygidial part, but without any line separating them; the axis is distinct, and is marked by furrows; the free cheek, if present, is narrow. The glabella usually reaches the front margin of the head. In the next or *meraspid* (B, E) stage a transverse suture separates the head from the pygidium. The pygidium increases in size by the addition of new segments in front of the last (or anal) segment; and the thoracic segments are gradually introduced between the head and the pygidium, and arise by the front segments of the pygidium becoming free (D, H). The eyes, which appear first at the margin, move backwards and inwards until they attain their adult position, and the free cheeks increase in size (H). The glabella may become rounded in front and relatively shorter; its furrows become more distinct, indicating the existence of five cephalic segments. In some cases (C, D) the facial suture appears first at the lateral margin of the head-shield; in others (G) at the anterior margin. The *holaspid* stage begins after the full number of thoracic segments has appeared. During this stage further growth in size takes place, and changes occur in the form of the thoracic segments and of the pygidium.

The youngest stages of some of the Olenellids are of interest since segmentation is shown on the cheeks by means of grooves which extend outwards from the glabella; in these forms the eye-lobe appears first as an outgrowth from the front segment of the glabella.

The possession of antennæ, and the biramous character of the other appendages, connect the Trilobites with the Crustacea. The great variability in the number of segments in the thorax and pygidium, the large hypostome, and the gnathobases on the thoracic appendages seem to indicate that the Trilobites are related to the Phyllopod group of the Branchiopoda (p. 377), and especially to *Apus* and *Branchipus*; other features in which the two groups agree have been furnished by the discovery of the median unpaired eye and the caudal fork in Trilobites. But the Trilobites differ from the Phyllopods in the trilobation of the body, in the occurrence of a facial suture, and in the posterior segments being fused together to form a pygidium. In the character of their appendages the Trilobites are more primitive than the Branchiopods or any other Crustacea, since all except the first pair are very similar in structure and show but little specialisation in different regions of the body, and all are deeply biramous. It is probable that all the trunk limbs served in swimming, feeding and respiration. Other primitive characters are seen in the indication of segmentation on the dorsal surface of the head, and in the presence of a pair of appendages on every segment of the body except the last. The Trilobites differ from other Crustacea in having only one pair of pre-oral appendages.

In the general form of the dorsal exoskeleton and in the dorsal position of the compound eyes many Trilobites show a resemblance to the Xiphosura (p. 411); this may be due to adaptation to a similar mode of life rather than to any close relationship, since the essential morphological features of the two groups are distinct.

The dorso-ventrally flattened body, and the position of the eyes at the summits of the cheeks make it probable that most of the Trilobites were benthonic. The general similarity in form to *Limulus*, especially in such genera as *Dalmanites* and *Homolonotus*, suggests that they were able to burrow in the sand or mud in search of food in the same way that *Limulus* does. A few, like *Cyclopyge* (fig. 175) and *Deiphon* (fig. 188), were adapted for swimming. Some, like *Acidaspis*, possessed numerous long spines which enabled the animal to float.

Agnostus. Body small, head-shield and pygidium similar in form and size; eyes and facial suture absent; glabella does not reach the anterior border of the head, and has a small lobe at each of the posterior angles. Thorax formed of 2 segments, axis wide, pleuræ grooved. Segmentation not shown on the lateral parts of the pygidium. *Olenellus* Beds to Bala Beds. Ex. *A. pisiformis*, Lingula Flags.

Microdiscus. Similar to *Agnostus* but with from 2 to 4 segments in the thorax, and axis of pygidium with numerous distinct segments. *Olenellus* Beds to Lingula Flags. Ex. *M. punctatus*, Lingula Flags.

Trinucleus. Head-shield large, with long genal spines, and a broad flat, ornamented border; glabella inflated, pyriform, furrows sometimes absent. Eyes generally absent. Facial suture absent or indistinct. Thorax with 6 segments, pleuræ grooved, straight, but slightly curved near their extremities. Pygidium short, triangular, margin entire. Arenig to Bala Beds. Ex. *T. concentricus*, Bala Beds.

Ampyx. Similar to *Trinucleus*. Head-shield triangular, without a border, and with a long straight spine given off from the front of the glabella; facial sutures near the external margin, not continuous in front; free cheeks very narrow. Arenig to Wenlock Beds (chiefly Ordovician). Ex. *A. nudus*, Llandeilo Beds.

Olenellus. Head-shield large, semicircular, with a border and genal spines; glabella of nearly the same width throughout, the front lobe longer than the others; facial sutures not visible; eyes large, elongate, curved, joined to the front segment of the glabella. 14 segments in the thorax; pleuræ grooved and produced into backwardly-curved spines; the third segment larger than the others and with longer spines. Pygidium elongate, spine-like, without lateral lobes. Lower Cambrian. Ex. *O. thompsoni*.

Mesonacis. Similar to *Olenellus*. Thorax elongated, tapering posteriorly, consisting of 15 anterior segments, behind which are 10 shorter segments with the pleuræ less well developed; the axis of the 15th segment bears a long backwardly-directed spine. Pygidium small, plate-like. Lower Cambrian. Ex. *N. vermontana*.

Holmia (fig. 182). Similar to *Olenellus*. A spine at the posterior margin of the head-shield between the glabella and the genal spine. Thorax of 16 segments, the third not enlarged; pleuræ produced into narrow, separated spines. A row of spines extends down the axis of the body from the neck-ring nearly to the pygidium. Pygidium small, plate-like, with indications of segments on the axis. Lower Cambrian. Ex. *H. kjerulfi*.

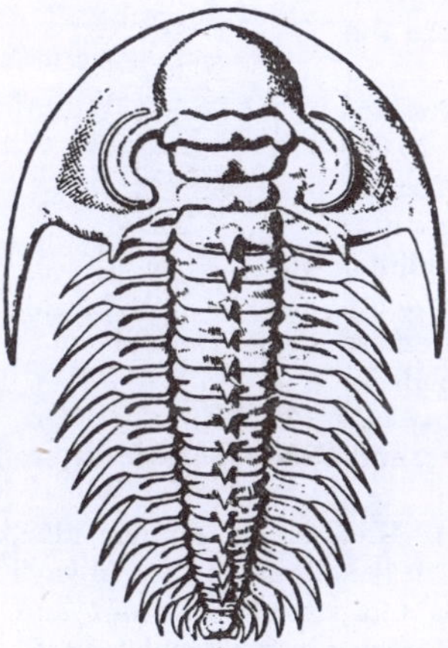

A

Fig. 182.

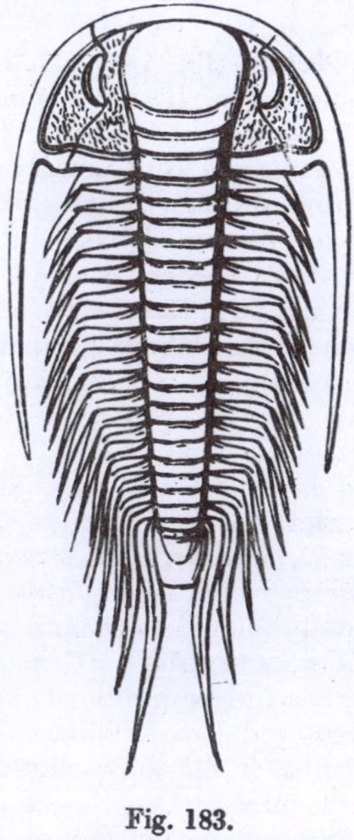

Fig. 183.

Fig. 182. *Holmia kjerulfi*, Lower Cambrian. *A*, pygidium. Natural size. (After Holm.)

Fig. 183. *Paradoxides davidis*, from the Menevian Beds. $\times \frac{1}{4}$.

Callavia. Similar to *Holmia*. Glabella narrow, especially in front. Pleuræ produced into broad spines. A long spine from the neck-ring. Lower Cambrian. Ex. *C. bröggeri*.

Paradoxides (fig. 183). Body large, elongated, narrowed posteriorly. Head-shield broad, semicircular, with a border, and long genal spines; glabella broad in front, with 2 to 4 furrows on each side, some of which are continuous across. Facial sutures extend from the posterior to the anterior border. Eyes large and arched. Thorax long, of 16 to 20 segments; pleuræ grooved and produced into long backwardly-directed spines. Pygidium very small, plate-like, its axis with 2 to 8 segments. Middle Cambrian. Ex. *P. davidis*, Menevian; *P. bohemicus*, Cambrian.

Olenus (fig. 184). Body oval; head-shield larger than the pygidium, with a narrow border, and with genal spines; glabella not reaching the anterior border, and not expanding in front, usually with three pairs of furrows; facial sutures extend from the posterior margin (near the genal angle) to the front border; eyes a little in front of the middle of the cheeks, and united to the front of the glabella by an eye-line. Thorax of from 12 to 15 (typically 14) segments; axis narrow, pleuræ with short points. Pygidium small, with 3 or 4 segments indicated on the axis, and with entire border. Lingula Flags to Tremadoc Beds. Ex. *O. gibbosus*, *O. cataractes*, Lingula Flags. *Parabolina*, *Peltura*, *Parabolinella*, *Leptoplastus*, *Eurycare*, and *Sphærophthalmus* are closely related to *Olenus*.

Fig. 184. *Olenus cataractes*, from the Lingula Flags. Natural size.

Conocoryphe (= *Conocephalites*). Head-shield semicircular, with a furrow inside the border, with genal spines (not always preserved); axial furrows deep, glabella narrow in front and with 3 or 4 backwardly-directed furrows and a well-marked neck-furrow; free cheeks narrow; eyes absent. Facial sutures begin just within the genal angles, and cut the front margin. Hypostome convex, formed of a central oval portion surrounded by a narrow border. Thorax with 14 or 15 segments; pleuræ grooved. Pygidium small, margin entire, axis with from 2 to 8 segments. Lower Cambrian to Tremadoc Beds. Ex. *C. lyelli*, *C. sulzeri*, Lower Cambrian.

Angelina. Body oval. Head-shield with long genal spines, glabella parabolic, without furrows; eyes small, near the middle of the cheeks. Thorax with 14 or 15 segments, pleuræ faceted. Pygidium short, margin provided with two teeth, axis of 4 or 5 segments. Tremadoc Beds. Ex. *A. sedgwicki*.

Calymene (figs. 174, 176). Head-shield semicircular, genal angles rounded, occasionally pointed; glabella inflated, broadest behind, with three pairs of lateral furrows separating three globular lobes on each side. Eyes small, prominent. Facial sutures extending from the genal angles to the anterior border, where they are connected by a transverse suture below the margin. Thorax of 13 segments, axis prominent, pleuræ grooved and faceted. Pygidium with 6 to 11 segments, margin entire. Arenig to Upper Ludlow. Ex. *C. tuberculata*, Wenlock Limestone.

Homalonotus (fig. 185). Body large, elongated, with indistinct trilobation. Head-shield broad, genal angles rounded, furrows on the glabella indistinct or absent. Eyes small. Facial suture passing from the genal angles to the front margin, and often continuous in front. Thorax with 13 segments; axis wide, not well marked. Pygidium triangular, axis with 10 to 14 segments. Arenig to Devonian. Ex. *H. delphinocephalus*, Wenlock Beds; *H. bisulcatus*, Ordovician.

Ogygia. Body oval, nearly flat. Head-shield large, semicircular, with a flattened border; glabella distinct, wider in front, with 4 or 5 lateral furrows. Eyes large. Facial sutures pass from the posterior border to the front margin, and are generally continuous at the margin. Free cheeks large. Hypostome not notched. Thorax of 8 segments, axis narrow, distinct; pleuræ grooved, usually with pointed ends. Pygidium large, semicircular, margin entire, axis of numerous segments. Tremadoc to Llandeilo Beds. Ex. *O. buchi*, Llandeilo Beds.

Asaphus (figs. 177, 186). Body oval, surface smooth or with striæ. Head-shield large, semicircular with a flattened border, genal angles rounded or spinose; glabella indistinctly defined, wide in front, with indistinct lateral furrows. Eyes large. Facial sutures pass from the posterior to the anterior margin and are generally continuous at the front margin. Free cheeks large. Hypostome notched posteriorly. Thorax formed of 8 segments, axis rather broad, pleuræ obliquely grooved, with rounded extremities. Pygidium of about the same size as the head, rounded, formed of numerous segments; margin entire. Tremadoc to Bala Beds. Ex. *A. powisi*, *A. tyrannus*, Llandeilo. Sub-genus *Asaphellus*: hypostome not notched. Tremadoc Beds. Ex. *A. homfrayi*.

Illænus. Body oval, convex. Head-shield large, semicircular; glabella indistinctly limited except near the posterior end, with-

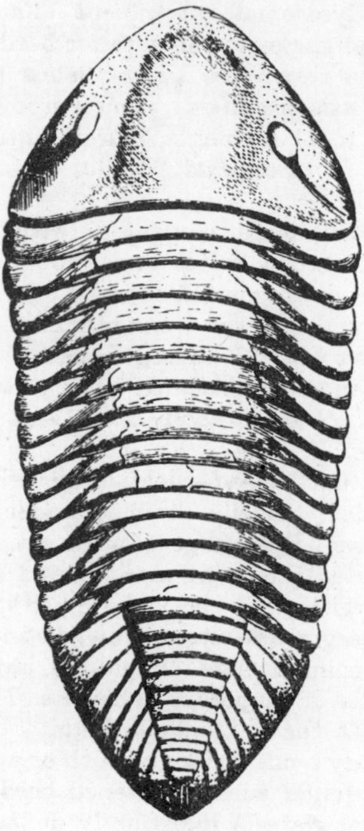

Fig. 185.

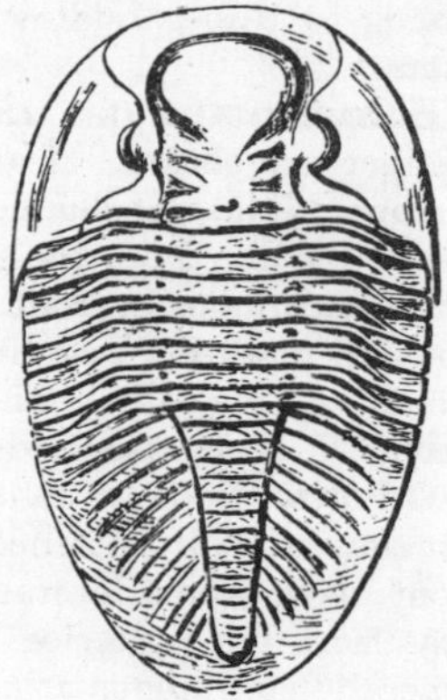

Fig. 186.

Fig. 185. *Homalonotus delphinocephalus*, Silurian. Natural size. (From Nicholson.)

Fig. 186. *Asaphus tyrannus*, from the Llandeilo Beds. × ½.

out furrows externally. Eyes remote from one another. Facial sutures commence on the posterior border, cut the anterior border in front of the eye, and unite on the inferior surface. Free cheeks small. Thorax with usually 10 segments, axis broad, pleuræ neither grooved nor ridged. Pygidium large, semicircular, axis indistinct, segments not visible externally. Arenig to Wenlock. Ex. *I. davisi*, *I. bowmanni*, Bala Beds.

Cyclopyge (= *Æglina*) (fig. 175). Head-shield large; glabella large, convex, projecting beyond the margin in front. Cheeks narrow; eyes very large, occupying nearly all the free cheeks. Facial sutures discontinuous, close to the glabella. Thorax with 5 or 6 segments, axis broad, pleuræ grooved. Pygidium rounded, axis short. Arenig to Bala Beds. Ex. *C. binodosa*, Arenig Beds.

Bronteus (= *Goldius*). Head-shield large, semicircular, genal angles pointed. Glabella expanding rapidly in front, with 3 lateral furrows in some species, none in others. Facial sutures start from the posterior border and are discontinuous in front. Free cheeks large; eyes crescentic, placed near the posterior border. Thorax with 10 segments, pleuræ ridged. Pygidium very large, fan-shaped; axis very short; lateral lobes large, with radiating grooves. Bala Beds to Devonian. Ex. *B. flabellifer*, Devonian.

Harpes. Form similar to *Trinucleus*, but border of head-shield broader, finely punctate, and extended posteriorly to near the end of the thorax instead of bearing narrow genal spines. Glabella short, convex, not expanded in front. Eyes consist of 2 or 3 lenses, and are usually joined to the front part of glabella by an eye-line. Thorax with 22 to 29 segments; axis narrow, pleuræ long, grooved. Ordovician to Devonian. Ex. *H. ungula*, Ordovician.

Phacops. Head-shield nearly semicircular; glabella prominent, broadest in front, with 3 or 4 furrows, which are sometimes indistinct; facial sutures commencing on the lateral borders of the cheeks in front of the genal angle, and continuous in front of the glabella. Eyes generally large, formed of large distinct lenses. Thorax with 11 segments, pleuræ grooved. Pygidium variable. Ordovician to Devonian.

Phacops, as defined above, may be divided into:

Phacops (restricted): glabella inflated and expanded in front, with the two anterior furrows obscure. Eyes large. No genal spines. Silurian and Devonian. Ex. *P. stokesi*, Silurian.

Trimerocephalus: glabella furrows obscure or absent. Eyes small. No genal spines. Devonian. Ex. *T. lævis.*

Acaste: glabella not much expanded in front, all the furrows distinct. Ordovician and Silurian. Ex. *A. downingiæ*, Silurian

Chasmops: glabella greatly expanded in front, two anterior furrows large, two posterior very small. With genal spines. Ordovician. Ex. *C. conophthalmus*, Bala Beds.

Dalmanites: glabella not much expanded in front, all the furrows distinct. Eyes large. Genal spines long. Pleuræ often produced into spines. Silurian. Ex. *D. caudatus*, Silurian.

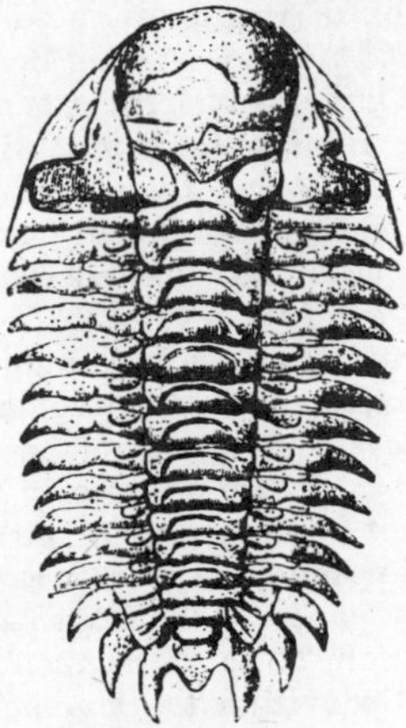

Fig. 187.

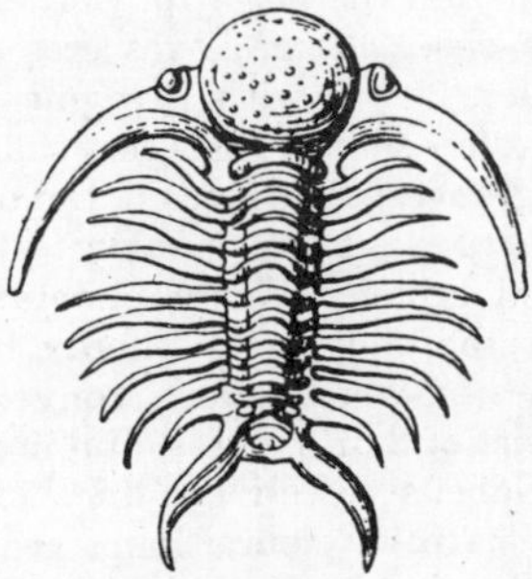

Fig. 188.

Fig. 187. *Cheirurus insignis.* Silurian. Natural size. (From Nicholson after Barrande.)

Fig. 188. *Deiphon forbesi*, Wenlock Shales. ×2. (After Whittard.)

Cheirurus (fig. 187). Head-shield semicircular, genal angles pointed or with spines; glabella convex, oblong or ovoid, with three pairs of furrows which are sometimes continuous across, the last pair uniting with the neck-furrow. Facial sutures continuous in front and ending on the external margins. Free cheeks small; eyes prominent. Thorax with usually 11 segments, pleuræ grooved, and produced into spines. Pygidium small, with 4 segments, lateral lobes with backwardly-directed spines. Tremadoc to Devonian. Ex. *C. articulatus*, Devonian; *C. bimucronatus*, Bala to Ludlow Beds; *C. juvenis*, Bala Beds.

Deiphon (fig. 188). Glabella globular or ovoid without furrows. Fixed cheeks forming two long curved spines. Free cheeks small, subtriangular. Thorax with 9 segments; pleuræ in the form of free spines. Pygidium short, 5 segments, prolonged into two spines on each side. Llandovery and Wenlock. Ex. *D. forbesi.*

Sphærexochus. Glabella large, spheroidal, with 3 pairs of furrows—the two anterior indistinct, the posterior curving backwards and joining the deep neck-furrow. Cheeks small; eyes small, near the axial furrow; facial suture starts from the genal angle. Thorax with 10 segments; pleuræ without grooves, with rounded ends. Pygidium small, with 3 segments. Ordovician and Silurian. Ex. *S. mirus*, Wenlock Limestone.

Staurocephalus. Glabella with a spherical lobe projecting in front of the cheeks; the remainder of the glabella narrow and cylindrical with 2 pairs of furrows and a deep neck-furrow. Cheeks very convex, with a flat border. Facial suture starts from the lateral margin and cuts the front margin. Eyes on stalks. Thorax with 10 segments; pleuræ ridged, produced into spines. Pygidium small, of 4 segments, with pleuræ produced into spines. Bala to Wenlock Limestone. Ex. *S. murchisoni*, Wenlock Limestone.

Encrinurus. Head-shield covered with tubercles; with a flat border, and pointed genal angles; glabella pyriform, confluent with the border in front, its furrows indistinct or absent; eyes small, on short peduncles. Facial sutures continuous in front, ending just in front of the genal angles. Free cheeks narrow. Thorax with 11 similar segments, pleuræ ridged. Pygidium narrow, triangular, with many segments in the axis, with 6 to 12 pleuræ bent backwards and diverging from the axis. Bala to Upper Ludlow. Ex. *E. punctatus*, Wenlock Limestone.

Cybele. Similar to *Encrinurus.* Three pairs of more distinct glabella furrows; border continuous in front of the glabella; genal angles usually rounded; facial sutures continuous in front. Thorax with 12 segments; pleuræ of the first 5 with blunt ends, those of the remaining 7 produced into spines. Pygidium with 4 or 5 pleuræ which bend sharply backwards and converge towards the axis. Ordovician. Ex. *C. verrucosa*, Bala Beds.

Lichas. Test covered with tubercles. Head-shield convex, relatively small, with genal spines. Glabella broad, with a central raised part, furrows directed backwards. Facial sutures pass from the posterior to the anterior border. Cheeks and eyes small. Thorax with 9 or 10 segments; pleuræ grooved, ending in rather long spines. Pygidium large, showing 2 or 3 segments, lateral parts produced into spines. Llandeilo to Wenlock. Ex. *L. anglicus*, Wenlock.

Acidaspis (fig. 189). Head-shield broad, its trilobation not well marked, with genal spines, and usually with spines at the margin of the head; glabella with a pair of longitudinal furrows parallel to the axial furrows, and with two or three lateral furrows. Facial sutures start from the posterior margin just within

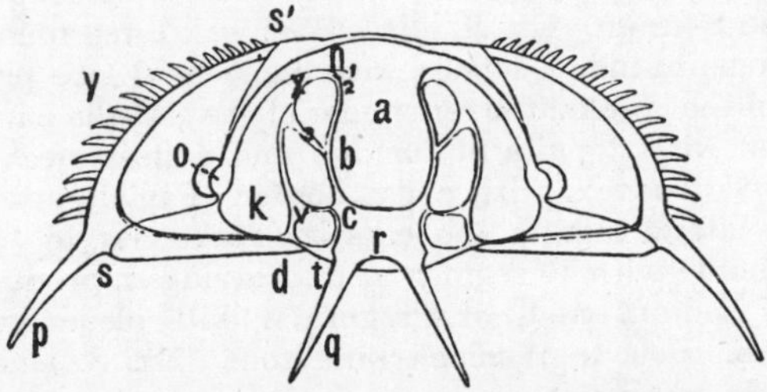

Fig. 189. *Acidaspis prevosti*, from the Silurian. Head-shield. (After Barrande.) 1, 2, 3, first, second, and third glabella furrows (the first usually indistinct); *a*, central part of the glabella; *c—b—n*, inner furrow of glabella; *c—v*, neck-furrow; *d—v—x*, axial furrow; *k—x*, fixed cheek; *o*, eye; *o—n*, eye-line; *p*, genal spines; *q*, spines from neck-ring; *r*, neck-ring; *s—s'*, facial suture; *y*, spines. Enlarged.

the genal angle and cut the front margin. Free cheeks large. Eyes connected with the glabella by an eye-line. Thorax with 9 or 10 segments, pleuræ with ridges produced into long spines. Pygidium small, with long spines. Llandeilo Beds to Devonian. Ex. *A. barrandei*, *A. brighti*, Wenlock.

Phillipsia (fig. 190 B—E). Body oval; glabella with nearly parallel sides, with 3 or 4 narrow lateral furrows, of which the posterior one curves backwards and joins the deep neck-furrow, thus cutting off a basal lobe. Facial sutures cut the posterior border obliquely, and the anterior border in front of the eye. Free cheeks large; eyes large, reniform. Thorax with 9 segments, pleuræ grooved. Pygidium semicircular, with 12 to 18 segments, margin entire. Devonian to Permian. Ex. *P. derbiensis*, Carboniferous.

Proetus. Closely allied to *Phillipsia* but with fewer segments in the pygidium. Ordovician to Permian, chiefly Devonian. Ex. *P. fletcheri*, Wenlock.

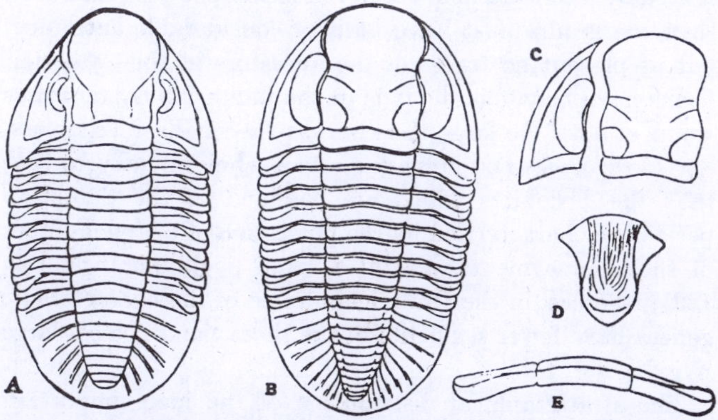

Fig. 190. A, *Griffithides globiceps*, Carboniferous Limestone. B—E, *Phillipsia derbiensis*, Carboniferous Limestone; D, hypostome; E, thoracic segment. (From Nicholson, after Woodward.) × 1½.

Griffithides (fig. 190 A). Body oval; glabella with inflated basal lobes cut off by the posterior furrow, and without other lateral furrows; main part of glabella pyriform; eyes rather small. Thorax with 9 segments. Pygidium rounded, with about 13 segments. Carboniferous Limestone. Ex. *G. seminiferus*.

Distribution of the Trilobita

The Trilobites are confined to the Palæozoic period, and form one of the most important and striking features in the faunas of the Lower Palæozoic deposits. They occur first in the Lower Cambrian Beds, and reach their maximum in the Ordovician. In the Silurian, Trilobites are still abundant, but become less important in the Devonian. Only one family survives from the Devonian into the Carboniferous and is represented by four or five genera only. Trilobites have been found in the Permian of Sicily, the Crimea, the United States, China, Timor and Western Australia.

Already in the Cambrian period the Trilobites were represented by a considerable variety of forms, showing that even then the group must have been of considerable antiquity, but at present no traces of the ancestors of the Cambrian forms have been found. It is in the Cambrian System that we meet with the largest, as well as the smallest Trilobites, *e.g. Paradoxides* and *Agnostus*. As a whole, it may be said that the Trilobites which are confined to the Cambrian period are characterised by the possession of a large number of thoracic segments, and of a small pygidium (figs. 182, 183); whereas, in the Ordovician, most of the characteristic genera have fewer segments in the thorax and possess large pygidia (fig. 186).

The stratigraphical distribution of the more important genera is shown below.

Lower Cambrian. Characterised especially by *Olenellus* and its allies (*Mesonacis, Holmia, Callavia*). *Agnostus, Microdiscus, Redlichia*, etc.

Middle Cambrian. Distinguished by *Paradoxides*. Other common forms are *Agnostus, Microdiscus, Solenopleura, Centropleura, Conocoryphe, Arionellus, Sao, Ellipsocephalus, Ogygiopsis.*

Upper Cambrian. Characterised by *Olenus* and its allies (*Parabolina, Parabolinella, Sphærophthalmus, Peltura, Ctenopyge, Triarthrus*). *Dikelocephalus, Niobe, Anacheirurus, Angelina, Asaphellus, Orometopus, Shumardia.*

Ordovician. *Agnostus, Ampyx, Trinucleus, Ogygia, Asaphus, Illænus, Cyclopyge, Chasmops, Calymene, Cybele, Lichas. Ogygia, Asaphus, Trinucleus* and *Ampyx* are abundant.

Silurian. *Calymene, Homalonotus, Illænus, Phacops, Dalmanites, Acaste, Cheirurus, Deiphon, Sphærexochus, Encrinurus, Acidaspis, Proetus, Lichas. Calymene* and *Phacops* are particularly abundant.

Permian and are also represented in the Lias and the Upper Jurassic. Examples belonging to modern families occur in Tertiary deposits.

Trichoptera (caddis-flies) are represented by primitive types in the Lias and Purbeck Beds. Genera belonging to modern groups are found in the Eocene of Wyoming, the Oligocene amber, and in the Miocene of Colorado.

Lepidoptera. Butterflies and moths are very rare as fossils. A few occur in the Middle and Upper Jurassic rocks, *e.g.* *Palæontina oolitica* from the Stonesfield Slate. The Order is better represented, although still uncommon, in the Tertiary Beds; examples have been found in the Oligocene of the Isle of Wight, the Oligocene amber of the Baltic, and in the Miocene of Colorado.

The *Coleoptera* (beetles) first appear in the Permian; they are more numerous in the Trias and Upper Jurassic, and are well represented in some of the Tertiary Beds. Examples have been found in the Lias, the Stonesfield Slate, the Solenhofen Limestone, the Purbeck Beds, the Lower Chalk of Bohemia, the Oligocene amber, and in the Miocene of Oeningen and Colorado.

The *Diptera* include flies, fleas, gnats, and mosquitoes. A few forms are found in the Lias, the Solenhofen Limestone, and the Purbeck Beds; the Order is represented in Tertiary deposits by numerous forms belonging to modern families. Forerunners of this group are found in the Permian and Trias.

The *Hymenoptera* include ants, bees, wasps, saw-flies, etc. The earliest examples are found in the Jurassic (Solenhofen Limestone and Purbeck beds), and the Order shows a considerable development in the Cretaceous. A large number of forms are met with in the Tertiary, where most of the important modern families are represented. Ants, wasps and bees are common in the Miocene of Colorado; saw-flies in the Oligocene amber, etc. Hymenoptera have been found in the Oligocene of the Isle of Wight. Insects allied to this group occur in the Permian (Protohymenoptera).

CLASS V. ARACHNIDA

Scorpions (fig. 220), spiders, and mites are common forms of the Arachnida. In the members of this Class the anterior segments of the body are fused together, forming a *prosoma* or cephalothorax which is covered by a carapace. This region usually bears six pairs of appendages, of which one pair is in front of the mouth. Antennæ are absent, and no pair of appendages is modified to serve exclusively as jaws. The first pair, known as *chelicerœ*, are pre-oral; the second pair, the *pedipalps*, are behind the mouth and serve partly as jaws; the four remaining pairs are long limbs, used for locomotion and to some extent as jaws. The trunk may or may not be segmented; in some groups it is divided into an anterior and a posterior region (*mesosoma* and *metasoma*), each of which consists typically of six segments. The first segment of the mesosoma bears the genital pore. The metasoma bears no appendages, and those on the mesosoma are never in the form of locomotory limbs, but are connected with respiration; in the primitive aquatic arachnids they are plate-like and bear lamellar gills; in the terrestrial forms the gills are replaced by lung-books or by tracheæ.

The Arachnida are divided into two sub-classes: (1) Mero stomata, (2) Euarachnida.

SUB-CLASS I. MEROSTOMATA

The Merostomata are aquatic Arachnids which breathe by means of gills borne on the plate-like appendages of the mesosoma. There are two Orders: (1) Xiphosura, (2) Eurypterida.

ORDER I. XIPHOSURA

The only living representative of the Xiphosura is the king-crab, *Limulus* (figs. 211, 212), found on the eastern shores of North America and Asia, and in the Malay Archipelago and the Indian Ocean. The body of *Limulus* is covered by a chitinous exoskeleton, and consists of the prosoma (figs. 211 A; 212, 1), and the trunk or *opisthosoma* (figs. 211 B; 212, 2), formed of the mesosoma and metasoma fused together. At the end of the body, behind the anus, is a long, movable tail-spine (fig. 212, 3).

The prosoma is covered dorsally by a large crescentic or nearly semicircular carapace (fig. 211, 1), which is very convex above and carries on its upper surface two pairs of eyes, one compound and lateral (5), the other simple and median (4). The large compound eyes are near the middle of the lateral parts of the carapace; the small simple eyes are close together in the middle line, near the anterior margin. The carapace is continued on to the under surface of the prosoma as a marginal rim. The trunk is more or less hexagonal in outline and is movably articulated with the prosoma; both have two longitudinal furrows on the dorsal surface, dividing a narrow axial part from a broad lateral portion on each side, thus giving a superficial resemblance to a Trilobite. The mesosoma forms the main part of the trunk and consists of six fused segments, the segmentation being shown by grooves on the dorsal surface, and by the six movable spines borne on each side. The small posterior part of the trunk without grooves represents the greatly reduced metasoma.

The prosoma carries six pairs of appendages concealed in the concavity of its under surface; the anterior pair (fig. 211, 1) (*chelicerœ*) only are in front of the mouth and

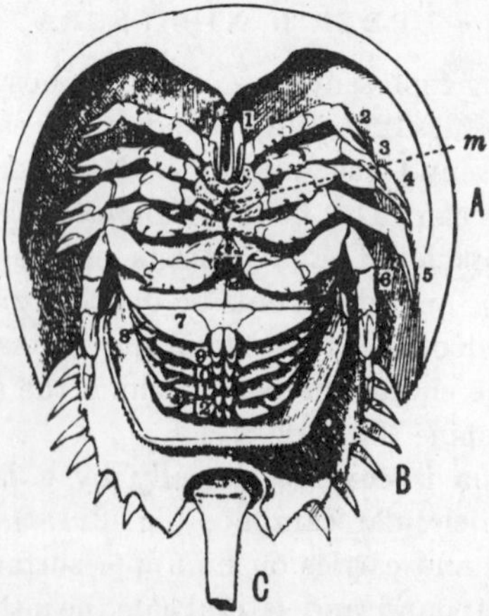

Fig. 211. *Limulus polyphemus*, Recent. Ventral surface. A, cephalothorax or prosoma; B, trunk (opisthosoma); C, portion of the tail-spine. 1–6, appendages of the prosoma; 1, chelicera; 2–6, ambulatory legs—behind the mouth are the small chilaria; 7–12, appendages of the mesosoma; 7, genital operculum; 8–12, lamellar appendages bearing gills; *m*, mouth. Reduced.

are small, three-jointed appendages with chelæ. The other five pairs (2–6) are the long, six-jointed walking-legs placed just behind the mouth; most of them (except the last pair) end in chelæ, and their basal joints (except in the sixth pair) are spinose and function as gnathobases. Behind the mouth are a pair of small unjointed processes, the *chilaria*, which represent the appendages of a pre-genital segment. The

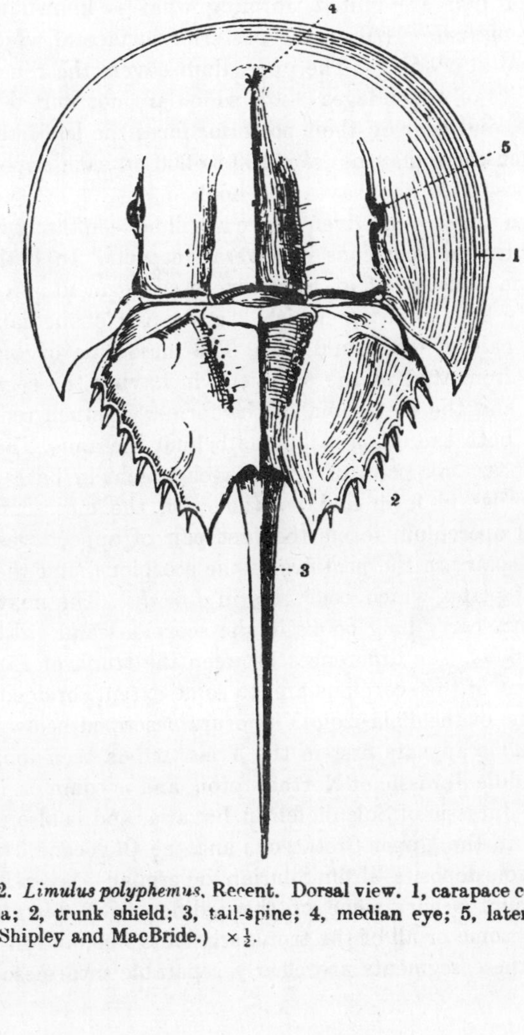

Fig. 212. *Limulus polyphemus*, Recent. Dorsal view. 1, carapace covering prosoma; 2, trunk shield; 3, tail-spine; 4, median eye; 5, lateral eye. (From Shipley and MacBride.) ×½.

mesosoma carries six pairs of plate-like appendages; the anterior pair are united, forming what is known as the *genital operculum* (7), on the posterior surface of which are the genital openings. The operculum covers the remaining five pairs of appendages (8–12), which are not united in the middle, and bear on their posterior faces the leaf-like gills, of which there may be from 150 to 200 on each appendage superposed like the leaves of a book.

From the account given above it will be seen that *Limulus* resembles the scorpions in several respects. In both, the prosoma consists of at least six fused segments, covered dorsally by a carapace which bears a pair of median eyes and a pair of compound eyes. The mesosoma of *Limulus* differs from that of the scorpions in having the segments fused, and the metasoma of the former is much reduced; but in both there is a tail-spine behind the anus. The prosoma bears six pairs of appendages which, in both cases, are similar in form and position. On the mesosoma the genital operculum forms the first pair of appendages; the second pair are the pectines of the scorpions, and the first pair of plates which bear gills in *Limulus*. The next four segments carry lung-books in the scorpions and gill-books in *Limulus*. The differences between the trunk of *Limulus* and that of the scorpions are, to some extent, bridged over by some of the Palæozoic Xiphosura described below.

Limulus appears first in the Trias; it has been found in the Middle Jurassic of Northampton, and is common in the Upper Jurassic of Solenhofen in Bavaria, and is also represented in the Upper Cretaceous and the Oligocene. In the Palæozoic deposits—from Silurian to Permian—several other Xiphosura occur; some of these differ from *Limulus* in having some or all of the trunk segments free, and in some cases these segments are clearly separable into mesosoma and metasoma. Thus in *Neolimulus* and *Hemiaspis* (fig. 213 A) all the segments are free, and in the latter they are clearly separable into mesosoma and metasoma. In *Belinurus* (B) the segments of the mesosoma are free, but those of the metasoma are fused together. *Prestwichianella*, *Euproöps* (C) and *Paleolimulus* approach *Limulus* in having all the segments of the trunk fused, and the metasoma reduced. Those genera in which the trunk segments are free approach

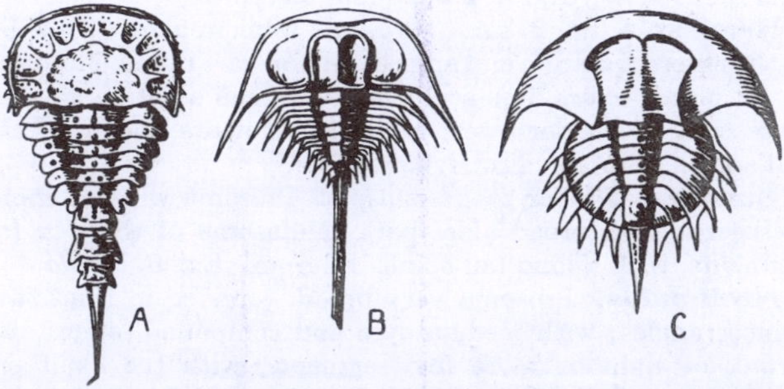

Fig. 213. A. *Hemiaspis limuloides*, Upper Silurian. $\times \frac{3}{4}$. (After Woodward.) B, *Belinurus reginæ*, Coal Measures, $\times 1\frac{1}{2}$. (After Woodward.) C, *Euproöps danæ*, Carboniferous. $\times \frac{1}{2}$. (After Packard.)

both the Eurypterida and the Scorpionida more nearly than does *Limulus*. Some of the appendages have been found in *Paleolimulus* but in most of the Palæozoic specimens they are not preserved. The examples found in the Coal Measures may perhaps have lived in fresh water.

Belinurus (fig. 213 B). Form similar to *Limulus*. Prosoma semicircular, with a flat border and long spines from the posterior angles; median part raised, with compound eyes at the sides and median eyes at the front. Mesosoma of five free segments, with the lateral parts produced into spines. Metasoma small, formed of three fused segments with a long tail-spine. Upper Old Red Sandstone and Coal Measures. Ex. *B. reginæ*.

Prestwichianella (=*Prestwichia*). Prosoma semicircular, continued into spines at the posterior angles; median raised part ('glabella') broad, with the compound eyes at the anterior lateral angles. Trunk segments (probably seven) fused, with a flat marginal part produced into spines, and a tail-spine. The axial part of the trunk is narrow. Coal Measures. Ex. *P. rotundata.*

Euproöps (fig. 213 C). Similar to *Prestwichianella*, but the median raised part of the prosoma is quadrangular, and the compound eyes are more anterior in position. Coal Measures. Recorded from the Upper Devonian of Pennsylvania and Permian of Kansas. Ex. *E. danæ*, Coal Measures.

Hemiaspis (fig. 213 A). Prosoma semicircular, with spines at the external margin and angles; central part raised. Mesosoma of six broad, short, free segments, with the axial part raised; metasoma much narrower, of three segments and a pointed tail-spine. Silurian. Ex. *H. limuloides.*

Bunodes. Similar to *Hemiaspis*. Prosoma without spines. Mesosoma with broad axial part. Metasoma of three or four segments, with a long tail-spine. Silurian. Ex. *B. lunula.*

Neolimulus. Prosoma very broad, rounded in front, with spinose angles; with median eyes and compound lateral eyes. Trunk of eight or more free segments, with the axial part tapering rapidly backwards; apparently not differentiated into mesosoma and metasoma. Silurian. Ex. *N. falcatus.*

Distribution of the Xiphosura

Fossil Xiphosura are rare, except in the Solenhofen Limestone (Upper Jurassic). The earliest forms which show affinities to the Xiphosura are *Aglaspis* from the Cambrian of Wisconsin and *Beckwithia* from the Cambrian of Utah. The chief genera are:

Silurian. *Hemiaspis, Neolimulus, Bunodes, Pseudoniscus.*
Devonian. *Belinurus, Protolimulus, Weinbergina.*
Carboniferous. *Belinurus, Euproöps, Prestwichianella.*
Permian. *Euproops* and *Paleolimulus* in Kansas.
Trias to Oligocene and Recent. *Limulus.*

ORDER II. EURYPTERIDA

The Eurypterids are found only in the Palæozoic rocks and are remarkable for the large size which they often attain; one form (*Pterygotus anglicus*) reaches a length of six feet and is the largest Arthropod known. The Eurypterids have a scorpion-like appearance; but, unlike the scorpions, they were all aquatic animals. The body is compressed dorso-ventrally, and is protected by a chitinous exoskeleton (fig. 214) which is covered with small scale-like markings.

The prosoma consists of the six anterior segments fused together, and is usually quadrate, semicircular or semi-oval in outline. The carapace, which covers the dorsal surface of the prosoma, bears a pair of small, simple eyes near its centre (fig. 214, *e*) and a pair of large, compound eyes—one at each of the outer front margins (*d*) or on the dorsal surface at some little distance from those margins.

Behind the prosoma come the twelve free and movable segments of the trunk or abdomen. In some genera (fig. 214) these segments gradually decrease in width in passing from the anterior to the posterior end, but in other cases (fig. 219) they are divisible into two groups—the anterior segments being short and broad, whilst the posterior are longer and narrower. The six anterior segments bear appendages and form the mesosoma (fig. 215, I–V; fig. 219, VII–XII); the six posterior segments form the metasoma (fig. 215, 7–12; fig. 219, XIII–XVIII), at the end of which is the post-anal tail-plate or spine (*g*); this may be spine-like (fig. 216), or triangular, or in the form of an oval plate which may be produced into a median spine as in *Slimonia* (fig. 219), or divided at the end into two lobes as in some species of *Pterygotus* (fig. 214). Each segment of the mesosoma is covered by a broad, slightly convex dorsal shield (or tergum), and by a ventral cuticle (or sternum), and the tergum of each segment overlaps the one next behind. In

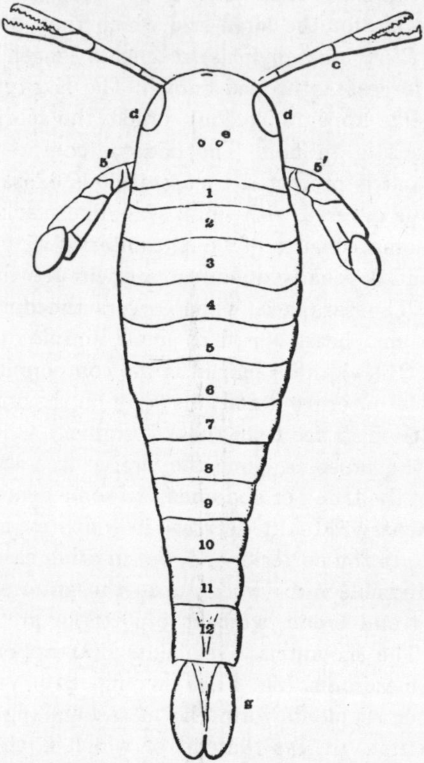

Fig. 214. Dorsal surface of *Pterygotus osiliensis*, from the Upper Silurian, Rootziküll. *c*, first pair of appendages (cheliceræ); *d*, compound eyes; *e*, simple eyes; *g*, tail-plate; 5′, sixth pair of appendages of prosoma; 1–6, segments of the mesosoma; 7–12, segments of the metasoma. Reduced. (After Schmidt.)

the metasoma each segment is surrounded by a continuous chitinous sheath.

The mouth is in a central position on the under surface of the prosoma (figs. 216, 219). In front of the mouth there

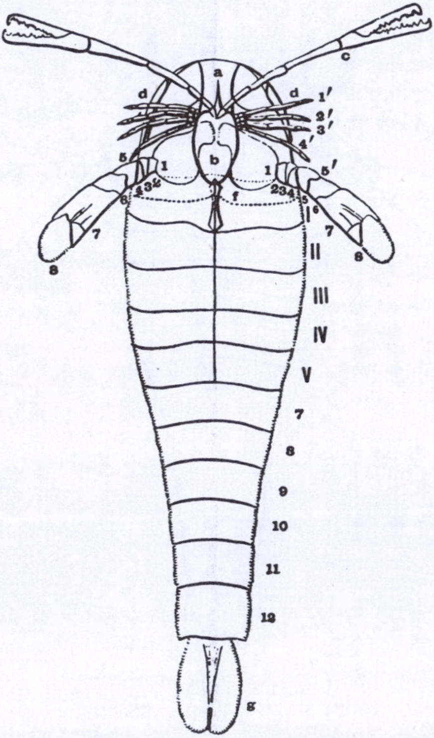

Fig. 215. Ventral surface of *Pterygotus osiliensis*, from the Upper Silurian, Rootziküll. *a*, epistome; *b*, metastoma; *c*, first pair of appendages (cheliceræ), consisting of three joints only (not as shown in the figure), a long basal joint, and two shorter joints forming the chela; *d*, compound eyes; *f*, I, genital operculum; *g*, tail-plate; 1′–5′, second to sixth pairs of appendages; II–V, ventral plate-like appendages of segments 3 to 6 of the mesosoma; 7–12, segments of the metasoma. Reduced. (After Schmidt.)

is one pair of appendages only (fig. 216, 1; 219, 1) which end in chelæ and are usually small; each consists of a basal

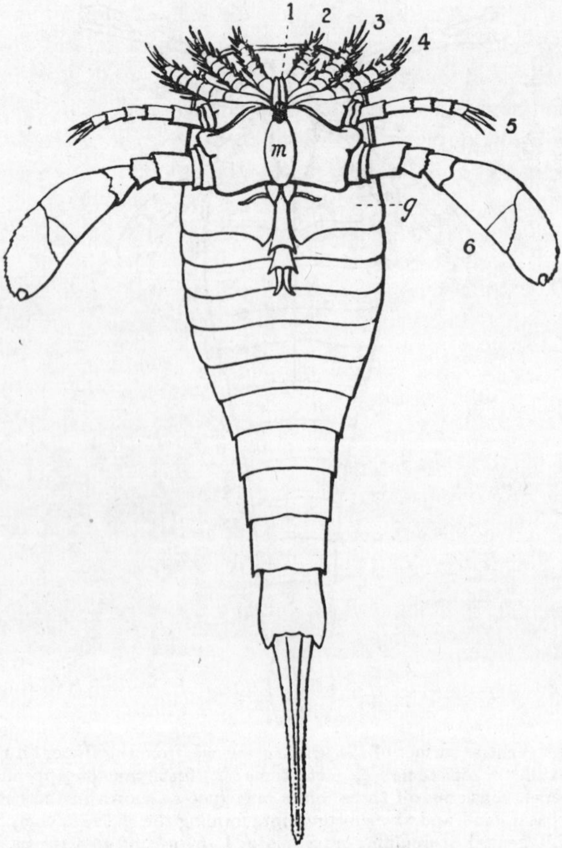

Fig. 216. ***Eurypterus remipes***, Silurian, New York. Restoration of ventral surface. 1–6, appendages of prosoma; *g*, genital operculum; *m*, metastoma. (After Ruedemann.) $\times \frac{1}{2}$.

joint (coxa) and two others which form the chela. The other five pairs of appendages (fig. 219, II–VI) are at the sides of the elongate mouth and usually increase in size from front to back; they consist of from six to eight joints each, and are not chelate; they functioned in locomotion, and also in mastication since the inner margins of the basal joints (or coxæ) are provided with tooth-like processes; the posterior pair (VI), except in *Stylonurus* and *Mixopterus*, are much larger than the others and have a very large basal joint. Placed just behind the mouth, in the median line, is an oval or heart-shaped plate, the *metastoma* (*b*), which covers the inner parts of the basal joints of the sixth pair of appendages. The metastoma represents the pair of chilaria of *Limulus* (p. 412); the presence in some cases of a notch in front, and a median longitudinal groove on the surface, supports the view that the metastoma originated from a pair of appendages. Although attached to the prosoma the metastoma is believed to represent the appendages of the first segment of the mesosoma which is generally of smaller size than the other segments (fig. 214, I)

Just as in *Limulus* (fig. 211) the carapace is continued on to the ventral surface, where it forms a marginal rim or 'doublure' which is separated from the dorsal part of the carapace by a marginal suture. The rim may be divided into two halves by a single suture in front (fig. 217 A); or by a pair of sutures which separate a median plate, the epistome (B, *e*), from the lateral parts; or there may be another suture on each side dividing each lateral part into an anterior and posterior part (C).

The six segments of the mesosoma bear on their ventral surfaces five pairs of plate-like appendages (fig. 215, I–V; fig. 219, VII–XII), each of which overlaps the one behind like the tiles on a roof. The first pair of plates form the genital operculum, and are divided in the middle by a median process, which often extends beyond the posterior margin of the operculum on to the next pair of appendages; the shape and size of the median process differ in the two sexes. The genital operculum covers the ventral surfaces of both the first and second segments of the mesosoma (fig. 219, VII, VIII). The other four pairs of appendages (fig. 215, II–V) are attached only near the front margin of each segment, and bear leaf-like gills (fig. 219, *c*) on their inner (or dorsal) surfaces. The segments of the metasoma

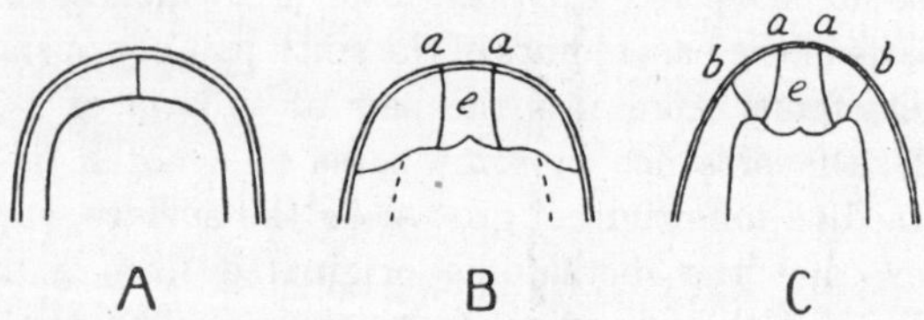

Fig. 217. Part of ventral surface of the prosoma, showing the marginal rim or 'doublure'. A, *Eurypterus*. B, *Pterygotus*. C, *Hughmilleria*. *a*, *b*, sutures; *e*, epistome. Reduced. (After Störmer.)

(figs. 215, 7–12; 219, XIV–XVIII) are protected by continuous chitinous rings and bear no appendages.

In the larval stages (fig. 218) the prosoma is relatively larger than in the adult owing to the fact that all the trunk segments are not yet developed. There is no clear distinction between mesosoma and metasoma, and in this respect the young forms agree with the adults of the more primitive types of Eurypterids. The large size of the compound eyes, and the prominence of the ocelli (fig. 218 B) are probably adaptations for planktonic life.

In many respects the Eurypterids resemble the Scorpions. The number of segments in each of the three regions of the body is the same, and the two pairs of eyes are similar in character and position. In both Eurypterids and Scorpions the prosoma bears six pairs of appendages, of which the first are pre-oral and chelate, and the remaining five agree in position and in general form; but in the Eurypterids the number of joints in the walking legs varies, and the basal segments of all serve as jaws, whereas in the Scorpions the last two pairs function only in locomotion; also in the Eurypterids the last leg and the genital operculum are much

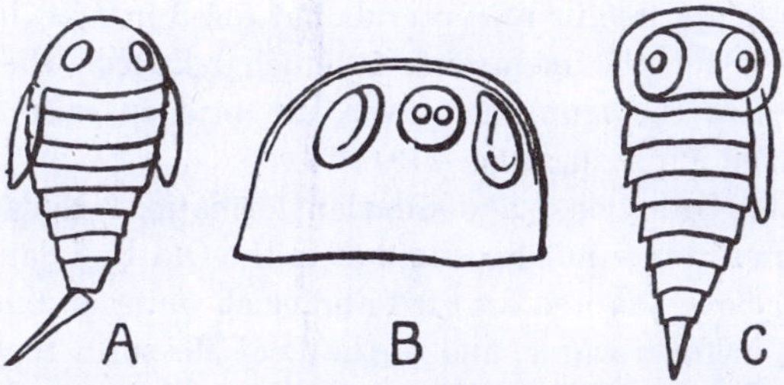

Fig. 218. Young stages of Eurypterids, Silurian. A, B, *Eurypterus maria*; A, ×6; B, prosoma, ×12; C, *Stylonurus myops*, ×20. (After Ruedemann.)

larger relatively than in the Scorpions. One of the characteristic features of the Eurypterids is the large metastoma. The pectines are absent in the Eurypterids, except perhaps in *Glyptoscorpius* from the Carboniferous. The lung-books of the Scorpions are represented by the leaf-like gills of the Eurypterids, but the plate-like appendages of the mesosoma are absent in the Scorpions. In both groups the segments of the metasoma are free and without appendages and at the posterior end is a tail-spine. The differences between the Eurypterids and recent Scorpions are to some extent bridged over by *Palæophonus*, a Silurian Scorpion (see p. 429).

The Eurypterids agree in many respects with *Limulus*. The principal points of difference are: (1) only the first pair of appendages are chelate in Eurypterids, whereas in *Limulus* all the walking-legs except the last, and the first in the male, may be chelate; (2) the last pair of legs are larger in Eurypterids than in *Limulus* and their basal joints assist in mastication; (3) the large, single plate forming the metastoma in Eurypterids is represented by the pair of small chilaria of *Limulus*; (4) the second segment of the mesosoma in Eurypterids is without appendages and is covered by the genital operculum; (5) in the trunk all the segments are free in Eurypterids but fused in *Limulus*, and in the latter the metasoma is much reduced—these differences in the trunk, however, are bridged over by the Palæozoic Xiphosura (fig. 213).

In the Ordovician and Silurian formations Eurypterids are found in marine deposits, but in the Old Red Sandstone they became adapted for life in brackish water and, in some places, in fresh water, and in the Coal Measures they seem to have lived in fresh water only. The broad flattened forms, with the compound eyes on the dorsal surface, and a tail-spine were probably benthonic and able to burrow in mud and sand in search of food in the same way that *Limulus* does at the present day. The narrower and more convex forms, with relatively smaller prosoma, lateral eyes, stream-lined body and broad tail-plate were probably active swimmers belonging to the necton.

Eurypterus (figs. 216, 218 A, B). Prosoma quadrate, the anterior angles rounded; the compound eyes are a little in front of the median lateral point on each side. The tail-spine is long, narrow, and pointed. The pre-oral appendages are small and consist of a basal joint and a chela; the second appendage consists of seven joints, the remaining four pairs of eight joints; all these five pairs of appendages are without chelæ. The second, third and fourth pairs are similar in structure and bear spines; the fifth pair are longer than the preceding and without spines;

and the sixth pair are much longer and also larger, with a large quadrate basal joint. The metastoma is oval. The median process of the genital operculum is short in the male, long in the female. Ordovician to Permian. Ex. *E. fischeri*, Upper Silurian.

Stylonurus. General form similar to *Pterygotus*. Second, third, and fourth pairs of appendages with spines; the two posterior pairs very long and slender. Compound eyes near the middle of the prosoma. Tail-spine long, pointed. Body sometimes nearly 5 feet long. Upper Silurian and Old Red Sandstone. Recorded from the Ordovician of New York. Ex. *S. powriei*, Upper Silurian and Old Red Sandstone.

Pterygotus (figs. 214, 215). Prosoma semi-oval, rounded in front; the compound eyes are at the margins. The tail-plate is oval and either bilobed or pointed at its extremity. The pre-oral appendages are long and chelate; the second, third, fourth and fifth pairs are similar to each other in size and structure; the sixth pair long and stout. Metastoma oval. The examples of this genus are often of enormous size, *P. anglicus* sometimes reaching a length of

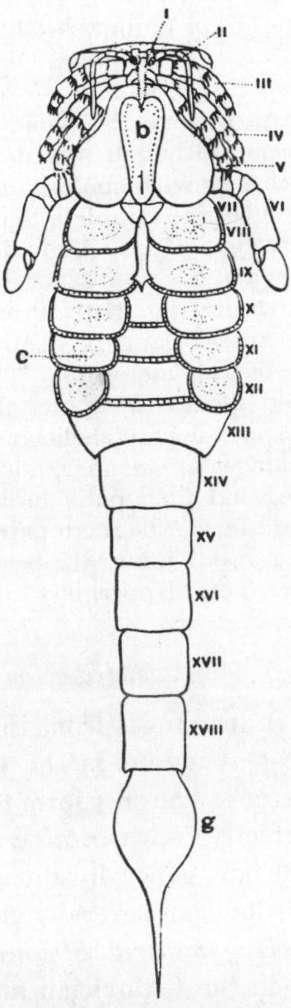

Fig. 219. *Slimonia.* Restoration of the under surface by M. Laurie. *b*, metastoma; *c*, leaf-like gills seen through the ventral plate-like appendages of the mesosoma; *g*, tail-plate; I–VI, appendages of the prosoma; VII–XII, segments of the mesosoma; XIII–XVIII, segments of the metasoma; VII–VIII, genital operculum. Reduced.

6 feet. Lower Ludlow to Old Red Sandstone. Ordovician of New York. Ex. *P. anglicus*, Old Red Sandstone; *P.* (*Erettopterus*) *bilobus*, Upper Silurian.

Hughmilleria. Of small size. Similar to *Pterygotus*, but the chelicerae, although well developed, are much shorter; second to fifth legs with spines; metastoma cordate; eyes not always at the margin; tail-plate lanceolate, in this respect approaching *Eurypterus*. Upper Ordovician to Old Red Sandstone. Ex. *H. socialis*, Silurian.

Slimonia (fig. 219). Prosoma quadrate; the compound eyes at the anterior angles. Segments of the mesosoma broader than those of the metasoma. The tail-plate is oval, ending in a pointed process or spine. Metastoma heart-shaped. The pre-oral appendages (chelicerae) are small; the second pair of appendages are slender, and composed of six joints; the third, fourth, and fifth pairs have seven joints, and are similar in size and form; the sixth pair are longer and have a large retort-shaped basal joint. Upper Ludlow and Passage Beds. Ex. *S. acuminata*, Uppermost Silurian.

Distribution of the Eurypterida

This Order ranges from the Cambrian to the Permian, but is most abundant in the Upper Silurian and the Old Red Sandstone. The only form recorded from the Cambrian is the imperfectly known *Strabops*,[1] from Missouri. Although Eurypterids are generally uncommon and poorly preserved in the Ordovician several genera have been recognised. *Pterygotus*, *Hughmilleria*, *Stylonurus*, *Mixopterus*, and *Eurypterus* begin in the Ordovician and continue into the Silurian, in which *Slimonia*, *Drepanopterus* and some others appear. In the Old Red Sandstone *Pterygotus*, *Slimonia*, *Stylonurus* and *Eurypterus* are the chief forms. In the Carboniferous and Permian the number of genera is reduced, the chief being *Eurypterus*.

[1] Perhaps more nearly related to the Xiphosura.

SUB-CLASS II. *EUARACHNIDA*

The Euarachnids breathe air by means of either pulmonary sacs or tracheæ, and the mesosoma is without plate-like appendages. The principal Orders are: (1) Scorpionida, (2) Pedipalpi, (3) Araneida, (4) Pseudoscorpionida, (5) Phalangida, (6) Acarina.

ORDER I. SCORPIONIDA

The Scorpions (fig. 220) have a long, narrow body, in which three regions are clearly marked. In front, the *prosoma* or cephalothorax consists of six fused segments, covered dorsally by a chitinous carapace which bears a pair of simple eyes near its centre, and a group of simple eyes at each of the two outer front margins. The middle region of the body—the *mesosoma* or pre-abdomen (7–12)—is formed of six free segments, which are short and broad; the chitinous sheath of each segment consists of a dorsal plate or *tergum* and a ventral plate or *sternum*. The posterior portion of the body is the *metasoma* or post-abdomen (13, 14), and is formed of six segments, each being encased in a complete chitinous cylinder, and all, except the first (13), are narrow; at the end of the last segment is the tail-spine (15), which bears the poison glands. The anal opening is on the last segment.

The prosoma bears six pairs of appendages: (1) the *chelicerœ* (fig. 220, 1) are small three-jointed limbs with chelæ, placed just in front of the mouth; (2) the *pedipalps* (2) are the largest appendages and are at the sides of the mouth; they consist of six joints, ending with chelæ, and the basal joints function in mastication; next come four pairs of seven-jointed walking legs (3–6) which end in claws, instead of chelæ; the basal joints of the third and fourth

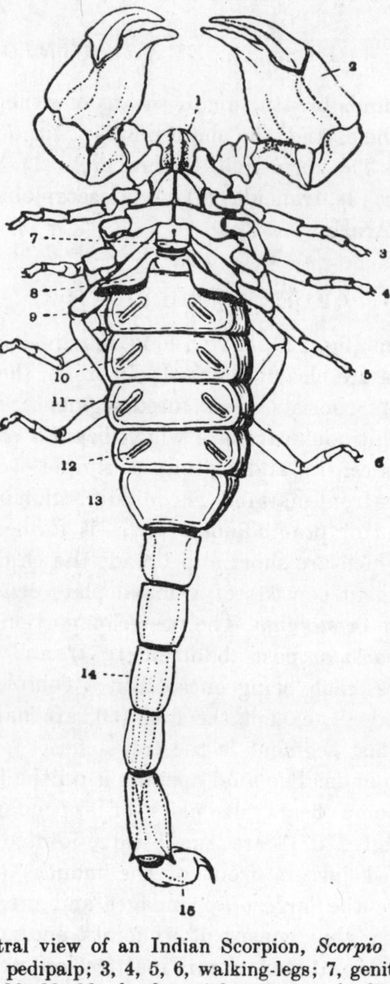

Fig. 220. Ventral view of an Indian Scorpion, *Scorpio swammerdami*. 1, chelicera; 2, pedipalp; 3, 4, 5, 6, walking-legs; 7, genital operculum; 8, pectines; 9, 10, 11, 12, the four right stigmata leading to the lung-books; 13, first segment of metasoma; 14, fourth segment of metasoma; 15, tail-spine. (From Shipley and MacBride.) $\times \frac{2}{3}$.

pairs assist in mastication. Between the bases of the last two pairs of legs, and immediately in front of the genital operculum, is a small plate, the *metasternite*, which represents fused sterna corresponding to these limbs.

On the seventh segment of the body (the *first* of the mesosoma) there is a small rounded plate—the *genital operculum* (fig. 220, 7). The eighth segment bears the *pectines* (8), which are tactile organs and consist of a stem with a row of short processes like the teeth of a comb. On segments nine to twelve, there are, in the adult, no proper appendages; but a pair of oblique, slit-like openings—the *stigmata*—occur on each of these segments, and lead into pulmonary sacs which contain the lung-books. The metasoma (segments 13 to 18) has no appendages.

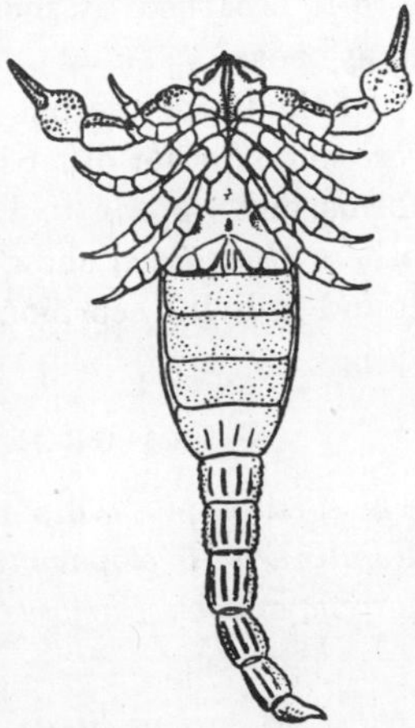

Fig. 221. *Palæophonus caledonicus* from the Upper Silurian of Lesmahago, Lanarkshire. Restoration of ventral surface by R. I. Pocock. × 1½.

Although this Order is of great antiquity, it has but few fossil representatives. *Palæophonus* (fig. 221) occurs in the Silurian rocks of Gotland and Lanarkshire; *Proscorpius* in the Silurian of North America. *Eoscorpius*, *Archæoctonus* and *Anthracoscorpio* are found in the Carboniferous. Imperfect specimens of scorpions have been obtained from the Trias of Warwickshire. One form (*Tityus*) is known from the Oligocene beds.

In some of its characters *Palæophonus* (fig. 221) is more primitive than later scorpions; the walking legs consist of nearly equal-sized joints and seem to be without claws;

the basal joints of all these legs could serve to some extent as jaws and in this respect resemble the walking legs of *Limulus* and still more those of the Eurypterida. *Palæophonus*, unlike later scorpions, seems to have been aquatic, since it is found associated with marine fossils, and, moreover, stigmata appear to have been absent—probably therefore it breathed by means of branchial lamellæ instead of lung-books.

Of the Carboniferous genera some (*Archæoctonus*, *Anthracoscorpio*) do not differ in any important respect from living forms and appear to have been as highly organised, but others (*Eobuthus*) show some morphological characters not found in living scorpions.

ORDER II. PEDIPALPI

The Pedipalpi ('whip scorpions', etc.) are represented by *Geralinura*, *Protophrynus* and *Græophonus* in the Carboniferous, and by *Phrynus* in the Tertiary rocks.

ORDER III. ARANEIDA

Spiders belonging to the genera *Protolycosa*, *Arthrolycosa*, etc., are found in the Coal Measures. In the Oligocene—especially in the amber of Prussia—a large number of forms occur. Others are found in the Eocene of Wyoming, and the Miocene of the Florissant, Colorado.

ORDER IV. PSEUDOSCORPIONIDA (CHERNETIDEA)

This order includes the 'book scorpions' (*Chelifer*) and others. Various forms, belonging to existing genera, occur in the Oligocene amber, *e.g. Chelifer*, *Chernes*.

ORDER V. PHALANGIDA (OPILIONINA)

Examples of this Order ('harvest-men', etc.) have been found in the Oligocene amber. A few forms found in the Carboniferous may belong to this Order.

ORDER VI. ACARINA

This Order comprises the mites and ticks. A mite (*Protacarus*) has been found in the Old Red Sandstone (Rhynie chert); various forms, belonging chiefly to living genera, occur in the Oligocene amber and other Tertiary deposits.

ORDER VII. ANTHRACOMARTI

This is an extinct Order, found in the Old Red Sandstone (Rhynie chert) and the Carboniferous, and appears to be related to the Pedipalpi and Phalangida. The principal genera are *Brachypyge*, *Anthracomartus*, *Kreischeria*, *Eophrynus*, *Anthracosiro*.

LIST OF PALÆONTOLOGICAL WORKS

ABBREVIATIONS

A.J.S.	American Journal of Science.
A.M.N.H.	Annals and Magazine of Natural History.
G.M.	Geological Magazine.
Q.J.G.S.	Quarterly Journal of the Geological Society.

GENERAL

Zittel, K. A. von. (1) Handbuch der Palaeontologie. 1876–93. (Also in French.) (2) and **Eastman, C. R.** Text-book of Palæontology. Ed. 2. 1913.

Swinnerton, H. H. Outlines of Palæontology. Ed. 2. 1930.

Bather, F. A. (1) Fossils and Life. Rep. Brit. Assoc. (1921), p. 61. (2) Biological Classification. Q.J.G.S. LXXXIII. (1927), p. lxii.

Depéret, C. The Transformation of the Animal World. 1909.

Osborn, H. F. Palæontology. Encyc. Brit., ed. 11, xx. 1911, p. 579.

McCoy, F. (1) British Palæozoic Fossils. [Pp. 1–184, 1851; pp. 185–406, 1852; pp. 407 to end, 1855.] (2) Carboniferous Limestone Fossils of Ireland. 1844. (3) Silurian Fossils of Ireland. 1846.

Murchison, R. I. (1) The Silurian System. 1839. (2) Siluria. Ed. 5. 1872. [For Lower Palæozoic Fossils.]

Whidborne, G. F. Devonian Fauna of the South of England. 3 vols. 1889–1907. (Palæont. Soc.)

Roemer, F. and **Frech, F.** Lethæa geognostica. (1) Palæozoica. 1876–1902. (2) Das Mesozoicum. 1903–13. (3) Das Cainozoicum. 1903–4.

Koninck, L. G. de. Faune du Calcaire Carbonifère de Belgique. 6 parts. 1878–87.

Davies, A. M. Tertiary Faunas. Vol. I. 1935.

Bassler, R. S. Index to American Ordovician and Silurian Fossils. Smithson. Inst., Bull. U.S. Nat. Mus. 92 (1915).

Etheridge, R. Fossils of the British Islands. (Palæozoic.) 1888.

Morris, J. Catalogue of British Fossils. Ed. 2. 1854.

Sherborn, C. D. Index Animalium. 1902–33.

Fossilium Catalogus. I. Animalia. 1913– (In course of publication.)

FORAMINIFERA

Brady, H. B. (1) Foraminifera (Challenger Report). 1884. (2) Carboniferous and Permian Foraminifera. 1876. (Palæont. Soc.)

Carpenter, W. B. Introduction to the Foraminifera. 1862.

Chapman, F. The Foraminifera. 1902.

Ciry, R. Les Fusulinidés de Turquie. Ann. Paléont., XXIX., XXX., 1941–43

Cushman, J. A. (1) Monograph of the Foraminifera of the North Pacific Ocean. Bull. U.S. Nat. Mus., 71 (1910–17). (2) Atlantic Ocean. *Ibid.* 104 (1918–24). (3) Foraminifera. Ed. 2. 1933. (4) Illustrated Key to the genera of Foraminifera. 1933.

Douvillé, H. (1) Évolution des Nummulites. C.R. Acad. Sci. Paris, CLXVIII. (1919), p. 651. (2) Revision des Lepidocyclines. Mém. Soc. géol. France, N.S. II. (1924–25), p. 1.

Gamble, F. W. Radiolaria (Lankester's Treatise on Zoology, I. 1, p. 94). 1909.

Gubler, J. Fusulinidés du Permian de l'Indochine. Mém. Soc. géol. France, N.S. XI. 26 (1935).

Jones, T. R., Parker, W. K. and **Brady, H. B.** Crag Foraminifera. 1866, 1895. (Palæont. Soc.)

Lemoine, P. and **Douvillé, R.** *Lepidocyclina.* Mém. Soc. géol. France, Paléont. XXXII. (1904).

Liebus, A. and **Thalman, H. E.** Foraminif. récent et foss. Foss. Catalogus, 49, 59, 60 (1931–33).

Lister, J. J. (1) The Foraminifera (Lankester's Zoology, I. 2). 1903. (2) Dimorphism of *Nummulites.* Proc. Roy. Soc. LXXVI. B (1905), p. 298.

Sherborn, C. D. (1) Bibliography of Foraminifera. 1888. (2) Index to Foraminifera. 1893–96. (Smithson. Misc. Coll.)

Staff, H. Anatomie und Physiologie der Fusulinen. Zoologica, XXII. (1910).

RADIOLARIA

Campbell, A. S. and **Clark, B. L.** (1) Eocene Radiolarian Faunas from Mt. Diablo Area, California. Geol. Soc. America, Special Paper 39, 1942. (2) Miocene Radiolarian Faunas from S. California. *Ibid.* 51, 1944.

Cayeux, L. Organismes dans le Terrain Pré-cambrien. Bull. Soc. géol France (3), XXII. (1894), p. 197. Also G.M. (1894), p. 419; and **Rauff, H.** Neues Jahrb. für Min. etc. I. (1896), p. 116.

Haeckel, E. (1) Die Radiolarien. 1862. (2) Report on the Radiolaria (Challenger Report). 1887.

Hinde, G. J. (1) Ordovician Radiolaria. A.M.N.H. (6), VI. (1890), p. 40. (2) Devonian and Carboniferous Radiolaria. Q.J.G.S. XLIX. (1893), p. 215; LI. (1895), p. 609; LV. (1899), pp. 38, 214.

Holmes, W. M. Chalk Radiolaria. Q.J.G.S. LX. (1899), p. 694. Also *ibid.* LI. (1895), p. 600; G.M. (1895), p. 345.

Rüst, D. Radiolarien. Palæontographica, (Jurassic) XXXI. (1885), p. 269; (Chalk) XXXIV. (1888), p. 181; (Trias and Palæozoic) XXXVIII. (1892), p. 107; (Jurassic and Chalk) XLV. (1898), p. 1.

PORIFERA

Dendy, A. (1) Sponges. Encyc. Brit., ed. 11, XXV. 1911, p. 715. (2) The Tetraxonid Sponge Spicule: a Study in Evolution. Acta Zoologica, II. (1921). (3) Orthogenetic Series of growth forms in Tetraxonid spicules. Proc. Roy. Soc. XCVII. B (1924), p. 243.

Gerth, H. Spongien aus dem Perm von Timor. Paläont. v. Timor, XVI. (1929).

Hall, J. and **Clarke, J. M.** Palæozoic Reticulate Sponges (Dictyospongidæ). New York. 1898.

Hinde, G. J. (1) Catalogue of Sponges in the Geological Department of the British Museum. 1883. (2) British Fossil Sponges. 1887–93. (Palæont. Soc.) (3) *Porosphara.* Journ. R. Micr. Soc. (1904), p. 1.

Hudson, R. G. S. *Erythrospongia.* Proc. Yorks. Geol. Soc. XXI. (1929), p. 181.

Kolb, R. Die Kieselspongien des schwäbischen weissen Jura. Palæontographica, LVII. (1910).

Minchin, E. A. Sponges (Lankester's Zoology, II.). 1900.

Moret, L. (1) Spongiaires siliceux du Miocène de l'Algérie. Mém. Soc. géol. France, N.S. Mém. 1 (1924). (2) Spongiaires siliceux du Crétacé supérieur français. *Ibid.* 5 (1926)

Okulitch, V. J. North American Pleospongia. Geol. Soc. America, Special Paper 48 (1943).

Parona, C. F. Le Spugne Permiana di Palazzo Adriano in Sicilia. Mem. Soc. geol. Ital. I. (1933).

Rauff, H. (1) Palæospongologie. Palæontographica, XL. (1893–94); XLI. (1895), p. 223. (2) Receptaculitidæ. Abhandl. der math.-phys. Classe d. k. bayer. Akad. XVII. (1892), p. 645.

Ridley, S. O. and **Dendy, A.** Monaxonida (Challenger Report). 1887.

Schrammen, A. Die Kieselspongien d. oberen Kreide N.W. Deutschland. Palæontographica, Suppl.-Bd. v. (1910–12); Mon. z. Geol. u. Palæont. I. 2 (1924).

Schultze, F. E. Hexactinellida (Challenger Report). 1887.

Sollas, W. J. Tetractinellida (Challenger Report). 1888.

Vinassa de Regny. P. Trias-spongien aus dem Bakony. 1901.

Walcott, C. D. Middle Cambrian Spongiæ. Cambr. Geol. and Pal. IV. 6 (1920).

GRAPTOLITHINA

Barrande, J. Graptolites de la Bohême. 1850.

Bulman, O. M. B. (1) British Dendroid Graptolites. (Palæont. Soc.) 1927– . (2) Graptolites prepared by Holm. Arkiv f. Zool. 24–28 (1932–36). (3) Programme-evolution in Graptolites. Biol. Reviews, VIII. (1933), p. 311. (4) Graptolithina. Handb. d. Paläozool. 1938. (5) Caradoc Graptolites from Ayrshire, 1945– . (Palæont. Soc.)

Eisenach, A. Mikrofossilien des baltischen Silurs; des böhmischen Silurs. Palæont. Zeitschr. XVI. (1934), p. 52.

Elles, G. L. and **Wood, E. M. R.** British Graptolites 1901–14. (Palæont. Soc.) **Elles,** Graptolite Faunas of the British Isles. Proc. Geol. Assoc. XXXIII. (1922), p. 168.

Hall, J. Graptolites of the Quebec group. 1865.

Holm, G. (1) On *Didymograptus*, *Tetragraptus*, and *Phyllograptus*. G.M. (1895), pp. 433, 481. (2) Gotlands Graptoliter. Bih. k. Svenska Vet. Akad. XVI. 7 (1890).

Hopkinson, J. (1) Morphology of Rhabdophora. A.M.N.H. (5), IX. (1882), p. 54. (2) Reproduction. *Ibid.* (4), VII. (1871), p. 317.

Kozlowski, R. Graptolithes du Tremadoc de la Pologne. Ann. Mus. Zool. Polonici, XIII. (1938), p. 183.

Kraft, P. Ontogenetische Entwickelung und Biologie von *Diplograptus* und *Monograptus*. Palæont. Zeitschr. VII. (1926), p. 207.

Lapworth, C. (1) Distribution of Rhabdophora. A.M.N.H. (5), III. (1879), pp. 245, 449; IV. (1879), pp. 333, 423; V. (1880), pp. 45, 273, 358; VI. (1881), pp. 16, 185. (2) Classification of Rhabdophora. G.M. X. (1873), pp. 500, 555.

Nicholson, H. A. and **Marr, J. E.** Phylogeny of Graptolites. G.M. (1895), p. 529.

Perner, J. Graptolites de la Bohême. Parts I.–III. 1894–99.

Ruedemann, R. (1) Development and Mode of Growth of *Diplograptus*. 14th Ann. Rep. State Geol. New York for 1894 (1895), p. 219. Also A.J.S. (3), XLIX. (1895), p. 453. (2) Graptolites of New York, Part I. 1904; Part II. 1908. (N. York State Mus. Mem. 7.)

Stubblefield, C. J. Early British Graptolites. G.M. (1929), p. 268.

Törnquist, S. L. (1) Structure of Diprionidæ. Särtryck af Konl. Fysiogr., Handl. Ny Följd, IV. (1893). (2) Graptolites of *Phyllo-Tetragraptus* Beds. Konl. Fysiogr. Sällsk. Handl. XII. (1901).

Walther, J. Die Lebensweise der Graptolithen. Zeitschr. der deutsch. geol. Gesellsch. XLIX. (1897), p. 238.

Wiman, C. (1) Diplograptidæ and Monograptidæ. Bull. Geol. Inst. Upsala, I. (1894), pp. 97, 113. (2) Über die Graptoliten. *Ibid.* II. (1895), p. 1.

STROMATOPOROIDEA

Heinrich, M. Structure and Classification of Stromatoporoidea. Journ. Geol. XXIV. (1916), p. 57.

Hickson, S. J. *Gypsina* and the systematic position of the Stromatoporoids. Q.J. Micr. Sci. LXXVI. (1934), p. 433.

Kuhn, O. Hydrozoa. Foss. Catalogus, 36 (1928).

Nicholson, H. A. British Stromatoporoidea. 1886–92. (Palæont. Soc.)

Parks, W. A. Niagara Ordovician and Silurian Stromatoporoids. Univ. Toronto Studies, Geol. Series, 5, 6, 7. 1908–10.

Steiner, A. Stromatopores secondaires. Mém. Soc. Vaudoise Sci. nat. IV. (1932), p. 105.

Törnquist, A. Ueber mesozoische Stromatoporiden. Sitzung. k. preuss. Akad. Wissensch. Berlin (1901), p. 1115. Also **Bakalow,** Neues Jahrb. für Min. etc. I. (1906), p. 13; **Deninger,** *ibid.* p. 61.

Tripp, K. Über den Skelettbau von Hydractinien zu einer vergleichenden Betrachtung der Stromatoporen. Neues Jahrb. für Min. etc. Beil.-Bd. LXII. B (1929), p. 467.

SCYPHOMEDUSÆ

ieslinger, A. (1) Medusæ fossiles. Foss. Catalogus, 26 (1924). (2) Eine Meduse aus der alpinen Trias. Neues Jahrb. für Min. etc. Beil.-Bd. LI. (1926), p. 494.

aas, O. Ueber Medusen aus dem Solenhofer Schiefer und der Kreide. Palæontographica, XLVIII. (1902), p. 297.

alcott, C. D. (1) Fossil Medusæ. Mon. U.S. Geol. Survey, XXX. (1898). (2) Cambr. Geol. and Pal. II. (1911), p. 55.

ANTHOZOA (ACTINOZOA)

cher, C. E. (1) Development of a Palæozoic Poriferous Coral. Trans. Connecticut Acad. VIII. (1893), p. 207. (2) Development of Favosi- idæ. *Ibid.* p. 215. **Girty,** Amer. Geol. XV. (1895), p. 131.

ard, H. M. *Alveopora* and the Favositidæ. Journ. Linn. Soc. Zool.), XXVI. (1898), p. 495.

ne, G. C. (1) The Anthozoa (Lankester's Zoology, II.). 1900. 2) *Heliopora,* etc. Phil. Trans. Roy. Soc. 186 (1895), p. 455.

k, G., Bernard, H. M. and **Matthai, G.** Catalogue of Madreorarian Corals in the British Museum. I.–VII. 1893–1928.

n, T. C. Development of *Streptelasma.* A.J.S. (4), XXIII. (1907), 277.

Carruthers, R. G. (1) Septal Plan of the Rugosa. A.M.N.H. (7), XVIII. (1906), p. 356. (2) Revision of Carboniferous Corals. G.M. (1908), pp. 20, 63. (3) Evolution of *Zaphrentis delanouei*. Q.J.G.S. LXVI. (1910), p. 523.

Dana, J. D. Zoophytes (Wilkes Expedition). 1848.

Duerden, J. E. (1) Morphology of Madreporaria. Septal Sequence. Biol. Bull. VII. (1904), p. 79; IX. (1905), p. 27. (2) Relationships of Rugosa to Zoantheæ. A.M.N.H. (7), IX. (1902), p. 381; XVIII. (1906), p. 226.

Duncan, P. M. (1) British Fossil Corals. 1866–72. (Palæont. Soc.) (2) Revision of the Families and Genera of the Madreporaria. Journ. Linn. Soc. (Zool.), XVIII. (1885), pp. 1–204.

Dybowski, W. N. Monographie der Zoantharia Rugosa, etc. Arch. für Naturk. Liv-, Est- und Kurlands, V. (1874).

Edwards, H. Milne. (1) Histoire naturelle des Coralliaires. 1857–60. (2) and **Haime, J.** British Fossil Corals. 1850–54. (Palæont. Soc.)

Faurot, L. Affinités des Tétracoralliaires et des Hexacoralliaires. Ann. Paléont. IV. (1909), p. 69.

Felix, J. Die Anthozoën der Gosauschichten. Palæontographica, XLIX. (1903), p. 163.

Frech, F. (1) Die Korallenfauna der Trias. Palæontographica, XXXVII. (1890), p. 1. (2) Die Korallenfauna des Oberdevons in Deutschland. Zeitschr. der deutsch. geol. Gesellsch. XXXVII. (1885), pp. 21, 946.

Gerth, H. Anthozoën der Dyas von Timor. Paläont. v. Timor, IX. (1921), p. 67.

Gordon, C. E. Early Stages in Palæozoic Corals. A.J.S. (4), XXI. (1906), p. 109.

Grabau, A. W. Palæozoic Corals of China. Palæont. Sinica, B, II. 1, 2 (1922, 1928).

Gregory, J. W. Jurassic of Cutch. II. Corals. Palæont. Indica (1900).

Hickson, S. J. *Tubipora*. Q.J. Micr. Sci. XXIII. (1883), p. 556.

Hill, D. (1) Silurian Rugose Corals with acanthine septa. Phil. Trans. Roy. Soc. 226 B (1936), p. 189. (2) Carboniferous Rugose Corals of Scotland, 1938–41. (Palæont. Soc.)

Hinde, G. J. *Archæocyathus*, etc. Q.J.G.S. XLV. (1889), p. 125.

Huang, T. K. Permian Corals of S. China. Palæont. Sinica, B, VIII. 2 (1932).

Kiär, J. (1) Korallenfaunen des norwegischen Silursystems. Palæontographica, XLVI. (1899), p. 1. (2) Mittelsilurischen Heliolitiden. Skrift. Videns.-Selsk. i Christiana, I. (1903), p. 10.

Koby, F. (1) Polpiers jurassiques de la Suisse. (Mém. Soc. Pal. Suisse.) 1881–95. (2) Polypiers crétacés de la Suisse. (*Ibid.*) 1896–97.

Kunth, A. Foss. Korallen. Zeitschr. der deutsch. geol. Gesellsch. XXI. (1869), p. 647.

Lang, W. D. (1) Trends in Carboniferous Corals. Proc. Geol. Assoc. XXXIV. (1923), p. 120. (2) and **Smith, S.** Rugose Corals in Murchison's 'Silurian System'. Q.J.G.S. LXXIII. (1927), p. 448. (3) and **Smith, S.** and **Thomas, H. D.** Index of Palæozoic Coral Genera. 1940.

Lindström, G. Heliolitidæ. Bih. k. Svenska Vet. Akad. Handl. XXXII. No. 1 (1899). **Gregory**, Proc. Roy. Soc. LXVI. (1900), p. 291.

Nicholson, H. A. (1) Tabulate Corals of the Palæozoic Period. 1879. (2) *Tubipora* and *Syringopora*. Proc. Roy. Soc. Edinb. XI. (1880–81), p. 219. (3) and **Thomson, J.** Study of the chief types of Palæozoic Corals. A.M.N.H. (4), XVI. (1875), pp. 305, 424; XVII. (1876), pp. 60, 123, 290, 451; XVIII. (1876), p. 68.

Ogilvie, M. M. (1) Microscopic and systematic study of Madreporarian Corals. Phil. Trans. Roy. Soc. 186 (1896), p. 83; Nature, LV. (1897), p. 280. (2) Q.J. Micr. Sci. LI. (1907), p. 473. (3) Korallen der Stramberger Schichten. Palæontographica, Suppl. III. (1897).

Ortmann, A. Die Morphologie des Skelettes der Steinkorallen. Zeitschr. für wiss. Zool. L. (1890), p. 278; Neues Jahrb. für Min. etc. II. (1887), p. 185.

Rominger, C. Fossil Corals. Geol. Survey, Michigan, III. (1876).

Sardeson, P. W. Ueber die Beziehungen der fossilen Tabulaten zu den Alcyonarien. Neues Jahrb. für Min. etc. X. (1896), p. 249. **Weissermel**, Zeitschr. der deutsch. geol. Gesellsch. XLIX. (1897), p. 368, and L. (1898), p. 54. **Janensch**, *ibid.* LV. (1903), p. 486.

Schindewolf, O. H. *Petraia.* Q.J.G.S. LXXXVIII. (1931), p. 630.

Schlüter, C. Anthozoën des rheinischen Mittel-Devon. Abhandl. der preuss. geol. Landes-Anst. VIII. (1889).

Smith, S. (1) *Aulophyllum.* Q.J.G.S. LXIX. (1913), p. 51. (2) *Lonsdaleia. Ibid.* LXXI. (1916), p. 218. (3) *Phillipsastræa, Orionastræa. Ibid.* LXXII. (1917), p. 280. (4) *Aulina.* A.M.N.H. (9), XVI. (1925), p. 485. (5) *Corwenia. Ibid.* XVII. (1926), p. 149. (6) *Tryplasma. Ibid.* XX. (1927), p. 305. (7) Calostylidæ. *Ibid.* (10), V. (1930), p. 257. (8) Valentian Corals. Q.J.G.S. LXXXVI. (1930), p. 291. (9) Upper Devonian Corals of the Mackenzie River Region. Geol. Soc. America, Special Paper 59, 1945.

Vaughan, T. W. (1) Eocene and Oligocene Coral Faunas of the United States. Mon. U.S. Geol. Survey, XXXIX. (1900). (2) Revision of the Scleractinia. Geol. Soc. America, Special Paper 44, 1943.

Wedekind, R. Zoantharia Rugosa von Gotland. Sverig. geol. Undersök., Ca. 19 (1927).

Wentzel, J. Zoantharia Tabulata. Denkschr. der k. Akad. Wissensch. Math.-nat. Classe (Wien), LXII. (1895), p. 479.

Yü, C. C. Lower Carboniferous Corals of China. Palæont. Sinica, B, XII. 3 (1933).

ECHINODERMA

General

Bather, F. A., Gregory, J. W. and **Goodrich, E. S.** The Echinoderma (Lankester's Zoology, III.) 1900. **Bather**, Encyc. Brit. ed. 11, VIII. 1910, p. 871.

Clark, W. B. and **Twitchell, M. W.** Mesozoic and Cenozoic Echinodermata of the United States. Mon. U.S. Geol. Surv. LIV. (1915).

Delage, Y. and **Hérouard.** Zoologie Concrète. III. Échinodermes. 1903.

Eleutherozoa

Agassiz, A. (1) Revision of the Echini. Mem. Mus. Comp. Zool. III. (1872–74). (2) Panamic Deep Sea Echini. *Ibid.* XXXI. (1904). (3) Echinoidea (Challenger Report). 1881.

Bather, F. A. (1) Triassic Echinoderms of Bakony. 1909. (2) *Bothriocidaris.* Palæont. Zeitschr. XIII. (1931), p. 55. (3) and **Spencer, W. K.** Ordovician Echinoid from Girvan. A.M.N.H. (10), XIII. (1934), p. 557.

Beurlen, K. Collyritidæ. Palæontographica, LXXX. A (1934), p. 41.

Broili, F. Eine Holothurie aus dem oberer Jura. Sitz. Bayerisch. Akad. (1926), p. 341.

Cotteau, G. (1) Paléont. Franç., Terrain Tertiaire. Échinides éocènes. 1885–94. (2) and **Triger**, Échinides de la Sarthe. 1855–69.

Croneis, C. and **McCormack, J.** Fossil Holothuroidea. Journ. Paleont. VI. (1932), p. 111.

Deecke, W. Echinoidea jurassica. Foss. Catalogus, 39 (1928).

Desor, E. (1) Synopsis des Échinides fossiles. 1858. (2) and **Loriol, P. de.** Échinologie Helvétique. Jurassique. 1868–72.

Duncan, P. M. (1) Structure of the Ambulacra of fossil Regular Echinoidea. Q.J.G.S. XLI. (1885), p. 419. (2) Revision of the Genera of the Echinoidea. Journ. Linn. Soc. (Zool.), XXIII. (1889), p. 1.

Etheridge, R., jun. Holothuroidea in the Carboniferous. Proc. Roy. Phys. Soc. Edinb. VI. (1880–81), p. 183.

Forbes, E. (1) Asteriadæ in British Strata. Mem. Geol. Survey. Organic Remains, dec. I. (1849). (2) Echinodermata of the British Tertiaries. 1852. (Palæont. Soc.)

Fourtau, R. Catalogue des Invertébrés fossiles de l'Égypte. Échinides Éocènes, 1913; Échinodermes Crétacés, 1914; Échinides Néogènes, 1920; Jurassiques, 1924.

Gordon, I. (1) Development of the test of *Echinus.* Phil. Trans. Roy. Soc. 214 B (1926), p. 259. (2) *Echinocardium. Ibid.* 215 B (1926), p. 255. (3) *Arbacia. Ibid.* 217 B (1929), p. 289.

Gregory, J. W. (1) British Cainozoic Echinoidea. Proc. Geol. Assoc. XII. (1891), p. 16. (2) Echinothuridæ. Q.J.G.S. LIII. (1897), p. 112. (3) *Lindstromaster*, etc. G.M. (1899), p. 341. (4) Palæozoic Ophiuroidea. Proc. Zool. Soc. (1896), p. 1028.

Hawkins, H. L. (1) Holectypoida. Proc. Zool. Soc. (1912), p. 440 and Phil. Trans. Roy. Soc. 209 B (1920), p. 377. (2) *Echinocystis* and *Palæodiscus*. Q.J.G.S. LXXXIII. (1927), p. 574. (3) Lantern and girdle of fossil and recent Echinoids. Phil. Trans. Roy. Soc. 223 B (1934), p. 617. (4) Evolution and Habit among the Echinoidea. Q.J.G.S. XCIX. (1943), p. lii.

Jackson, R. T. (1) Phylogeny of Echini, with a revision of Palæozoic species. Mem. Boston Soc. Nat. Hist. VII. (1912). (2) Palæozoic Echini of Belgium. Mém. Mus. R. hist. nat. Belg. XXXVIII. (1929). (3) Fossil Echini of the W. Indies. Carnegie Inst. Publ. 306 (1922).

Lambert, J. (1) Échinides de l'Infra-Lias et du Lias. Bull. Soc. Sci. de l'Yonne, LIII. (1900), p. 3. (2) and **Thiéry, P.** Essai de Nomenclature raisonnée des Échinides. 1909–12.

Loriol, P. de. Échinologie Helvétique. Crétacé. 1873.

Lovén, S. (1) Echinologiea. Bih. k. Svenska Vet. Akad. Handl. XVIII. 4 (1892). (2) Études sur les Échinoides. *Ibid.* XI. 7 (1874).

Lyman, T. Ophiuroidea (Challenger Report). 1882.

MacBride, E. W. and **Spencer, W. K.** *Aulechinus*, *Ectinechinus* and *Eothuria* from the Ordovician of Girvan. Phil. Trans. R.S., B 229 (1938), p. 91.

Mortensen, T. Monograph of the Echinoidea. I. Cidaroidea. 1928. II. Bothriocidaroida, etc. 1935.

Orbigny, A. d'. Paléontologie française. Terr. crét.: VI. Échinides irréguliers. 1855–60. Continued by **Cotteau**. VII. Échinides réguliers. 1862–67. Terr. jur.: IX. Échinides irréguliers. 1867–74. XI. Échinides réguliers. 1882–89.

Pomel, A. Classification des Échinides. 1883.

Ravn, J. P. J. Echinider i Dammarks Kridtaflej. Mém. Acad. R. Sci. et Lett. Danemark (8). XI. (1927). p. 307; (9), I. (1928), p. 1.

Rowe, A. W. *Micraster*. Q.J.G.S. LV. (1899), p. 494.

Sladen, W. P. (1) Asteroidea (Challenger Report). 1889. (2) and **Spencer, W. K.** British Fossil Echinodermata. II. Cretaceous Asteroidea and Ophiuroidea. 1891–1908. (Palæont. Soc.)

Smiser, J. S. Belgian Cretaceous Echinoids. Mém. Mus. R. hist. nat. Belg. LXVIII. (1935).

Sollas, W. J. (1) Silurian Echinoidea and Ophiuroidea. Q.J.G.S. LV. (1899), p. 692. (2) *Lapworthura*, Phil. Trans. Roy. Soc. 202 B (1912), p. 213.

Spencer, W. K. (1) Evolution of the Cretaceous Asteroidea. Phil. Trans. Roy. Soc. 204 B (1913), p. 99. (2) British Palæozoic Asterozoa. 1914– (Palæont. Soc.)

Stürtz, B. Beiträge zur Kenntniss palæozoischer Seesterne. Palæontographica, XXXII. (1886), p. 75; XXXVI. (1890), p. 203.

Wright, T. (1) British Oolitic Echinodermata. I. Echinoidea. 1857–78. II. Asteroidea and Ophiuroidea. 1863–80. (2) British Cretaceous Echinodermata. I. Echinoidea. 1864–82. (Palæont. Soc.)

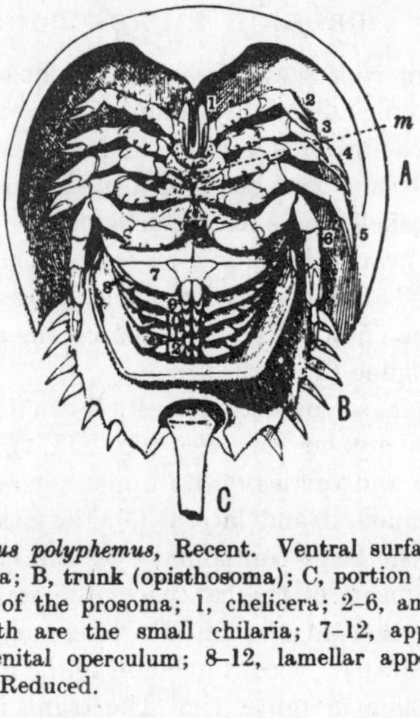

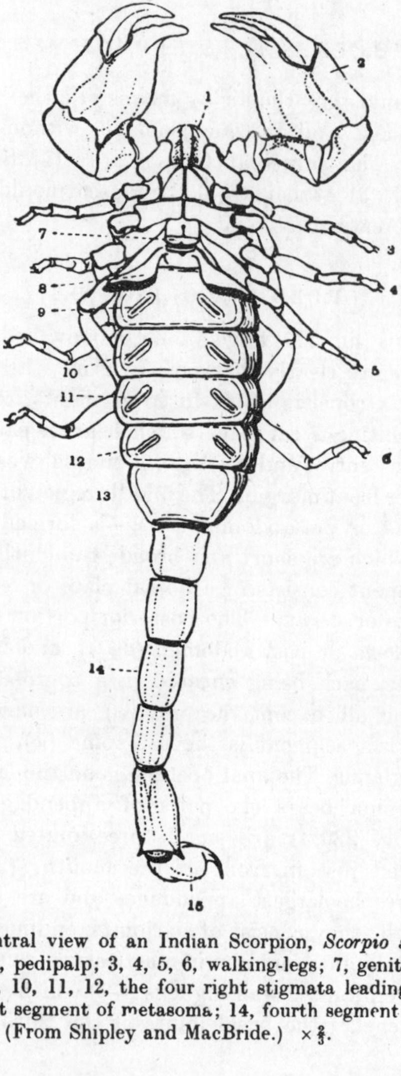

Pelmatozoa

Bassler, R. S. (1) Edrioasteroidea. Smithson. Misc. Coll. XCIII. 8 (1935); XCIV. 6 (1936). (2) Index of Paleozoic Pelmatozoan Echinoderms. Geol. Soc. America, Special Paper, 45, 1943.

Bather, F. A. (1) British Fossil Crinoids. A.M.N.H. (6), V. (1890), pp. 306, 373, 485; VI. (1890), p. 222; VII. (1891), pp. 35, 389; IX. (1892), pp. 189–202. (2) Terms in Crinoid Morphology. *Ibid.* IX. (1892), p. 51. (3) Crinoidea of Gotland. Part I. Bih. k. Svenska Vet. Akad. Handl. XXV. (1893). (4) *Uintacrinus*. Proc. Zool. Soc. (1895), p. 974. (5) *Petalocrinus*. Q.J.G.S. LIV. (1898), p. 401. (6) Studies in Edrioasteroidea, I.–IX. 1915, reprinted from G.M. (1898–1915). (7) Caradocian Cystidea from Girvan. Trans. Roy. Soc. Edinb. XLIX. (1913), p. 359. (8) *Cothurnocystis*: a study in adaptation. Palæont. Zeitschr. VII. (1925), p. 1. (9) *Saccocoma cretacea*. Proc. Geol. Assoc. XXXV. (1924), p. 111. (10) The Fossil and its environment. Q.J.G.S. LXXXIV. (1928), p. lxi. (11) Triassic Echinoderms of Timor. Paläont. v. Timor, XVI. (1929).

Billings, E. Cystideæ of the Lower Silurian Rocks of Canada. Geol. Survey of Canada: Organic Remains, dec. III. (1858).

Buch, L. von. Ueber Cystideën. Abhandl. d. k. Akad. d. Wiss. zu Berlin (1844), p. 89.

Carpenter, P. H. (1) Crinoidea (Challenger Report). 1884–8. (2) Oral and Apical Systems of Echinoderms. Q.J. Micr. Sci. XVIII. (1878), p. 351; XIX. (1879), p. 176. (3) Morphology of Cystidea. Journ. Linn. Soc. (Zool.). XXIV. (1894), p. 1. (4) and **Etheridge, R.** Catalogue of Blastoidea in the British Museum. 1886.

Clark, A. H. Monograph of existing Crinoids. Bull. U.S. Nat. Mus. 82 (1915–21).

Croneis, C. and **Geis, H. L.** Ontogeny of the Blastoidea. Journ. Paleont. XIV. (1940), p. 345.

Forbes, E. Cystidea of the Silurian Rocks of the British Islands. Mem. Geol. Survey, Vol. II. Part II. (1848).

Goldring, W. Devonian Crinoids of New York. N. York State Mus. Mem. 16 (1923).

Jaekel, O. Stammesgeschichte der Pelmatozoen. I. Thecoidea und Cystoidea. 1899.

Koninck, L. de and **Le Hon, H.** Crinoïdes du Terrain carbonifère de la Belgique. 1854.

Moore, R. C. and **Laudon, L. R.** Evolution and Classification of Paleozoic Crinoids. Geol. Soc. America, Special Paper 46, 1943.

Peck, R. E. Lower Cretaceous Crinoids from Texas. Journ. Paleont. XVII. (1943), p. 451.

Waagen, W. and **Jahn, J. J.** In Barrande's Système Silurien du centre de la Bohême. VII. 1, Cystidées, 1887; 2, Crinoïdes, 1899.

Wachsmuth, C. and **Springer, F.** (1) Revision of the Palæocrinoidea. Proc. Acad. Nat. Sci. Philadelphia (1879), p. 226; (1881), p. 177; (1885), p. 225; (1886), p. 64. (2) *Crotalocrinus. Ibid.* (1888), p. 364. (3) North American Crinoidea Camerata. Mem. Mus. Zool. Harvard, XX., XXI. (1897). (4) **Springer**, *Uintacrinus. Ibid.* XXV. (1901). (5) **Springer**, Crinoidea Flexibilia. Smithson. Inst. 2501 (1920). (6) **Springer**, American Silurian Crinoids. 1926.

Wanner, J. (1) Die permischen Krinoiden von Timor. Jaarb. Mijnw. Ned. O.-Indië, Verh. 3 (1921), 1923. (2) Die permischen Echinodermen. Paläont. v. Timor, VI. (1916); XIV. (1924).

Withers, T. H. Catalogue of Machæridia (Brit. Mus.). 1926.

Wright, J. (1) British Carboniferous Crinoids. G.M. (1926), p. 145; (1927), p. 353; (1928), p. 246; (1932), p. 337; (1933), p. 193; (1934), p. 241; (1935), p. 193, (1942), p. 269; (1943), pp. 81, 231; (1945), p. 114. (2) Scottish Carboniferous Crinoids. Trans. R. Soc. Edinb. LX. (1941), p. 1.

Yakovlev, N. Crinoidi Permiani di Sicilia. Palæont. Ital. XXXIV. (1933), 1935, p. 269.

BRACHIOPODA

Barrande, J. Système Silurien du centre de la Bohême. Vol. v. 1879.

Beecher, C. E. Development and Classification of Brachiopoda. (Studies in Evolution.) 1901, pp. 229–415.

Bittner, A. Brachiopoden der Alpinen Trias. Abhandl. d. kk. geol. Reichsanst. XIV. (1890); XVII. (1892).

Buckman, S. S. Brachiopoda of Namyau Beds, Burma. Palæont. Indica, N.S. III. (1917).

Davidson, T. (1) British Fossil Brachiopoda. 6 vols. 1851–86. (Palæont. Soc.) (2) Monograph of Recent Brachiopoda. Trans. Linn. Soc. (Zool.), (2), IV. (1886–88).

Deslongchamps, E. Sur le développement du deltidium. Bull. Soc. géol. France, (2), XIX. (1862), p. 409.

Fenton, C. L. Evolution in *Spirifer*. Publ. Wagner Free Institute, Philadelphia, II. (1931).

Fischer, P. Manuel de Conchyliologie. 1887. (Brachiopods, by Œhlert.)

Fischer, P. and **Œhlert, D. P.** Sur l'évolution de l'appareil brachial de quelques Brachiopodes. C.R. Acad. Sci. Paris, CXV. (1892), p. 749.

Friele, H. Development of the Skeleton in *Waldheimia*. Arch. Math. Nat. II. (1877), p. 380.

George, T. N. Carboniferous Reticulate Spiriferidæ. Q.J.G.S. LXXXVIII. (1932), p. 516.

Hall, J. and **Clarke, J. M.** (1) Introduction to the Palæozoic Brachiopoda. Geol. Survey New York, Palæontology, VIII., I. (1892); II. (1894). (2) Introduction to the study of the Brachiopoda. 1894.

Jones, O. T. *Plectambonites*, etc. Mem. Geol. Survey Palæont. I. 5 (1928).

Kozlowski, R. Les Brachiopodes Gothlandiens. Palæont. Polonica, I. (1929).

Muir-Wood, H. M. (1) British Carboniferous Producti. II. Mem. Geol. Survey, Palæont. III. 1 (1928). (2) Internal Structure of Mesozoic Brachiopoda. Phil. Trans. Roy. Soc. 223 B (1934), p. 511. (3) Brachiopoda of the Great Oolite. 1936. (Palæont. Soc.)

Reed, F. R. C. Ordovician and Silurian Brachiopoda of Girvan. Trans. Roy. Soc. Edinb. LI. (1917), p. 795.

Schuchert, C. (1) Classification of the Brachiopoda. Amer. Geol. XI. (1893), p. 141; XIII. (1894), p. 80. (2) Synopsis of American Fossil Brachiopoda. Bull. U.S. Geol. Survey (1897). (3) Palæogeographic and Geologic Significance of Recent Brachiopoda. Bull. Geol. Soc. Amer. XXII. (1911), p. 258. (4) and **Le Vene, C. M.** Brachiopoda. Foss. Catalogus, 42 (1929). (5) and **Cooper, G. A.** Orthoidea and Pentameroidea. Mem. Peabody Mus. IV. 1 (1932).

Thomas, I. (1) British Carboniferous Orthotetinæ. Mem. Geol. Survey Gt. Britain (1910). (2) British Carboniferous Producti. *Ibid.* (1914).

Walcott, C. D. Cambrian Brachiopoda. Mon. U.S. Geol. Survey, LI. (1912).

Wisniewska, M. Rhynchonellidés du Jurassique sup. de Pologne. Palæont. Polonica, II. 1 (1932).

CHÆTOPODA

Hinde, G. J. Annelid Jaws from the Palæozoic. Q.J.G.S. XXXV. (1879), p. 370; XXXVI. (1880), p. 368.

Stauffer, C. R. Ordovician Polychæta from Minnesota. Bull. Geol. Soc. Amer. XLIV. (1933), p. 1173.

POLYZOA

Bassler, R. S. (1) Early Palæozoic Bryozoa of the Baltic Provinces. Bull. U.S. Nat. Mus. 77 (1911). (2) Permian Bryozoa of Timor. Paläont. v. Timor, XVI. (1929). (3) Foss. Catalogus, 67, Bryozoa (1935).

Brydone, R. M. Chalk Polyzoa. G.M. (1906–18).

Busk, G. (1) Polyzoa (Challenger Report). 1884–86. (2) The Crag Polyzoa. 1859. (Palæont. Soc.)

Canu, F. (1) Bryozoaires éocènes de la Belgique. Mém. Mus. R. hist. nat. Belg. XXXIX. (1929). (2) Bryozoaires oligocènes. *Ibid.* 50 (1931).

Canu, F. and **Bassler, R. S.** Synopsis of American early Tertiary Cheilostome Bryozoa. Bull. U.S. Nat. Mus. 96 (1917); 106 (1920).

Canu, F. and **Lecointre, G.** Bryozoaires des Faluns de Touraine et d'Anjou. Mém. Soc. géol. France, N.S. IV. (1925–34).

Cumings, E. R. (1) Development of *Fenestella*. A.J.S. (4), XX. (1905), p. 169. (2) Development and systematic position of the Monticuliporoids. Bull. Geol. Soc. Amer. XXIII. (1912), p. 257. (3) and **Galloway, J. J.** Trepostomata. *Ibid.* XXVI. (1915), p. 349.

Davis, A. G. English Lutetian Polyzoa. Proc. Geol. Assoc. XLV. (1934), p. 205.

Gregory, J. W. (1) British Palæogene Bryozoa. Trans. Zool. Soc. XIII. (1893), p. 219. (2) Catalogue of Fossil Bryozoa in the British Museum: Jurassic. 1896. (3) Ditto: Cretaceous 1899, 1909.

Haime, J. Bryozoaires de la formation jurassique. Mém. Soc. géol. France (2), V. (1854), p. 156.

Hincks, T. History of the British Marine Polyzoa. 1880.

Lang, W. D. (1) Calcium Carbonate and Evolution in Polyzoa. G.M. (1916), p. 73. (2) Cribrimorph Cretaceous Polyzoa. A.M.N.H. (8), XVIII. (1917), pp. 81, 381. (3) Pelmatoporinæ. Phil. Trans. Roy. Soc. 209 B (1919), p. 191. (4) Catalogue of Fossil Bryozoa in the British Museum (Cretaceous). Vols. III., IV. 1921, 1922.

Lee, G. W. British Carboniferous Trepostomata. Mem. Geol. Survey, Palæont. I. (1912).

Nicholson, H. A. Structure and Affinities of *Monticulipora*, etc. 1881.

Orbigny, A. d'. Paléontologie française. Terr. crét. V. 1850–52.

Pergens, E. Revision des Bryozoaires du Crétacé figurés par d'Orbigny. 1. Cyclostomata. Bull. Soc. géol. Belg. Mém. III. (1889), p. 305.

Shrubsole, G. W. (1) Carboniferous Fenestellidæ. Q.J.G.S. XXXV. (1879), p. 275. (2) Silurian Fenestellidæ. *Ibid.* XXXVI. (1880), p. 241.

Ulrich, E. O. (1) Lower Silurian Bryozoa of Minnesota. Geol. and Nat. Hist. Survey, Minnesota, III. (1) (1895), p. 96. (2) and **Bassler, R. S.** Revision of Palæozoic Bryozoa. Smithson. Misc. Coll., XLV. (1903), p. 256; XLVII. (1904), p. 15.

Vine, G. R. Reports on Fossil Polyzoa. Rep. Brit. Assoc. (1880–92).

MOLLUSCA

1. *General*

Adams, H. and **A.** Genera of Recent Mollusca. 3 vols. 1858.

Böggild, O. B. Shell Structure of the Mollusks. Mém. Acad. R. Sci. et Lett. Danemark (9), II. (1930), p. 231.

Carpenter, W. Microscopic Structure of Shells. Rep. Brit. Assoc. for 1844 (1845), p. 24; for 1847 (1848), p. 93.

Cossmann, M. (1) Catalogue illustré des Coquilles fossiles de l'Éocène des environs de Paris. Vols. I.–V. 1886–92. (2) and **Pissaro, G.** Iconographie des Coquilles de l'Éocène de Paris. 2 vols. 1904–13.

Fischer, P. Manuel de Conchyliologie. 1887.

Morris, J. and **Lycett, J.** Great Oolite Mollusca. 1850–63. (Palæont. Soc.)

Newton, R. B. List of the Edwards Collection of British Oligocene and Eocene Mollusca in the British Museum. 1891.

Pelseneer, P. Mollusca. (Lankester's Zoology, v.) 1907.

Sowerby, J. Mineral Conchology of Great Britain. 7 vols. 1812–46.

Thiele, J. Handbuch der systematischen Weichtierkunde. 2 vols 1929–35.

Wood, S. V. Crag Mollusca. 4 vols. 1848–82. (Palæont. Soc.)

Woodward, S. P. Manual of the Mollusca. Edition 4 by Tate. 1880

2. *Lamellibranchia*

Amalitzky, W. Ueber die Anthracosien der Permformation Russlands. Palæontographica, XXXIX. (1892), p. 125.

Arkell, W. J. British Corallian Lamellibranchia. 1929–36. (Palæont. Soc.)

Barrande, J. Système Silurien du centre de la Bohême. Vol. VI. 1882.

Bernard, F. (1) Le développement et la morphologie de la coquille chez les Lamellibranches. Bull. Soc. géol. de France (3), XXIII. (1895), p. 104; XXIV. (1896), pp. 54, 412; XXV. (1897), p. 559. (2) La coquille des Lamellibranches. Ann. Sci. Nat. (Zool.), (8), VIII. (1898).

Beushausen, L. Die Lamellibranchiaten des rheinischen Devon. Abhandl. der preuss. geol. Landes-Anst. XVII. (1895).

Cossmann, M. and **Pissaro, G.** Iconographie des Coquilles de l'Éocène de Paris. I. Pélécypodes. 1904–6.

Dall, W. H. (1) The Hinge of Pelecypods. A.J.S. (3), XXXVIII. (1889), p. 445. (2) Classification of Pelecypoda. Trans. Wagner Inst. Sci. Philadelphia, III. (1895).

Davies, J. H. and **Trueman, A. E.** Non-marine Lamellibranchs of the Coal Measures. Q.J.G.S. LXXXIII. (1927), p. 210.

Dechaseaux, C. Pectinidés jurassiques. Ann. Paléont. XXV. (1936), p. 1.

Douvillé, H. (1) Classification des Lamellibranches. Bull. Soc. géol. France (4), XII. (1912), p. 419. (2) Les Rudistes et leur évolution. *Ibid.* (5), V. 1935 (1936), p. 319. (3) Les Lamellibranches cavicoles ou Desmodontes. *Ibid.* (4), VII. (1907), p. 96.

Ffab, L. Taxodonta des böhmischen Silurs. Palæontographica, LXXX. A (1934), p. 195.

Frech, F. Die devonischen Aviculiden Deutschlands. Abhandl. geol. Specialkarte von Preussen, IX. (1891).

Hind, W. (1) *Carbonicola, Anthracomya,* and *Naiadites.* 1894. (2) British Carboniferous Lamellibranchiata. 1896–1905. (Palæont. Soc.) (3) Silurian Lamellibranchs of Girvan. Trans. Roy. Soc. Edinb. XLVII. (1910), p. 479.

Isberg, O. Lamellibranchiaten des Leptænakalkes. 1934.

Jackson, R. T. Phylogeny of the Pelecypoda. The Aviculidæ and their allies. Mem. Boston Soc. Nat. Hist. IV. (1890), p. 277.

Lebkuchner, R. Die Trigonien des süddeutschen Jura. Palæontographica, LXXVII. (1932), p. 1.

Lycett, J. British Fossil Trigoniæ. 1872–79. (Palæont. Soc.)

McLearn, F. H. Palæontology of Silurian of Arisaig, Nova Scotia. Geol. Survey Canada, Mem. 137 (1924), pp. 91–140.

Neumayr, M. Morpholog. Eintheilung der Bivalven. Denkschr. der k. Akad. der Wissensch., Math.-nat. Classe (Wien), LVIII. (1891), p. 701.

Orbigny, A. d'. Paléontologie francaise. Terr. crét. III. Lamellibranches. 1843–47.

Popenhoe, W. P. and **Findlay, W. A.** Transposed Hinge Structures in Lamellibranchs. Trans. S. Diego Soc. Nat. Hist. VII. (1933), p. 299.

Stoliczka, F. Cretaceous Fauna of S. India. III. Pelecypoda. Palæontologia Indica (1870–71).

Vest, W. Ueber die Bildung und Entwicklung des Bivalven-Schlosses. Verhandl. u. Mittheil. siebenbürg. Vereins, XLVIII. (1898), p. 25.

Wood, S. V. Eocene Bivalves of England. 1861–71. (Palæont. Soc.)

Woods, H. Cretaceous Lamellibranchia of England. 2 vols. 1899–1908. (Palæont. Soc.)

Zittel, K. A. Die Bivalven der Gosaugebilde. Denkschr. der k. Akad. der Wissensch. XXIV. (2) (1865), p. 105; XXV. (2) (1866), p. 77.

3. *Amphineura and Gasteropoda*

Annandale, N. Evolution of Shell-Sculpture in the Viviparidæ. Proc. Roy. Soc. XCVI. B (1924), p. 60.

Cossmann, M. Essais de Paléoconchologie comparée. 1895–1925.

Donald, J. *Murchisonia*, etc. Q.J.G.S. LI. (1895), p. 210; XLIII. (1887), p. 617; LIV. (1898), p. 45; LV. (1899), p. 251; LVIII. (1902), p. 313; LXI. (1905), pp. 564, 567; LXII. (1906), p. 552; LXXIII. (1917), p. 59; LXXX. (1924), p. 408; LXXXII. (1926), p. 526; LXXXIX. (1933), p. 87.

Grabau, A. W. (1) Studies of Gastropoda. Amer. Nat. XXXVI. (1902), p. 917; XXXVII. (1903), p. 515; XLI. (1907), p. 607. (2) Phylogeny of *Fusus*, etc. Smithson. Misc. Coll. XLIV. (1904), p. 157.

Harmer, F. W. Pliocene Mollusca of Great Britain. 1914–25. (Palæont. Soc.)

Hudleston, W. H. (1) British Inferior Oolite Gasteropoda. 1887–96. (Palæont. Soc.) (2) and **Wilson, E.** Catalogue of British Jurassic Gasteropoda. 1892.

Knight, J. B. Gastropods of the St Louis Pennsylvanian Outlier. Journ. Paleont. IV. (1930), supplement; V. (1931), pp. 1, 177; VI. (1932), p. 189; VII. (1933), pp. 30, 359; VIII. (1934), pp. 139, 433.

Koken, E. (1) Ueber die Entwickelung der Gastropoden vom Cambrium bis zur Trias. Neues Jahrb. für Min. etc. VI. (1889), p. 305. (2) and **Perner, J.** Gastropoden des baltischen Untersilurs. Mém. Acad. Sci. Russie, (8), XXXVII. (1925).

Lindström, G. Silurian Gasteropoda and Pteropoda of Gotland. 1884

Longstaff, J. See under **Donald.**

Perner, J. Système Silurien de la Bohême. IV. Gastéropodes. 1903.

Quenstedt, W. Die Geschichte der Chitonen. Palæont. Zeitschr. XIV. (1932), p. 77.

Reed, F. R. C. British Ordovician and Silurian Bellerophontacea. 1920–21. (Palæont. Soc.)

Rochebrune, A. T. de. Monog. des espèces foss. des Polyplaxiphores. Ann. Sci. géol. XIV. (1883), p. 1.

Ulrich, E. O. and **Scofield, W. H.** Lower Silurian Gasteropoda of Minnesota. Rep. Geol. and Nat. Hist. Survey, Minnesota, III. 2 (1897).

Weir, J. British and Belgian Carboniferous Bellerophontidæ. Trans. Roy. Soc. Edinb. LVI. (1931), p. 767.

Wenz, W. Gastropoda. Handb. d. Paläozool. 6, 1938.

Wilson, E. British Liassic Gasteropoda. G.M. (1887), pp. 193, 258.

Wrigley, A. English Eocene and Oligocene Mollusca. Proc. Malacol. Soc. XVI. (1925), p. 232; XVII. (1927), p. 216; XVIII. (1929), p. 235; XIX. (1930), p. 91; XX. (1932), p. 127; XXI. (1934), p. 108, (1935), p. 356.

Zittel, K. A. Die Gastropoden der Stramberger Schichten. (Palæont. Mittheil.) 1873.

4. *Conularia, Hyolithes, etc.*

Bouček, B. Revise Českých Paleozoických Konularií. Pal. Bohemiæ, XI. (1928).

Holm, G. Sveriges Kambrisk-Siluriska Hyolithidæ och Conularidæ. Sveriges Geol. Undersök., Ser. C, No. 112 (Stockholm, 1893).

Kiderlin, H. Die Conularien. Neues Jahrb. für Min. etc. Beil.-Bd. B, 1937, p. 113.

Novák, O. Revision der palæozoischen Hyolithiden Böhmens. Abhandl. der böhm. Gesellschaft der Wissensch. IV. (1891).

Ruedemann, R. A sessile *Conularia*. Amer. Geol. XVII. (1896), p. 158; XVIII. (1896), p. 65.

Slater, I. L. British Conulariæ. 1907. (Palæont. Soc.)

Zázvorka, V. Hyolithi. Pal. Bohemiæ, XIII. (1930).

Zelizko, J. V. *Hyolithes.* Centralbl. für Min. etc. (1908), p. 362.

5. *Scaphopoda*

Gardner, J. S. Cretaceous Dentaliidæ. Q.J.G.S. XXXIV. (1878), p. 56.

Newton, R. B. and **Harris, G. F.** British Eocene Scaphopoda. Proc. Malacol. Soc. I. (1894), p. 63.

Richardson, L. Liassic Dentaliidæ. Q.J.G.S. LXII. (1906), p. 573.

6. *Cephalopoda*

Abel, O. Paläobiologie der Cephalopoden. Jena, 1916.

Arkell, W. J. Ammonites of the English Corallian. 1935– (Palæont. Soc.)

Barrande, J. Système Silurien du centre de la Boheme. II. (in 6 parts). Céphalopodes. 1867–70.

Blake, J. F. British Fossil Cephalopoda. Part I. Silurian. 1882.

Branco, W. Entwickelungsgeschichte der fossilen Cephalopoden. Palæontographica, XXVI. (1879), p. 19; XXVII. (1880), p. 17.

Buckman, S. S. (1) Inferior Oolite Ammonites. 1887–1907. (Palæont. Soc.) (2) Type Ammonites. 1909–30.

Bulow, E. v. Orthoceren und Belemnitiden der Trias von Timor. Paläont. v. Timor, IV. (1915).

Christensen, E. Neue Beiträge zum Bau der Belemniten. Neues Jahrb. für Min. etc. Beil.-Bd. LI. (1924), p. 118.

Crick, G. C. (1) Muscular Attachment in Ammonoidea. Trans. Linn. Soc. (Zool.), (2), VII. (1898), p. 71. (2) *Belemnites.* Proc. Malacol. Soc. II. (1896), p. 117; VII. (1907), p. 269.

Diener, C. Lebensweise und Verbreitung der Ammoniten. 1912.

Foerste, A. F. (1) Actinosiphonate, Trochoceroid and other Cephalopods. Journ. Sci. Lab. Denison Univ. XXI. (1926), p. 285. (2) and **Teichert, C.** Actinoceroids of E. Central N. America. *Ibid.* XXV. (1930), p. 201.

Foord, A. H. (1) Carboniferous Cephalopoda of Ireland. 1897–1903. (Palæont. Soc.) (2) and **Crick.** Catalogue of the Fossil Cephalopoda in the British Museum. Parts I–III. 1888–97.

Frech, F. Ueber devonische Ammoneen. Beitr. z. Geol. Österr.-Ung. u. d. Orients, XIV. (1902).

Grossouvre, A. de. Les Ammonites de la Craie supérieure. (Mém. explicat. Carte géol. de la France.) 1893.

Haug, E. Études sur les Goniatites. Mém. Soc. géol. France, Paléont. 18 (1898).

Holm, G. Organisation einiger silurischer Cephalopoden. Palæont. Abhandl. III. (1885).

Huxley, T. H. Structure of Belemnitidæ. (Mem. Geol. Survey.) 1864.

Hyatt, A. (1) Fossil Cephalopods. Bull. Mus. Comp. Zool. Harvard, III. no. 5 (1872). (2) Genesis of Arietidæ. Smithson. Contrib. XXVI. (1889). (3) Phylogeny (chiefly Nautiloidea). Proc. Amer. Phil. Soc. XXXII. (1894), p. 349.

Könen, A. v. Ammonitiden d. norddeutsch. Neocom. Abhandl. der preuss. geol. Landes-Anst. XXIV. (1902).

Miller, A. K. (1) Mixochoanitic Cephalopods. Univ. Iowa Stud. Nat. Hist. XIV. (1932). (2) and **Furnish, W. M.** Permian Ammonoids of the Guadalupe Mountain Region. Geol. Soc. America, Special Paper 26, 1940.

Mojsisovics von Mojsvár, E. (1) Die Cephalopoden der mediterranen Triasprovinz. Abhandl. d. kk. geol. Reichsanst. x. (1882). (2) Die Cephalopoden der Hallstätter Kalk. *Ibid.* vi. i. (1873); ii. (1893).

Næf, A. Die fossilen Tintenfische. Jena, 1922.

Neumayr, M. I. Jura Studien. Ueber Phylloceraten. Jahrb. d. kk. geol. Reichsanst. xxi. (1871), p. 297.

Neumayr, M. and **Uhlig, V.** Ueber Ammonitiden aus den Hilsbildungen Norddeutschlands. Palæontographica, xxvii. (1881), p. 135.

Newton, R. B. and **Harris, G. F.** British Eocene Cephalopoda. Proc. Malacol. Soc. i. (1894), p. 119.

Nötling, F. Untersuchungen ü. d. Bau der Lobenline von *Pseudosageceras*. Palæontographica, li. (1905), p. 155.

Orbigny, A. d'. Paléontologie française. Terr. crét. i. 1840–41. Terr. jur. i. (1842–49).

Pfaff, E. Ueber Form und Bau der Ammonitensepten. Jahresb. niedersächs. geol. Ver. (1911), p. 208.

Phillips, J. British Belemnitidæ. 1865–1909. (Palæont. Soc.)

Pia, J. *Oxynoticeras.* Abhandl. geol. Reichsanst. Wien, xxiii. (1914).

Pompeckj, J. F. Revision d. Ammoniten d. schwäbischen Jura. 1893, 1896.

Quenstedt, F. von. Die Ammoniten des schwäbischen Jura. 1883–89.

Reynès, P. Monographie des Ammonites (Lias). 1879.

Roman, F. Les Ammonites jurassiques et crétacées. 1938.

Ruedemann, R. Structure of some primitive Cephalopods. Bull. N. York State Mus. No. 80 (1905), p. 296.

Salfeld, H. Die Bedeutung der Konservativstämme für die Stammesentwicklung der Ammonoiden. 1924.

Schlüter, C. Cephalopoden der oberen deutschen Kreide. Palæontographica, xxi., xxiv. (1871–76).

Schmidt, H. Die carbonischer Goniatiten Deutschlands. Jahrb. preuss. geol. Landes-Anst. xlv. (1924), 1925, p. 489.

Sharpe, D. Mollusca in the Chalk of England. Cephalopoda. 1853–54. (Palæont. Soc.)

Smith, J. P. (1) Carbonif. Ammonoids. Mon. U.S. Geol. Survey, 42 (1903). (2) Middle Triassic Invertebrate Faunas. Prof. Paper 83, U.S. Geol. Survey (1914). (3) Upper Triassic Invertebrate Faunas. *Ibid.* 141 (1927). (4) Lower Triassic Ammonoids of N. America. *Ibid.* 167 (1932).

Spath, L. F. Development of *Tragophylloceras*. Q.J.G.S. lxx. (1914), p. 336. (2) Ammonoidea of the Gault. 1923–43. (Palæont. Soc.) (3) Jurassic Cephalopod Fauna of Kachh. Palæont. Indica, N.S., ix. (1927–33). (4) Evolution of the Cephalopoda. Biol. Reviews, viii. (1933), p. 418. (5) Catalogue of Fossil Cephalopoda in the British Museum. IV. Ammonoidea of the Trias. 1934. (6) Ammonites of the Liassic Family Liparoceratidæ. (Brit. Mus.) 1938.

Swinnerton, H. H. (1) Cretaceous Belemnites 1936– (Palæont. Soc.) (2) and **Trueman, A. E.** Morphology and development of the Ammonite Septum. Q.J.G.S. LXXIII. (1918), p. 26.

Teichert, C. Actinoceroiden Cephalopoden. Palæontographica, LXXVIII. (1933), p. 111.

Trauth, F. (1) Aptychienstudien. Ann. Nat. Mus. Wien, XLI. (1927), p. 171; XLII. (1928), p. 121; XLIV. (1930), p. 331; XLV. (1931), p. 17. (2) Die Anaptychien des Lias. Neues Jahrb. für Min. etc., Beil.-Bd. LXXIII. B (1935), p. 70.

Troedsson, G. T. (1) Ordovician Faunas of N. Greenland. Meddel. om Grönland, LXXI. (1929), p. 1. (2) Baltic Fossil Cephalopods. Lunds Univ. Årsskr. N.F. Avd. 2, N.F. 42 (1931); 43 (1932).

Ulrich, E. O., Foerste, A. F., etc. Ozarkian and Canadian Cephalopods. Geol. Soc. America, Special Papers 37 (1942), 49 (1943), 58 (1944).

Wedekind, R. Palæoammonoidea (Goniatiten). Palæontographica, LXII. (1918), p. 85.

Wright, T. Lias Ammonites. 1878–86. (Palæont. Soc.)

Würtenberger, L. Über die Stammesgeschichte der Ammoniten. 1880.

Zittel, K. A. Die Cephalopoden der Stramberger Schichten. 1868.

ARTHROPODA

1. *Crustacea*

A. *General*

Calman, W. T. Crustacea (Lankester's Zoology, VII. 3). 1909.

Edwards, H. Milne. Histoire naturelle des Crustacées. 1834–40.

Pruvost, P. (1) La Faune continentale du Terrain houiller du Nord de la France. 1919, pp. 35–92. (2) Faune continentale du terrain houiller Belgique. Mém. Mus. R. hist. nat. Belg. XLIV. (1930), p. 105.

Vogdes, A. W. Bibliography of Palæozoic Crustacea. Ed. 2. 1893. And Trans. S. Diego Soc. Nat. Hist. IV. 1925.

Woodward, H. Catalogue of British Fossil Crustacea. 1877.

Woodward, H. and **Salter, J. W.** Chart of Fossil Crustacea. 1865.

B. *Trilobita*

Barrande, J. Système Silurien du centre de la Bohême. Trilobites. 1852. Supplement. 1872.

Beecher, C. E. Structure and Development of Trilobites. (Studies in Evolution (1901), pp. 109–225.) G.M. (1902), p. 152.

Hall, J. and **Clarke, J. M.** Trilobites and other Crustacea, Lower Palæozoic. Geol. Survey New York, Palæontology, VII. (1888).

Henriksen, K. L. Segmentation of the Trilobite's Head. Medd. Dansk geol. Foren. VII. (1926).

Lake, P. British Cambrian Trilobites. 1906–46. (Palæont. Soc.)

Lindström, G. Visual organs of Trilobites. Bih. k. Svenska Vet. Akad. Handl. XXXIV. (1901).

Peach, B. N. *Olenellus* in N.W. Highlands. Q.J.G.S. XLVIII. (1892), p. 227; L. (1894), p. 661.

Raw, F. (1) Development of *Leptoplastus*, etc. Q.J.G.S. LXXXI. (1925), p. 223. (2) Ontogenies of Trilobites. A.J.S. XIV. (1927), pp. 1, 131. (3) Mesonacidæ of Comley. Q.J.G.S. XCII. (1936), p. 236.

Raymond, P. E. Appendages, Anatomy and Relationships of Trilobites. Mem. Connecticut Acad. Arts and Sci. VII. (1920).

Reed, F. R. C. Trilobites of Girvan. 1903–6. Supplements 1–3. 1914–35. (Palæont. Soc.)

Salter, J. W. British Trilobites. 1864–83. (Palæont. Soc.)

Shirley, J. British species of *Calymene*. Mem. Proc. Manchester Lit. Phil. Soc. LXXV. (1931), p. 1, and Q.J.G.S. XCII. (1936), p. 384.

Strand, T. Ontogeny of *Olenus*. Norsk geol. tidsskrift. IX. (1927), p. 320.

Stubblefield, C. J. Development of *Shumarclia*. Journ. Linn. Soc. (Zool.), XXXVI. (1926), p. 345.

Walcott, C. D. (1) Fauna of the Lower Cambrian or Olenellus zone. (Ann. Rep. U.S. Geol. Survey.) 1890. (2) Cambrian Geology and Palæontology. 5 vols. 1909–25. (Smithson. Misc. Coll.)

Warburg, E. Trilobites of the Leptæna Limestone. Bull. Geol. Inst. Upsala, XVII. (1925).

Weller, J. M. Permian Trilobite Genera. Journ. Paleont. XVIII. (1944), p. 320.

Woodward, H. British Carboniferous Trilobites. 1883–84. (Palæont. Soc.)

C. *Lower Crustacea*

Bassler, R. S. and **Kellett, B.** Bibliographic Index of Paleozoic Ostracoda. Geol. Soc. America, Special Paper, I. (1934).

Darwin, C. (1) Fossil Lepadidæ. 1851. (2) Fossil Balanidæ. 1854. (Palæont. Soc.)

Hopwood, A. T. The family Cyclidæ. G.M. (1925), p. 289.

Jones, T. R. (1) Fossil Estheriæ. 1862. (2) Tertiary Entomostraca. 1856, 1889. (3) Cretaceous Entomostraca. 1849, 1890. (Palæont. Soc.)

Jones, T. R., Kirkby, J. W. and **Brady, G. S.** British Entomostraca from the Carboniferous. 1874–84. (Palæont. Soc.)

Scourfield, D. J. Crustacean (*Lepidocaris*) from the Old Red Sandstone. Phil. Trans. Roy. Soc. 214 B (1926), p. 153.

Walcott. See under *Trilobita*.

Withers, T. H. Catalogue of Fossil Cirripedia (British Museum). I. Triassic and Jurassic. 1928. II. Cretaceous. 1935.

D. *Malacostraca*

Ammon, L. von. Beitrag zur Kenntniss der fossilen Asseln. Sitz. der KK. Akad. der Wissensch., math.-phys. Classe, IV. (1882), p. 507.

Assmann, P. Decapodenkrebse des deutschen Muschelkalks. Jahrb. preuss. geol. Landes-Anst. XLVIII. (1927), p. 332.

Bate, C. S. Crustacea Macroura (Challenger Report). 1888.

Beddard, F. E. Isopoda (Challenger Report). 1884, 1886.

Bell, T. Fossil Malacostracous Crustacea of Great Britain. 1858–1913. (Palæont. Soc.)

Beurlen, K. Die Decapoden des schwäbischen Jura. Palæontographica, LXX. (1928), p. 115.

Bill, P. C. Ueber Crustaceen a. d. Voltziensandstein des Elsasses. Mitt. geol. Landes-Anst. Elsass-Lothringen, VIII. (1914), p. 289.

Calman, W. T. (1) Syncarida from the Coal Measures. G.M. (1911), p. 488. (2) *Palæocaris* and *Uronectes*, A.M.N.H. (10), X. (1932), p. 537; XIII. (1934), p. 321.

Chilton, C. A fossil Isopod, *Phreatoicus*. Journ. Proc. Roy. Soc. N.S. Wales, LI. (1918), p. 365.

Fritsch, A. See under *Arachnida*.

Glaessner, M. F. (1) Crustacea decapoda. Foss. Catalogus, 41 (1929). (2) Stammesgeschichte der Dekapoden. Palæont. Zeitschr. XII. (1930), p. 25.

Huxley, T. H. *Pygocephalus* from Coal. Q.J.G.S. XIII. (1857), p. 363.

Jones, T. R. and **Woodward, H.** British Palæozoic Phyllocarida 1888–92. (Palæont. Soc.)

Miers, E. J. Brachyura (Challenger Report). 1886.

Oppel, A. Ueber jurassische Crustaceen. (Palæont. Mittheil.) 1862.

Peach, B. N. Higher Crustacea of the Carboniferous of Scotland. Mem. Geol. Surv. (1908).

Reiff, E. Isopoden aus dem Lias Schwabens. Palæont. Zeitschr. XVIII. (1936), p. 49.

Sars, G. O. Report on Phyllocarida (Challenger Report). 1887.

Scott, H. W. A Stomatopod from the Mississippian of Central Montana. Journ. Paleont. XII. (1938), p. 508.

Smith, G. Anaspidacea. Q.J. Micr. Sci. LIII. (1909), p. 489.

Straelen, V. van. (1) Crustacés Décapodes de la période jurassique. Mém. Acad. Roy. Belg. VII. (1925). (2) Crustacea Eumalacostraca, Foss. Catalogus, 48 (1931). (3) Isopodes Meso- et Cénozoiques. Mém. Acad. Roy. Belg. IX. (1928). (4) and **Schmitz.** Phyllocarida. Foss. Catalogus, 64 (1934).

Withers, T. H. A Liassic Crab, and the origin of the Brachyura. A.M.N.H. (10), IX. (1932), p. 313.

Woods, H. Fossil Macrurous Crustacea of England. 1925–31. (Palæont. Soc.)

Woodward, H. (1) *Pygocephalus*. G.M. (1907), p. 400. (2) *Præanaspides*. *Ibid.* (1908), p. 385.

2. *Myriapoda*

Peach, B. N. Myriapods from the Old Red Sandstone of Forfarshire. Proc. Roy. Phys. Soc. Edinb. VII. (1882), p. 77; XIV. (1899), p. 113.

Scudder, S. H. (1) Archipolypoda, a type of Carboniferous Myriapods. Mem. Boston Soc. Nat. Hist. III. (1882), p. 143. (2) Two new types of Carboniferous Myriapods. *Ibid.* III. (1884), p. 285.

Woodward, H. Carboniferous Myriapods. G.M. (1887), p. 1.

3. *Insecta*

Bolton, H. (1) Fossil Insects of the British Coal Measures. 1921–22. (Palæont. Soc.) (2) Insects from the Coal Measures of Commentry. (Brit. Mus.) 1925.

Brodie, P. B. Insects in the Secondary rocks of England. 1845.

Brongniart, C. L'histoire des Insectes des Temps Primaires. Bull. Soc. Industr. Min. St Étienne, Sér. 3, VII. (1893).

Handlirsch, A. (1) Die fossilen Insekten und die Phylogenie der rezenten Formen. 1906. (2) Revision of American Palæozoic Insects. Proc. U.S. Nat. Mus. XXIX. (1906), p. 661. (3) In C. Schröder, Handbuch der Entomologie, III. (1925), pp. 117–306.

Pruvost, P. La Faune continentale du Terrain houiller du Nord de la France. 1919, p. 93.

Scourfield, D. J. The oldest known fossil Insect. Nature, 145 (1940), p. 799.

Scudder, S. H. (1) Index to the Fossil Insects, Myriapods and Arachnids. Bull. U.S. Geol. Survey (1891). (2) Bibliography of Fossil insects. *Ibid.* 1890. (3) Palæodictyoptera. Mem. Boston Soc. Nat. Hist. III. (1885), p. 319. (4) Fossil Insects of North America. 1890.

Tillyard, R. J. (1) Mesozoic Insects of Queensland. Proc. Linn. Soc N.S. Wales, XLII. (1917), pp. 175, 676; XLIII. (1918), pp. 417, 568 XLIV. (1919), pp. 194, 358. 857; XLVI. (1921), p. 270; XLVIII. (1923), p. 481. (2) British Liassic Dragonflies. Brit. Mus. Nat. Hist. (1925). (3) The Panorpoid Complex in the Rhætic and Lias. *Ibid.* (1933). (4) Kansas Permian Insects. A.J.S. (1924–36).

4. *Arachnida*

Clarke, J. M. and **Ruedemann, R.** Eurypterida of New York. N. York State Mus. Mem. 14 (1912).

Dix, E. and **Pringle, J.** Xiphosura from the S. Wales Coalfield. Summ. Progress Geol. Survey (1928) 1929, p. 90, and A.M.N.H. (10), VI. (1930), p. 136.

Dunbar, C. O. *Paleolimulus*. A.J.S. (5), V. (1923), p. 443.

Fritsch, A. (1) Fauna der Gaskohle, etc. IV. 1901. (2) Palæozoische Arachniden. 1904.

Holm, G. *Eurypterus*. Mém. Acad. impér. Sci. St Pétersbourg. (8). VIII. (1898).

Huxley, T. H. and **Salter, J. W.** *Pterygotus*. Mem. Geol. Survey: Organic Remains. (1859.)

Laurie, M. Eurypterida. Trans. Roy. Soc. Edinb. XXXVII. (1892), p. 151, (1893). p. 509; XXXIX. (1899), p. 575.

Moore, P. F. Gill-like structures in the Eurypterida. G.M., 78 (1941), p. 62.

Peach, B. N. Scorpions from the Carboniferous of Scotland and the English Borders. Trans. Roy. Soc. Edinb. XXX. (1881). p. 397, (1882), p. 511.

Petrunkevitch, A. (1) Terrestrial Palæozoic Arachnida of N. America. Trans. Connecticut Acad. XVIII. 1913. (2) Amber Spiders. *Ibid.* XXXIV. (1942), p. 119.

Pocock, R. I. (1) Silurian Scorpion. Q.J. Micr. Sci. XLIV. (1901), p. 291. (2) Carboniferous Arachnida of Great Britain. 1911. (Palæont. Soc.)

Pruvost, P. See under *Insecta*.

Raasch, G. O. Cambrian Merostomata. Geol. Soc. America, Special Paper 19 (1939).

Schmidt, F. Die Crustaceenfauna der Eurypterusschichten von Rootziküll. Mém. Acad. impér. Sci. St Pétersbourg (7), XXXI. (1883), p. 28.

Störmer, L. (1) Merostomata from the Downtonian of Norway. Skrift. Norske Vidensk. Akad. Oslo, I. 10 (1934). (2) *Mixopterus* from Scotland. Summ. Progress Geol. Survey for 1934, II. (1936).

Wills, L. J. *Eobuthus*. Journ. Linn. Soc. (Zool.) XXXVI. (1925), p. 87.

Woodward, H. (1) The Xiphosura. Q.J.G.S. XXIII. (1867), p. 28; XXVIII. (1872), p. 46. (2) The Merostomata. 1866–78. (Palæont. Soc.)

INDEX

Names of genera are printed in italics.